Green Energy and Technology

Climate change, environmental impact and the limited natural resources urge scientific research and novel technical solutions. The monograph series Green Energy and Technology serves as a publishing platform for scientific and technological approaches to "green"—i.e. environmentally friendly and sustainable—technologies. While a focus lies on energy and power supply, it also covers "green" solutions in industrial engineering and engineering design. Green Energy and Technology addresses researchers, advanced students, technical consultants as well as decision makers in industries and politics. Hence, the level of presentation spans from instructional to highly technical. **Indexed in Scopus**.

More information about this series at http://www.springer.com/series/8059

Tomislav Pavlovic
Editor

The Sun and Photovoltaic Technologies

With contributions by Tomislav Pavlovic,
Aris Tsangrassoulis, Nikola Dj. Cekić, Plamen Ts. Tsankov,
Dragoljub Lj. Mirjanić, Ivana S. Radonjić Mitić

Springer

Editor
Tomislav Pavlovic
Department of Physics
University of Niš
Niš, Serbia

ISSN 1865-3529 ISSN 1865-3537 (electronic)
Green Energy and Technology
ISBN 978-3-030-22402-8 ISBN 978-3-030-22403-5 (eBook)
https://doi.org/10.1007/978-3-030-22403-5

This Springer imprint is published by the registered company Springer Nature Switzerland AG
The registered company address is: Gewerbestrasse 11, 6330 Cham, Switzerland

Contents

Solar Energy

Tomislav Pavlovic

Abstract In this chapter information about interior of the Sun, atmosphere of the Sun, origin of the solar energy, extraterrestrial and terrestrial solar radiation, measurements of sunshine duration, direct solar radiation intensity and intensity of global and diffuse solar radiation, solar radiation on an inclined surface, propagation of solar radiation through atmosphere and the Earth revolution are given.

1 Sun

In our galaxy, known as the Milky Way there are about 400 billion stars which include the Sun as well.

Age of the Sun is estimated at about five billion years. The Sun is halfway through its life cycle, which means that in the current way it will shine for another five billion years.

Since the Earth rotates around the Sun in an elliptical orbit, its distance from the Sun is changing. Average distance of the Earth from the Sun is $R =$ (149,597,870.5 $\pm$ 1.6) km (Fig. 1).

The Sun is plasma glaring ball diameter of $2R_S = 1.319 \times 10^6$ km. The surface temperature of the Sun is $t_s = 5500$–6000 °C, mean density is $\rho_s = 1409$ kg/m^3 and the total mass is $M_s = 2 \times 10^{30}$ kg. On the surface of the Sun a gravitational field is 27.9 times stronger than the gravitational field on the surface of the Earth. The total amount of energy (luminosity) which the Sun radiates into the surrounding area in a unit of time amounts to $L_s = 3.83 \times 10^{26}$ J/s.

The Sun rotates on its axis as a non-homogeneous sphere. Parts of the Sun closer to the Equator rotate at higher angular velocity relative to parts of the Sun, which are closer to the polar regions.

Chemical composition of the Sun is analysed by its spectral lines. Chemical composition of the Sun is shown in Table 1.

T. Pavlovic (✉)
Faculty of Sciences and Mathematics, University of Niš, Niš, Serbia
e-mail: pavlovic@pmf.ni.ac.rs

T. Pavlovic (ed.), *The Sun and Photovoltaic Technologies*,
Green Energy and Technology, https://doi.org/10.1007/978-3-030-22403-5_1

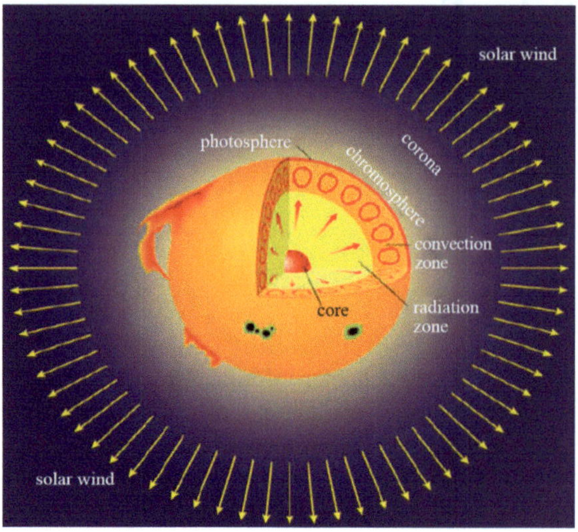

Fig. 1 Structure of the Sun [1]

Table 1 Chemical composition of the Sun [2]

Element	Percentage of total atoms	Percentage of total mass of the Sun
Hydrogen	92	73.4
Helium		25.0
Carbon	0.03	0.3
Nitrogen	0.008	0.1
Oxygen	0.06	0.8
Neon	0.008	0.1
Magnesium	0.002	0.05
Silicon	0.003	0.07
Sulfur	0.002	0.04
Iron	0.004	0.2

The Sun has its own magnetic field whose intensity and polarity varies depending on time and location on its surface. Magnetic field polarity is changing in cycles which on average last 22 years as a consequence of the ionized Sun fluid movement. In a calm period of the Sun's activity, magnetic field induction of the Sun is occurring more rarely 10^{-4} T.

1.1 *Interior of the Sun*

The Sun is composed of the interior and atmosphere. The interior of the Sun is composed of the core, radiative and convection zones (Fig. 2).

Due to the gravity and isotropy of the propagation of energy from the center of the Sun towards the surface, its features are dependent on the distance from the center. Distribution of temperature, density, mass and luminosity of the Sun, depending on the distance from its center are given in Table 2.

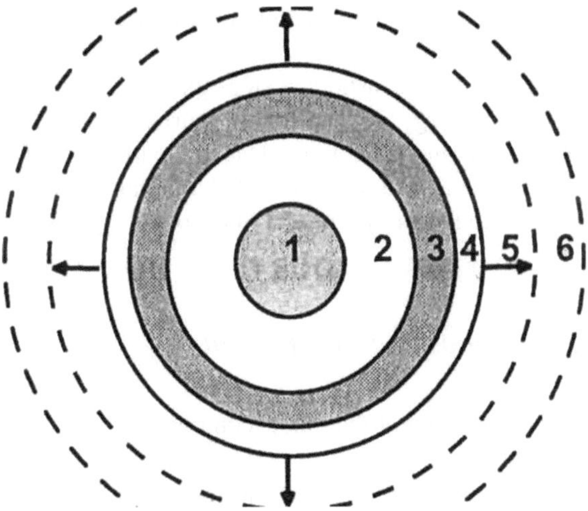

Fig. 2 Structure of the Sun: (1) core, (2) radiative zone, (3) convection zone, (4) photosphere, (5) chromosphere, (6) corona

Table 2 Distribution of temperature, density, mass and luminosity of the Sun, depending on r/R_S [2]

r/R_S	$t(r)$ (10^{6} °C)	$P(r)$ (kg/m^3)	$M(r)/M_S$	$L(r)/L_S$
1.0	0.006	0.0001	1.000	1.00
0.9	0.50	9	0.999	1.00
0.8	1.27	35	0.996	1.00
0.7	1.80	120	0.990	1.00
0.6	2.42	400	0.970	1.00
0.5	3.42	1300	0.920	1.00
0.4	4.74	4100	0.820	1.00
0.3	6.65	13,000	0.630	0.99
0.2	9.35	36,000	0.340	0.91
0.1	12.65	85,000	0.073	0.40
0.0	14.62	134,000	0.000	0.00

The core of the Sun's radius is up to $r = 0.25\,R_s$, and the core temperature is 10^7 °C, and pressure is about 10^{15} Pa. In the core solar energy is produced in thermonuclear process of converting hydrogen into helium.

The energy generated in the core in the form of gamma quantas, neutrinos and energy particles, is transmitted through the radiation layers ($0.25\,R_s < r < 0.85\,R_s$), towards the surface of the Sun, and from there out into the surrounding space.

Above the radiation zone there is a convection zone spanning the distance of $0.85\,R_s < r < 1\,R_s$. In the convective zone there are atoms and negative hydrogen ions which are very good absorbers of solar radiation. Due to the absorption of solar radiation in the convective zone there appears large negative gradient of temperature and convective instability (turbulent flow of ionized gases), according to which the layer is named.

In multiple inelastic collisions gamma quantas generated in the core of the Sun lose their energy, so that from the surface of the Sun, photons from the optical part of the spectrum are emitted. Due to the long-life of the excited states of atoms (10^5 years) and inelastic scattering of photons along their journey, the photons travel from the core to the surface of the Sun about 10^6 years [1–4].

1.2 Atmosphere of the Sun

The atmosphere of the Sun is composed of the photosphere, the chromosphere and the corona.

The photosphere is a shiny surface of the Sun thickness of ~200 km, which is seen from the Earth with the naked eye. This stems from the fact that the mean length of the photons' free path is approximately equal to the thickness of the photosphere, so that photons can directly come out and go into the surrounding space. Photosphere density is ~10^{-4} kg/m^3, the concentration of particles ~10^{23} atom/m^3, and the total mass M_f ~ 10^{20} kg. By comparison, the mass of all the oceans on the Earth (~10^{21} kg) is greater than the mass of the photosphere. The temperature of the photosphere is about 5527 °C. Above the photosphere there are other layers of solar atmosphere. In the lower layer—the chromosphere, the temperature is falling slightly, whereby on 1000 km above the photosphere it reaches a minimum of ~3727 °C. After that, the temperature rapidly rises so that in the top part of the solar atmosphere—the corona, in some places it reaches ~10^6 °C. From the photosphere solar radiation reaches the Earth in the form of an optical continuum, with discrete spectral lines of hydrogen, helium and other elements. The light from the photosphere is partially absorbed into the chromosphere and corona, which results in the appearance of the dark absorption lines which are known as *Fraunhofer* lines (Fig. 3).

The surface of the photosphere is not homogeneous, it consists of a series of light and dark granules, sunspots and many other phenomena (Fig. 4).

Granules are the pillars of gases erupting from the convection zone into photosphere (bright granules) and as cooled (dark granules) return to the convective zone. The temperature of the bright granules is by 100–200 °C higher than the temperature

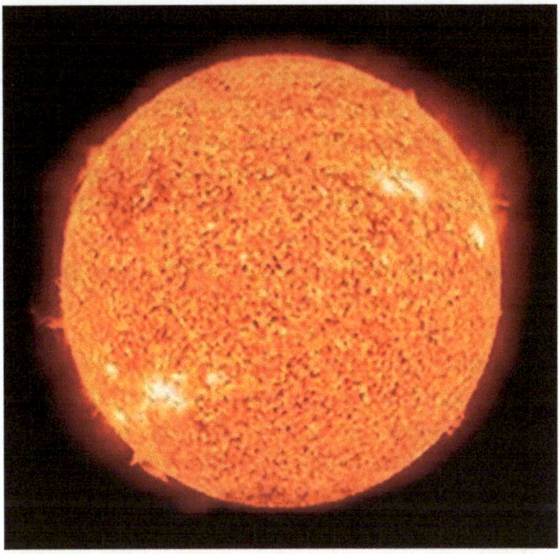

Fig. 3 The Sun photographed at 304 Å by the Atmospheric Imaging Assembly (AIA 304) of NASA's Solar Dynamics Observatory (SDO). This is a false-color image of the Sun observed in the extreme ultraviolet region of the spectrum [5]

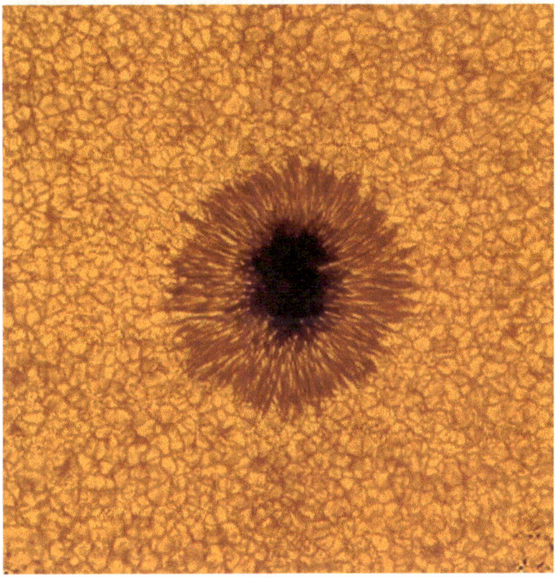

Fig. 4 Sunspot and granules on the photosphere [6]

of dark granules. Diameter of the granules is about 1000 km, and the time duration of about 10 min, after which they dissolve and disappear.

Sunspots were discovered by Galileo and his associates in 1609, and they are systematically counted from 1749 and on. They occur in almost equal numbers in the northern and southern parts of the photosphere. Sunspots consist of a central dark zone diameter of around 1800 km, and a brighter, wider field of half shadow diameter of, on average 37,000 km. Sunspots temperature is lower than the temperature of the surrounding photosphere and is around 3427 °C. Sunspots have magnetic fields induction up to 5 T. Sunspots are usually appearing in groups. Observations have led to the conclusion that the number of sunspots is periodically changed. Thus, there are years when for several weeks there are no spots on the Sun. This period is referred to as a period of minimum solar activity. Subsequently, there are periodic increases in the number of the sunspots. A number of sunspots is the biggest in the period of the maximum solar activity. The time interval between the minimum that is, the maximum number of spots is called a cycle of solar activity. It lasts for 11.2 years, and is twice shorter than the duration of the magnetic cycles. The lifetime of sunspots varies from one day to several months.

Chromosphere and *corona* represent the solar atmosphere which is above the photosphere and extends far through the interplanetary space. Chromosphere and corona can be seen during the solar eclipse when the Moon covers significantly brighter photosphere. Chromosphere, in the form of an orange-reddish ring, the thickness of 3000–7000 km, surrounds photosphere. In it, gas is under low pressure, has a low density and a temperature of about 10^4 °C. Chromosphere does not emit white light but one, more orange-reddish, consisting of several characteristic lines in the ultraviolet, violet, red and radiofrequency part of the spectrum. In the spectrum of the chromosphere, the most intense are the lines of hydrogen (red) and calcium (purple).

In the chromosphere, there are a large number of phenomena, but the most important are protuberances (Fig. 5).

Protuberances represent the reddish strands of gaseous substances discharged in the corona by chromosphere. They can be calm and eruptive (Fig. 6). Calm protuberances are significantly longer than the eruptive ones. Sometimes they can persist for several months until their dissipation. Protuberances are highly visible during solar eclipses. The width of the calm protuberances strands is ~6500 km (4000–15,000 km), height ~ 42,000 km (15,000–120,000 km) and the length of 2×10^5 to 1.1×10^6 km. The speed of the protuberances matter movement, in the direction of the chromosphere-corona and vice versa, is 10–20 km/s.

Eruptive prominences (eruptions) occur in the chromosphere in the form of fierce jets of gas that can reach 9×10^5 km (Fig. 7). They usually occur above large groups of sunspots. After sudden flashes, they last for 5–30 min and then go out. During the eruption, in addition to visible light, there is also an increased emission of ultraviolet, radiofrequency radiation and charged particles of different energies, some of which reach cosmic rays energy.

Eruptions, similar to the sunspots, appear in cycles of 11 years.

Fig. 5 Chromosphere [5]

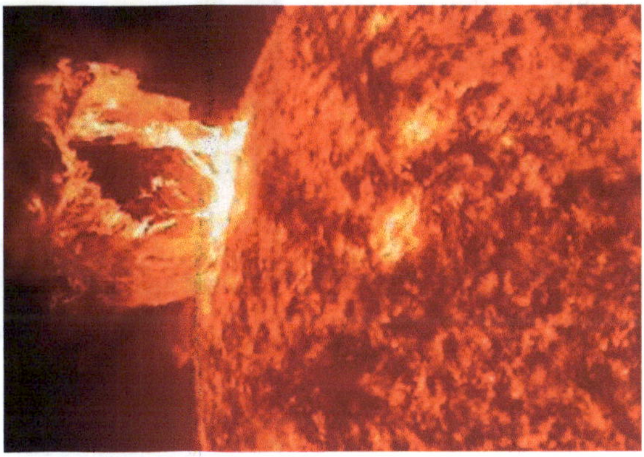

Fig. 6 Calm protuberances [7]

Corona extends above the chromosphere in the range of 1–3 R_s and even further, in the interplanetary space. Medium density of the corona is $\rho_k \sim 10^{14}$ kg/m^3, and the total mass in 1–3 R_s is $M_k \sim 5 \times 10^{14}$ kg. The temperature of the corona is the order of 10^6 °C, which is much higher than the temperature of the chromosphere and photosphere. For the time being, there is no theory that would satisfactorily explain such a high temperature of the corona. It is assumed that this is due to the absorption of the magneto-hydrodynamic waves in the corona, originating from the top of the convective zone (Fig. 8).

At the time of the greatest solar activity corona is of a circular shape, and at the time of the least activity, in some places, it has strands. Due to the high temperature, in the corona, individual particles have sufficient energy to overcome the Sun‘s gravity

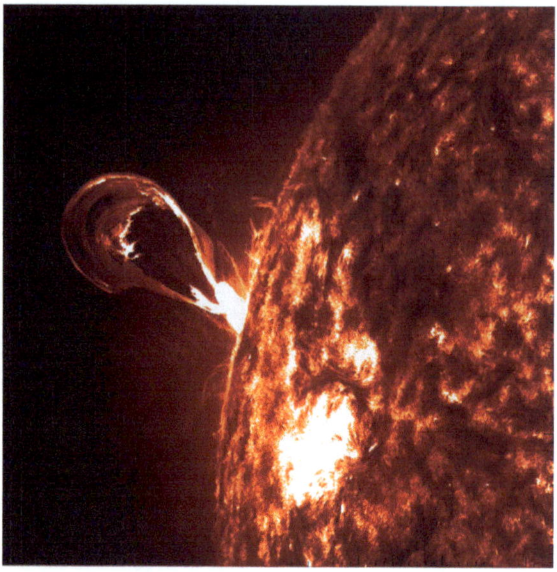

Fig. 7 Eruptive protuberances [8]

Fig. 8 Corona [7]

and go into the surrounding space. The flow of particles from the corona is known as the solar wind, which passes by the Earth at a speed of around 400–800 km/s. Solar wind represents plasma which is made up of free electrons and protons [1–7, 9–12].

1.3 Origin of the Solar Energy

An origin of the solar energy was revealed and explained by Hans Bethe in 1938. He showed that energy is generated in the Sun, in the process of the thermonuclear fusion of hydrogen into helium. Thermonuclear reactions take place in the Sun's core. This is the zone where in the fusion reactions of light nuclei, heavier atomic nuclei are obtained. On this occasion, nucleons transit from a state with less in a state with a higher binding energy, which is accompanied by the emission of the energy of the connection.

Basic fusion reactions in the Sun's core are carried out so that finally four hydrogen nuclei produce helium nucleus. Such thermonuclear reactions are exothermic and in them, on account of the defect of mass, energy is released, which the Sun emits into the surrounding area. These reactions take place in the so-called *proton-proton cycle* (*p-p* cycle) in several stages.

In the first phase, a collision of two protons (hydrogen nucleus) of precisely determined energies, and their unification into a single core occur.

The resulting nucleus composed of two protons, is very unstable and one proton from it enters the neutron releasing at the same neutrino (small elementary lepton particle) and positron (the antiparticle of the electron). In this way, two protons produce deuteron, a positron and a neutrino with the release of energy of 1.44 MeV:

$$ {}_1H^1 + {}_1H^1 \rightarrow {}_1D^2 + \beta^+ + \nu + 1.44\,\text{eV} \tag{1} $$

The major part of the released energy takes away a positron, and a smaller part a neutrino. Positron later reacts with some free-electron thus anihilating itself with the release of two gamma photons.

In the second phase deuterium nucleus, which was created in the first phase, rapidly reacts with another proton (hydrogen atom), building a core of a very rare isotope "helium 3", which is in an excited state:

$$ {}_1H^1 + {}_1D^2 \rightarrow \left({}_2He^3\right)^* \tag{2} $$

In the third phase, a newly founded core isotope "helium 3" tranforms from the excited state into the basic state, by releasing gamma photon:

$$ {}_1H^1 + {}_1D^2 \rightarrow \left({}_2He^3\right)^* \rightarrow \left({}_2He^3\right) + \gamma \tag{3} $$

In the last phase isotope "helium 3" collides with another identical isotope "helium 3" thus building an ordinary atom of helium with two protons released:

$$ {}_1H^1 + {}_2He^3 \rightarrow {}_2He^4 + 2{}_1H^1 \tag{4} $$

The resulting protons start a new local reaction, with the release of the energy, and thus one variant of *p-p* cycle is ending.

Schematic of *p-p* cycle is shown in Fig. 9.

The core ${}_2He^3$ can also enter reaction: ${}_2He^3 + {}_2He^4 \rightarrow {}_4Be^7 + \gamma$.

The chain of reaction of this variant of the *p-p* cycle ends in the following way:

$$ {}_4Be^7 + e^- \rightarrow {}_3Li^7 + \nu_4 Be^7 + {}_1H^1 \rightarrow {}_1B^2 + \gamma \quad (5) $$

or

$$ {}_3Li^7 + {}_1H^1 \rightarrow 2\,{}_2He^4\,{}_5B^8 \rightarrow {}_4B^8 + e^- + \nu \quad (6) $$

$$ {}_4B^8 \rightarrow 2\,{}_2He^4 \quad (7) $$

Apart from *p-p* cycle, hydrogen in the Sun can fusion burn also with the participation of the atoms of the heavier elements: carbon, oxygen, etc. These cores occur in the reactions as catalysts. The most important chain of such fusion reactions takes place with the participation of the nuclei of carbon ${}_6C^{12}$ isotope. It is a well known *carbon-nitrogen cycle*. Unlike the stars with greater mass, the carbon-nitrogen cycle is less present in relation to the *p-p* cycle.

In both cycles of fusion reactions approximately the same amount of energy of around 26.72 MeV per nucleus formed ${}_2He^4$ is released. The largest part of this energy goes to the electromagnetic radiation of continuous spectrum, with a peak in the shortwave gamma and X-ray wavelength area [1–4, 9, 13].

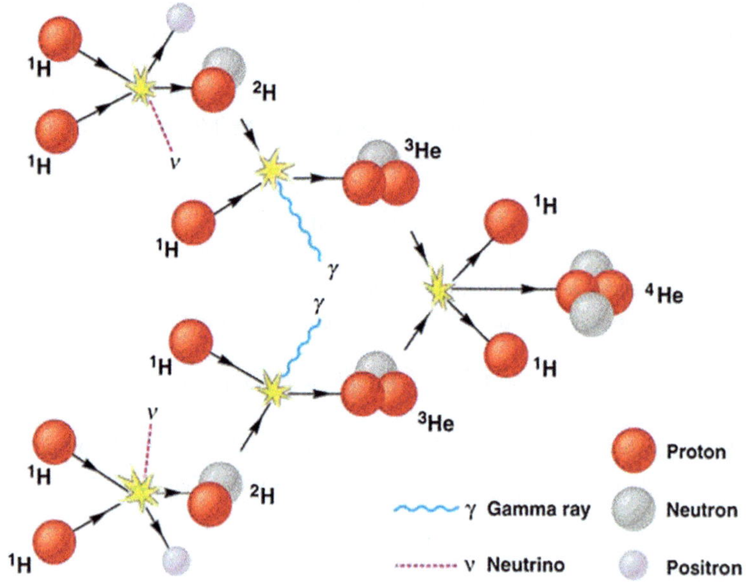

Fig. 9 Schematic of *p-p* cycle [13]

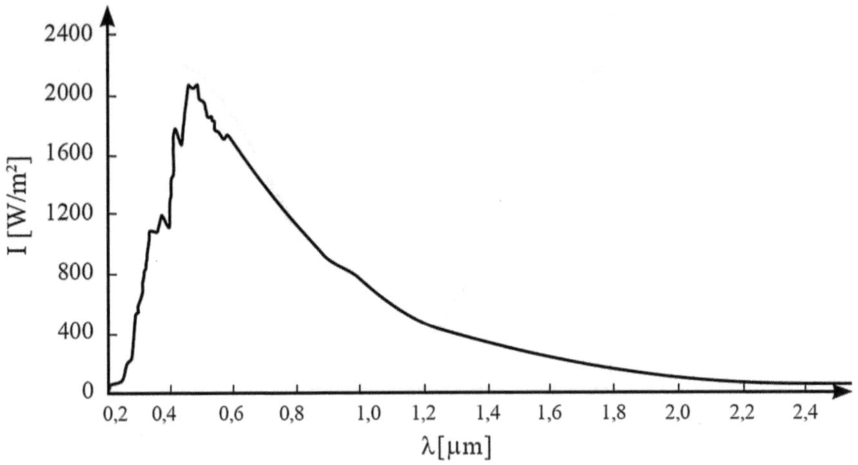

Fig. 10 Spectrum of extraterrestrial solar radiation [9]

1.4 Solar Radiation

1.4.1 Extraterrestrial Solar Radiation

The energy generated in the core of the Sun reaches the Earth in the form of the electromagnetic waves. Solar radiation at the entry into the Earth's atmosphere is known as the *extraterrestrial radiation*. Due to changes in the distance of the Earth from the Sun and solar activity, the intensity of the extraterrestrial radiation changes in the range of 1307–1393 W/m^2 (see Fig. 10).

The solar radiation spectrum before entering the Earth's atmosphere is in the wavelength range from 0.015 to 1000 μm.

The largest part of the extraterrestrial solar radiation is in the wavelength range of 0.704–1000 μm, less so in the area of 0.405–0.904 μm, and the least in the area of 0.015–1000 μm (Fig. 11).

1.4.2 Terrestrial Solar Radiation

Terrestrial radiation denotes solar radiation which after passing through the Earth's atmosphere reaches the Earth. Terrestrial radiation is generally in the range of wavelengths from 0.29 to 2.5 μm.

Of all radiation that reaches the Earth about 3% is in ultraviolet, about 42% in the visible and about 55% in the infrared part of the electromagnetic radiation spectrum. About 97% of the solar radiation reaches the Earth in the wavelength of 0.29–2.5 μm, and about 3% in wavelengths larger than 2.5 μm.

The spectral distribution of the extraterrestrial and terrestrial radiation intensity is given in Fig. 12.

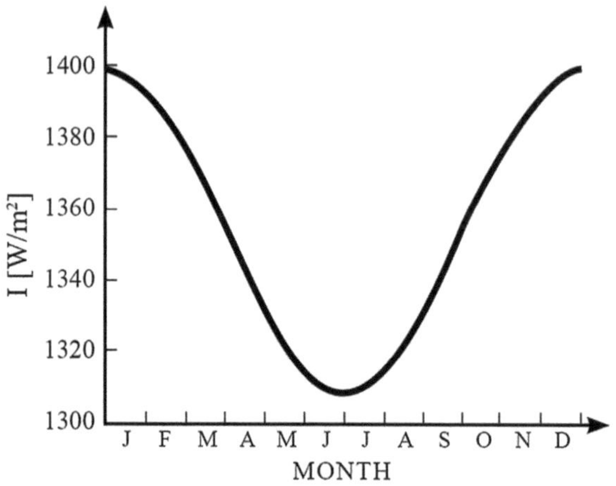

Fig. 11 Annual change in the intensity of extraterrestrial solar radiation [9]

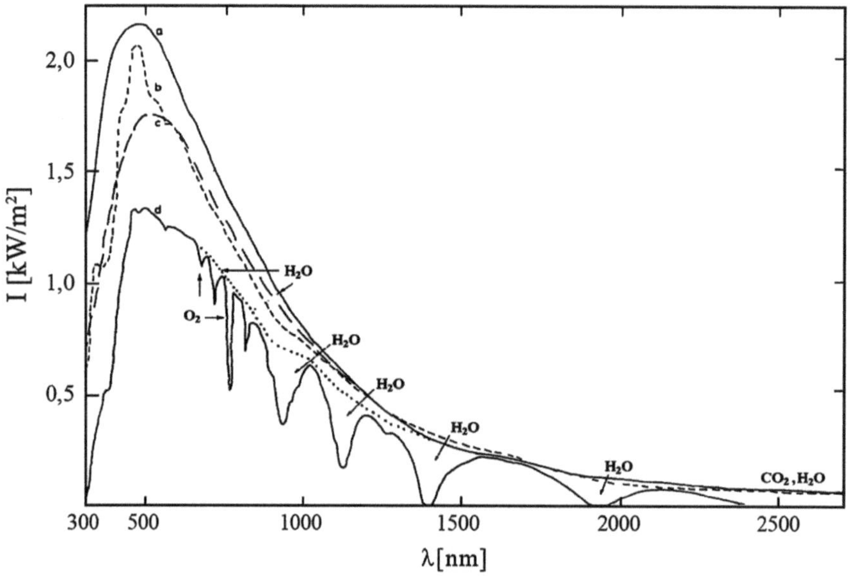

Fig. 12 Spectral distribution of solar radiation intensity: **a** radiation of black body at 5727 °C, **b** extraterrestrial radiation, **c** radiation of black body at 5357 °C and **d** terrestrial radiation [10]

Direct Solar Radiation

Because of the great distance of the Earth and the Sun it can be considered that solar radiation, before entering the Earth's atmosphere, consists of a beam of the parallel electromagnetic waves. Solar radiation can be absorbed, reflected or it can more or less freely pass through the atmosphere (transmit).

When passing through the Earth's atmosphere solar radiation weakens due to scattering and absorption on atoms, molecules and ions of present gasses. The degree of attenuation depends on the path length of the solar radiation through the Earth's atmosphere and its physical and chemical properties. This reduction in energy can be described by Bouquerel–Lambert–Beer law:

$$I = I_0 \times e^{-km} \tag{8}$$

where: I—is the energy of solar radiation which in a time unit falls normally on a square meter of the Earth's surface, I_0—is the energy of the extraterrestrial radiation which in a time unit falls normally on a square meter of surface, k—is a coefficient of solar radiation attenuation in the Earth's atmosphere, which depends on the composition and changes in the atmosphere and m—is an optical air mass, which depends on the angle of the incidence of solar radiation (Fig. 13).

In solar energy, an optical air mass represents the ratio of the path length of the Sun's rays through the atmosphere and the path length of the Sun's rays through the Earth's atmosphere when the Sun is at its zenith:

$$m = \frac{1}{\cos\alpha} \tag{9}$$

where α is the angle between the incident solar radiation and the normal on the surface of the Earth.

For solar radiation above the Earth's atmosphere it is assumed that the optical air mass is equal to zero, where the spectral energy distribution of that radiation is denoted with AM0.

If at zero altitude, solar radiation falls vertically to the Earth ($\alpha = 0$), an optical air mass is denoted with $M = 1$, and spectral distribution of solar radiation energy with AM1.

When solar radiation makes an angle $\alpha = 60°$ with the normal to the surface of the Earth, an optical air mass is $M = 2$, and the spectral distribution of solar radiation energy is denoted by AM2, etc. (Fig. 14).

Depending on the geographic location of a given location and the position of the Sun, that is, on the optical air mass, in the literature one can find various information on the spectral distribution of solar radiation energy.

The position of the Sun in the sky can at any moment be described by the altitude and azimuth of the Sun. A simple way of displaying the movement of the Sun across the sky is the solar diagram which can be represented in polar and cylindrical coordinate system.

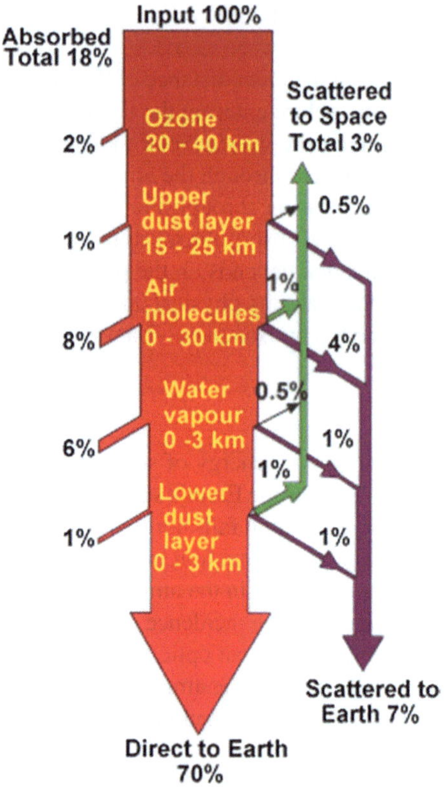

Fig. 13 Influence of the atmosphere on incident solar radiation [14]

In a cylindrical solar diagram, in a rectangular coordinate system the movement of the Sun shows as it appears to the observer who is facing south. On this diagram one can also draw the contours of the surrounding obstacles to predict how these obstacles might block (dazzle) the Sun during the year.

Polar solar diagram is a projection of the solar movement on a horizontal plane with the viewer in the center of the plane. In the polar solar diagram, it is easier to determine the compass direction of the Sun (Figs. 15 and 16).

Diffuse Radiation

Diffuse radiation is generated by scattering solar radiation on atoms and molecules of the gasses and particles of impurities in the air layer of the Earth. With increasing cloudiness and water vapor and aerosols concentration in the air, the proportion of diffuse in the global solar radiation increases, as well. Diffuse radiation for any place on the Earth can be calculated using the formula:

$$I_d = C \cdot I \cdot F_s \tag{10}$$

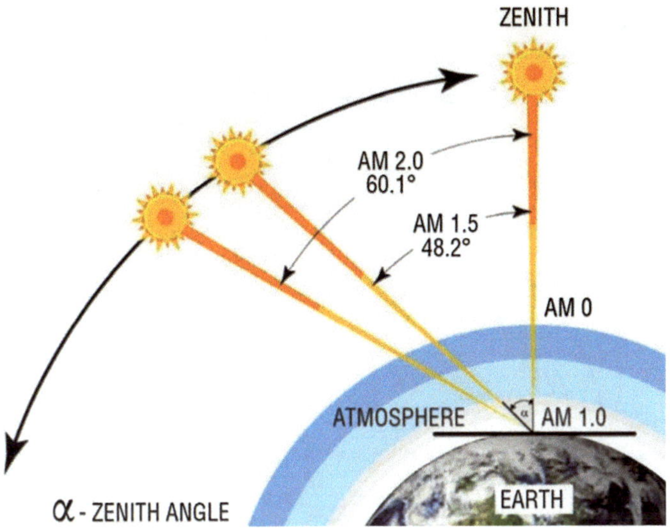

Fig. 14 With the explanation of the spectral distribution of solar radiation energy depending on the angle of incidence of solar radiation and optical air mass: (1) AM0, (2) $\alpha = 0$, AM1 and (3) $\alpha = 60°$, AM2 [15]

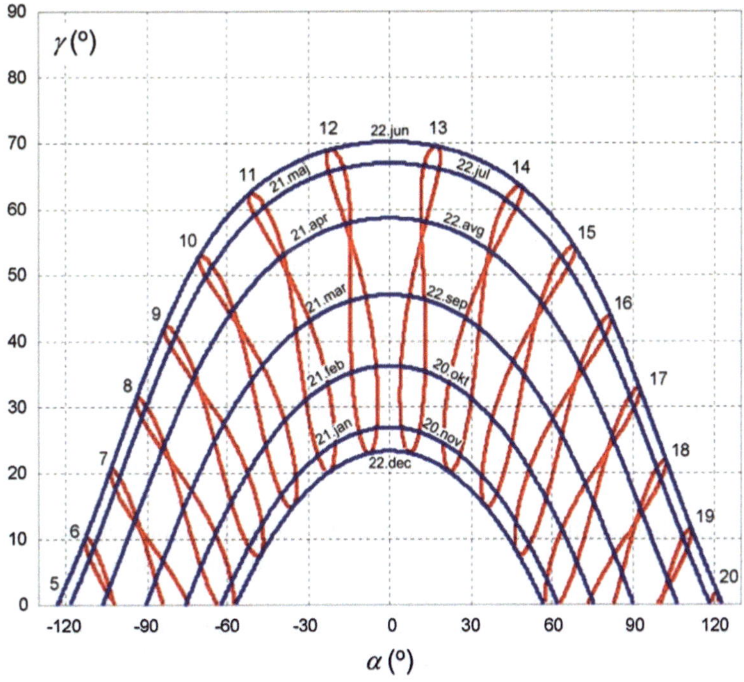

Fig. 15 Example of a cylindrical solar diagram without contours of the surrounding obstacles 43° 19′ 51″N and 21° 53′ 30″E [16]

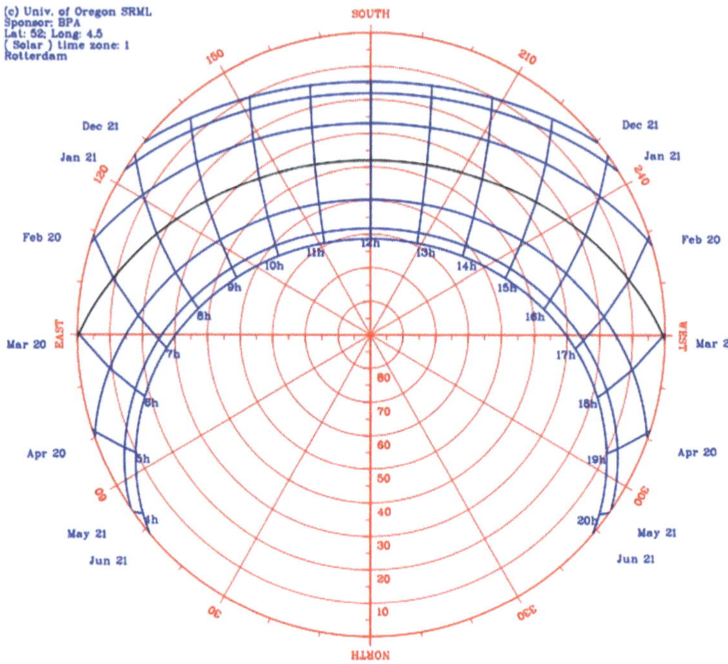

Fig. 16 Example of a polar solar diagram [17]

where: C—is a diffusion radiation factor, I—is an intensity of incident solar radiation and F_s—is an angular factor.

The angle factor is calculated using the formula:

$$F_s = \frac{1}{2}(1 + \cos\beta) \tag{10}$$

where β is the angle between the horizontal plane and a given surface.

When solar radiation on its way reaches the gas and particles molecules, it excites them to the oscillation and radiation, thus turning them into a source of the electromagnetic radiation of a certain wavelength. In this way, the received energy is transmitted unequally in all directions, depending on the properties of the gases or the particles. Energy is no longer expanding in one direction as before entering the atmosphere, but on all sides. The effect of diffusion is twofold: on the one hand, it reduces the intensity of direct solar radiation, and, on the other hand, it causes a scattered radiation of the sky. One part of the solar radiation goes back into the interplanetary space and it is lost for the processes in the atmosphere. The excited molecules or particles do not emit electromagnetic energy with the same spectral distribution of energy, but they change a relative share of any given single different wavelengths.

Diffuse solar radiation contains more a short-wave than a long-wave solar radiation energy. In low optical paths of radiation, in mid-day, diffuse radiation contains most shortwave solar radiation energy, making sky blue. By the Sun lowering scattering of the short-wave radiation increases, it becomes increasingly weak and its relative share in the total scattered radiation decreases. At low sun angles (at low altitudes of the Sun over horizon), at sunrise and sunset, the blue part of the spectrum is completely absorbed, thus only the yellow and red radiation remain. The absorption weakens the intensity of only certain wavelengths of the solar radiation.

Of all the gases that make up the atmosphere, certain wavelengths absorb significantly more oxygen, carbon dioxide, ozone and nitrogen, and in negligible amounts, nitrogen oxides, carbon monoxide and methane. Nitrogen absorbs only radiation with a wavelength of less than 0.2 mm, in the region of the spectrum where the intensity of solar radiation is negligible. Oxygen (O_2) significantly absorbs radiation in two areas, one between 0.76 and 9.8 mm, where it absorbs 8.9% of the radiation of that interval, and in another area with a maximum absorption of 0.69 mm wavelength. Carbon dioxide (CO_2) absorbs certain wavelengths in the infrared part of the spectrum of 1.4–15 mm and negligibly, some wavelengths in the visible part of the spectrum. Ozone (O_3) absorbs certain wavelengths in the ultraviolet (0.2–0.36 mm), visible (0.43–0.75 mm) and infrared (3–5 mm) part of the spectrum. The absorption depends on the thickness of the ozone layer which changes during the year. Atmospheric gases absorb only in a strictly limited area of the spectrum photons of a particular wavelength therefore, such absorption is called a *selective absorption*.

Reflected Radiation

On the inclined surface in relation to the Earth falls a direct and diffuse solar radiation from the sky, and a reflected radiation from the Earth and the surrounding objects.

After passing through the atmosphere, solar radiation encounters the ground or surface water (the seas, lakes, rivers). Depending on the properties of the substrate, larger or smaller part of the radiation will be reflected. There are three processes of reflection. *Mirror-like (specular) reflection* on the flat surfaces (in the nature, on flatwaters), when the surface roughness is smaller than the wavelength of the solar radiation (from 0.4 to 2 mm). If the surface roughness is comparable to the wavelength of radiation, a *scattered reflection* is possible, consisting of more mirror reflections in all the elementary levels from which the surface is made. *Bulk reflection* occurs when the radiation penetrates through the surface and reflects from different strata beneath the surface. The total reflection represents the sum of the specular, diffuse and volume reflections (Fig. 17).

The substrate property to reflect the radiation can be expressed by the *reflection, or albedo coefficient*. Albedo represents the ratio of the reflected radiation intensity to the total incident radiation, in relation to the observed body. Completely white body has an albedo of 1.0 because it would completely reflect the radiation, and completely black body has an albedo of zero. If the body has an albedo of 0.5 it means that it reflects half of the radiation falling on it. The bodies in nature have

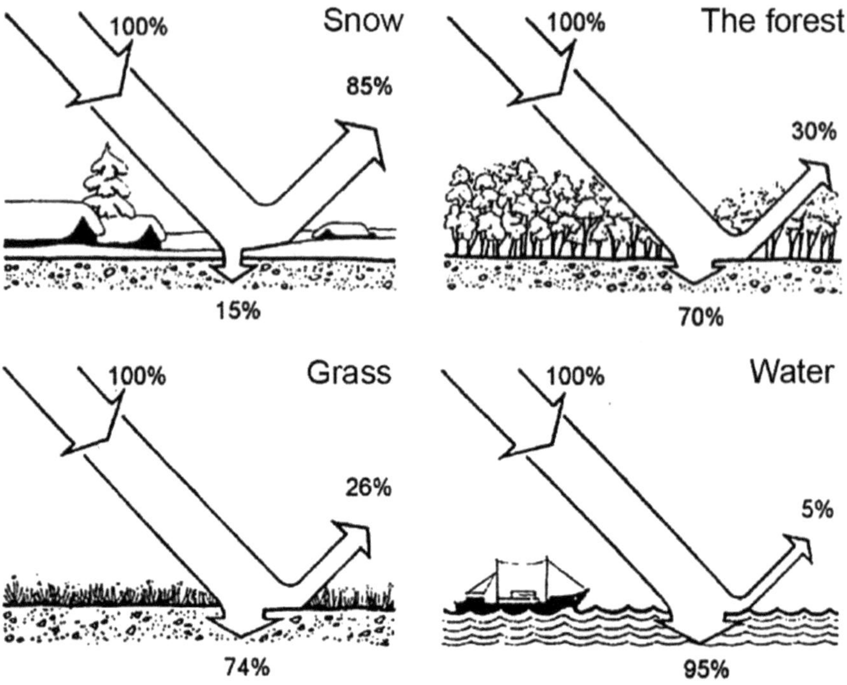

Fig. 17 Solar radiation absorption and reflection on the Earth [18]

very different albedos. Vegetation, as a rule, has a relatively low albedo, because a large part of the radiation is absorbed by plant pigments (chlorophyll, carotene, xanthophyll). Due to its structure consisting of the rare distributed small ice crystals, which cause multiple reflections on the contact between the ice and the light, fresh snow is one of the natural surfaces with the lowest albedo.

Reflected solar radiation as recommended by the World Meteorological Organization (WMO) is measured at a height of 1–2 m, preferably above mowed grassland. For the areas where winter snow lingers long, it is necessary to incorporate the instrument into the mechanism that regulates the same distance from the surface, regardless of the height of the snow. The instrument carrier itself must not affect the measurement and should be placed on the north side of the instrument.

The amount of solar radiation energy, fallen on a surface of the Earth depends on the location of the receiving surface, the slope of the surface in relation to the horizontal plane, orientation of the surface relative to the sides of the world, the time of the year, the conditions of the atmosphere, the size of the receiving surface, the characteristics of the receiving surface and irradiation time.

Comparative overview of the maps of the extraterrestrial and terrestrial solar radiation is given in Figs. 18 and 19, respectively [1–4, 9, 10, 14–19].

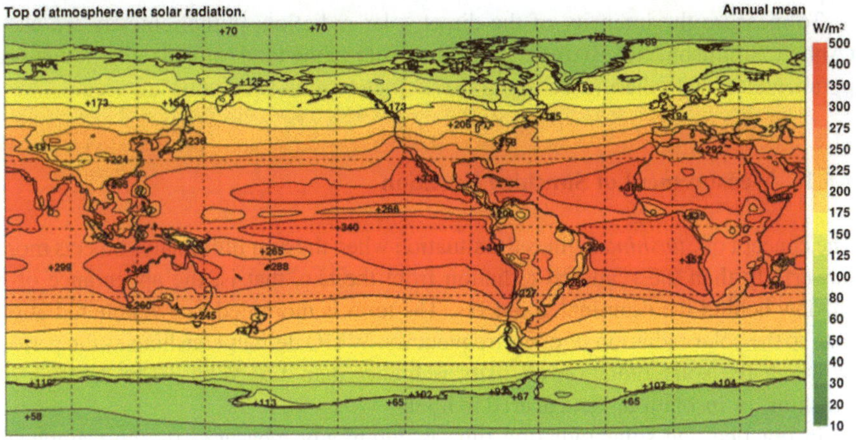

Fig. 18 Map of extraterrestrial solar radiation [20]

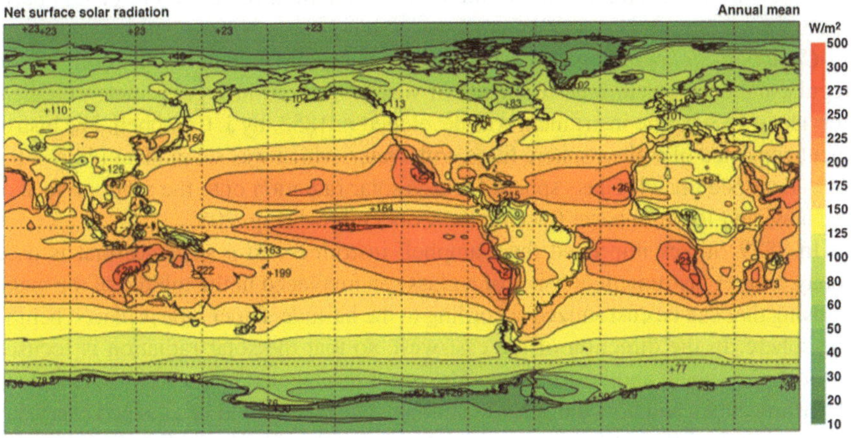

Fig. 19 Map of terrestrial solar radiation [20]

1.5 Solar Radiation Measurements

Solar radiation measuring belongs to a special branch of meteorology known as *actinometry*. For the practical use of solar energy, following actinometric information is important: duration of sunshine (insolation) and the energy of the total and diffuse solar radiation, falling on a horizontal surface.

To measure total solar radiation time, heliographs are used (Campbell-Stokes, Jordan, Maurer, etc). To measure the energy of solar radiation, radiometers are used including pyrheliometers, pyranometers, solarimeters, etc.

To measure the intensity of the direct solar radiation, pyrheliometers are used (Abbott, Smithson, etc). To measure the intensity of the total (global) solar radiation, pyranometers with thermo-cells and pyranometers with solar cells are used.

1.5.1 Measurement of Sunshine Duration

The concept of *Sunshine* defines the situation when the Sun illuminates objects more powerfully than the scattered radiation from the sky, that is, the appearance of a shadow behind the illuminated objects. The term is more tied to a visible radiation than to the solar radiation of other frequencies. WMO defines sunshine as the period in which the solar radiation intensity is higher than 120 W/m^2. The duration of sunshine, or insolation, is measured in hours.

In practice, the solar radiation time is commonly measured by the Campbell-Stokes heliograph, or instrument which registers the sunshine by the combustion of a special tape placed behind a glass sphere. Glass sphere focuses solar radiation onto the tape where, depending on the intensity of radiation, remain more or less burnt spots.

Campbell-Stokes heliograph is perhaps the oldest instrument which is still kept in regular meteorological measurements (it was introduced in regular meteorological service in 1880). The instrument was developed in 1853, by J. Campbell. His instrument consisted of a glass sphere filled with water, which was set in the middle of the carved wooden bawl. Glass sphere focuses solar radiation onto the inner surface of the bawl and leaves a burnt trace on the wooden bowl (Fig. 20).

In 1879 Stokes improved the Campbell instrument to a structure that is used today. Stokes instrument consists of a glass sphere placed in the middle of the metal housing that can be customized according to the latitude. Paper tape is placed behind the sphere, in the direction of the east-west, so that the solar radiation is focused through the sphere and burns a tape. The paper has labeled hours so it is possible to determine when and how the sun was shining. The tape should be changed every day and usually is different for summer, winter and spring (autumn). For a tape to start registering solar radiation a certain solar radiation intensity is required, depending on the instrument, between 80 and 280 W/m^2. If the tape is wet (which is the case in the winter mornings), before measuring, it is necessary to dry it because a wet tape begins to register the radiation later, as compared to dry one. The ends of the metal shell cover the opposite side of the tape at a specific time of the sunrise and sunset. Reading of the data is done visually. The very structure of the instrument does not provide automated data collection, and the WMO recommends abandoning the measurement of solar radiation by the Campbell-Stokes heliograph.

Heliographs are placed at the places with a free horizon, without any obstacles in the direction of the sunrise and sunset. For our latitude it means openness of the horizon for the sunrise from NE to SE (50°–140°) and the sunset from SW to NW (220°–310°). If obstacles still exist, they will shorten a registration of the heliograph, if larger than the height of the Sun at which heliograph starts or stops the registration. If obstacles are less, their impact will be covered by the instrumental error of the

Fig. 20 Campbell-Stokes heliograph *Courtesy* Kipp and Zonen

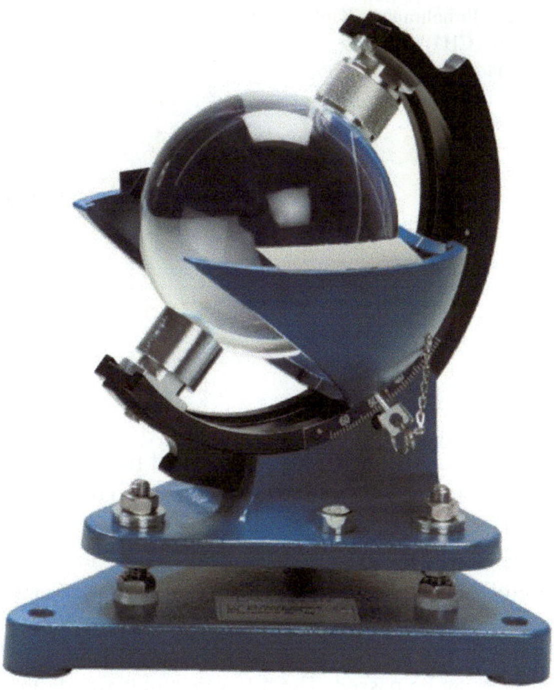

heliograph, and can be ignored. World Meteorological Organization recommends a maximum height of the obstacles to 3° where the lost part of the registration will not be significant. This horizon is considered quasi ideal.

Solarimeter is the instrument by which the intensity of solar radiation is easily and simply measured and read in W/m^2 [1, 2, 21–25].

1.5.2 Measurement of Direct Solar Radiation Intensity

Measurement of the intensity of direct solar radiation is certainly one of the most complex measurements in determining the potential of the solar energy. The intensity of direct solar radiation is measured by *pyrheliometer*, an instrument consisting of thermocouples at the bottom of the narrow collimating tube. Total collimator angle in most pyrheliometers amounts to 4°–5°, while the angle of the solar disk seen is at about 5°. Such a geometry allows the registration of only radiation that comes from a narrow band around the solar disk.

Reception area of the instrument should always be perpendicular to the sun's rays, so that pyrheliometers must follow the sun across the sky with the angular error of less than 0.75° (Kipp & Zonnen) or 1.5° (Epply), which requires a complex and precise mechanical system to track the movement of the Sun. Such systems for dual-axis Sun tracking are for an order of magnitude more expensive than the measuring

Fig. 21 Pirheliometar Kipp & Zonen CH1. *Courtesy* Kipp and Zonen

instrument. The measured data should be normalized to the mean distance of the Earth from the Sun, and the instrument should be calibrated according to the higher class instrument, according to ISO 9060: 1990 E standard (Fig. 21).

The standard instrument for measuring the direct solar radiation is Ångström compensation pyrheliometer. In this instrument, sensor part consists of two equal, black colored Manganin strips (dimensions of 20 × 2 × 0.01 mm). One of the strips is exposed to a direct solar radiation, while the other is shaded. Through the strip, current can be passed. The difference in temperature of the sensor foils is measured by a thermocouple. At the beginning of the measurement, foil exposed to the Sun is warmer than the shaded one. Through the shaded foil current is then gradually increased, until the temperatures of the foils become equal. Then, the power of solar radiation, falling on one foil, equals the power of the shaded foil heating. By measuring electrical power, at the same time, solar radiation power falling on the illuminated surface is measured.

Abott pyrheliometer consists of a movable supporting structure and collimating tube whose interior houses thermo cells and thermometer. This pyrheliometer can be directed towards the Sun manually or automatically [3].

1.5.3 Measurement of Intensity of Global and Diffuse Solar Radiation

The phenomena in the atmosphere and biological processes are highly influenced by the overall (global) solar radiation, coming from the direct and diffuse radiation emitted by the overall atmosphere. The instruments measuring the intensity of the global and diffuse radiation, are called pyranometers. Pyranometer measures the intensity of radiation of a wide wavelengths interval, coming from the entire celestial hemisphere, that is, the space angle of 2π srad. The sensor part of the pyranometer has a shape of the flat surface covered with a hemisphere of quartz glass. Pyranometers may have thermoelectric, photoelectric, pyroelectric or bimetallic elements as sensors.

World Meteorological Organization and the International Organization for Standards (ISO 9060: 1990) define three classes of pyranometers. Best Class Pyranometers (secondary standard) are used for the accurate meteorological measurements, instruments of the first class for regular meteorological measurements, and second class for operational measurements and monitoring of the photovoltaic and thermal solar systems.

The intensity of the scattered (diffuse) solar radiation can be measured by pyranometer as well, if the solar disk is so shaded that a direct solar radiation cannot reach the instrument. This can be achieved in several ways. Most often, for sheltering, a semicircular or circular metal strip is used, diameter of 0.5–1.5 m, oriented in the direction east-west, so that it obscures the solar disk from the sunrise to sunset, with a viewing angle of cover sufficient to completely block the solar disk (Figs. 22 and 23).

The strip is covered with a layer of black color of very small reflectivity in order to prevent reflection of the air from the strip. The ratio of the strip width and diameter is between 0.09 and 0.35. Producing strip shaped as U-profile, constancy of visual angle within ±2% can be achieved. As the Sun's declination changes during the year, it is necessary to move the strip every few (usually two) days. The tape shades pyranometer much more than it is necessary to block direct solar radiation. Because of the anisotropy of the scattered radiation with a peak near the solar disk, the loss of radiation, due to the shady part of the sky, can be considerable. Therefore, it is necessary to calculate the correction factor, which takes into account the area of the sky being shaded and the amount of the scattered radiation, which comes precisely from that part of the sky.

Often, instead of measuring backscattered (reflected) radiation, one measures albedo, that is, the ratio between the total radiation falling on the plane and the radiation that plane reflects. An instrument that measures the albedo, albedometer, consists of two pyranometers. Upper pyranometer (facing up) measures the total solar radiation, and the bottom (facing down) measures the solar radiation reflected

Fig. 22 Kipp & Zonen pyranometer for global solar radiation measurement. *Courtesy* Kipp and Zonen

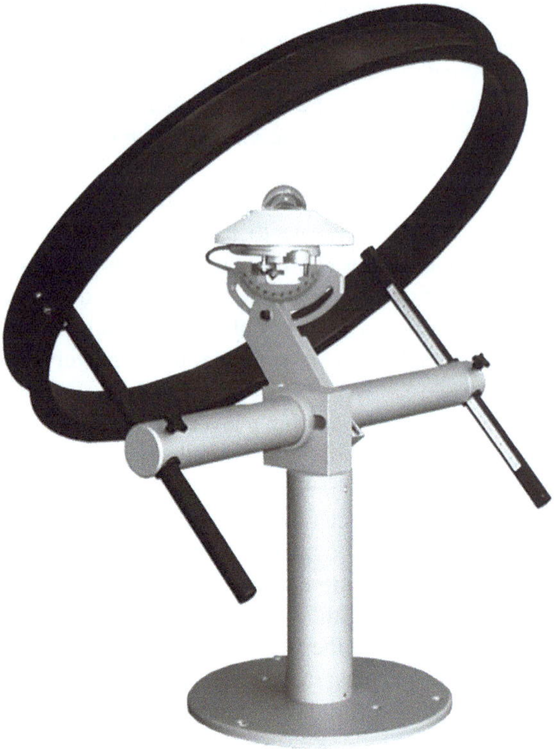

Fig. 23 Kipp & Zonen pyranometer with shadow ring for diffuse solar radiation intensity measurement. *Courtesy* Kipp and Zonen

from the ground. Although the literature emphasizes the importance of the use of the measured albedo for each location, albedo is very rarely measured (Fig. 24).

The albedo value of 0.2 is the most widely used when the measured albedo is not available, and is based on the works of Liu and Jordan [1–4, 21–34].

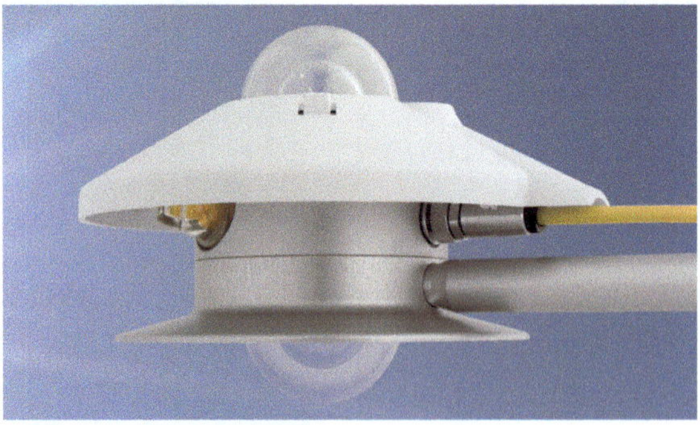

Fig. 24 CMA11 Kipp & Zonen Albedometer. *Courtesy* Kipp and Zonen

1.6 Solar Radiation on an Inclined Surface

Besides direct and diffuse radiation, on the PV module the radiation that is reflected from the Earth and the surrounding objects on the Earth falls (Fig. 25).

When one knows the values of the solar radiation intensity that reaches the horizontal surface (I_h), then the values of the solar radiation intensity reaching the surface oriented towards the south, and set at an angle relative to the horizontal plane (I_m), can be calculated using the following expression:

$$I_m = \frac{I_h \cdot \sin(\alpha + \beta)}{\sin \alpha} \tag{12}$$

where α is the elevation angle of the Sun (the height of the Sun), and β is the angle at which the solar module is tilted relative to the horizontal surface (Fig. 26).

Elevation angle α is calculated using the relation:

$$\alpha = 90 - \varphi + \delta \tag{13}$$

where φ is latitude and δ is the angle of the Sun declination expressed by relation:

$$\delta = 23.45^o \cdot \sin\left[\frac{360}{365}(284 + d)\right] \tag{14}$$

where d is the number of the days in a year [19, 24, 25].

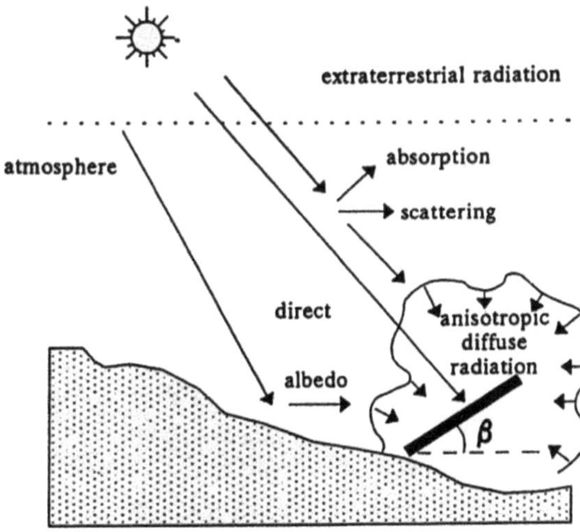

Fig. 25 Radiation falling on the PV module

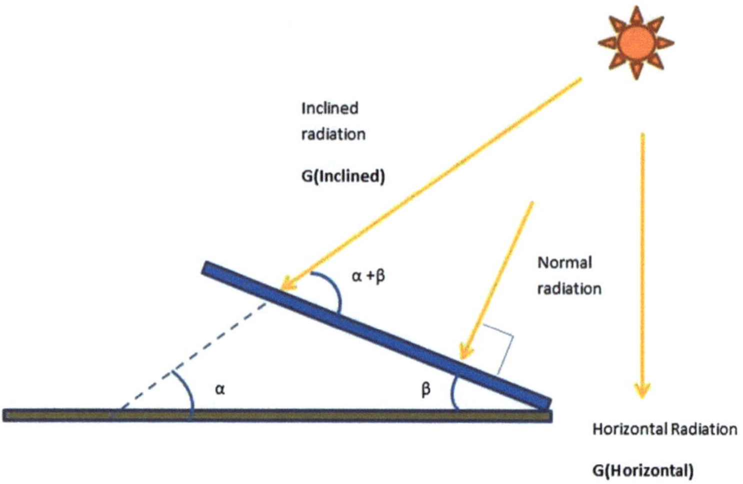

Fig. 26 Solar radiation reaching the surface oriented towards the south and set at an angle relative to the horizontal plane [19]

2 Propagation of Solar Radiation Through Atmosphere

2.1 Atmosphere

Atmosphere denotes the Earth's air layer which is located up to an altitude of 3000 km. Based on the physical characteristics and air composition atmosphere can be divided into homosphere (up to 94 km) and heterosphere (from 94 to 3000 km).

In terms of temperature changes, atmosphere can be divided into 5 main layers (spheres) and 4 transition layer (pauses). The interim layers of atmospheres include tropopause, stratopause, mesopause and thermopause. The main layers of the atmosphere and the altitudes to which they spread are given in Table 3.

Those altitudes are not fixed but are dependent on the latitude, time of day, season, etc.

In the atmosphere, at an altitude of 10–60 km, there is a layer known as the ozonosphere in which ozone (O_3) is formed. The highest concentration of ozone is at an altitude of 20–25 km from the Earth.

Table 3 Atmosphere basic layers [35]

Atmosphere layers	Altitude (km)
Troposphere	To 11
Stratosphere	From 11 to 40
Mesosphere	From 40 to 80
Thermosphere (ionosphere)	From 80 to 800
Exosphere	From 800 to 3000

The physical limits of the atmosphere are at the altitude of 21,644 km above the poles and 35,711 km above the Earth's equator.

Atmosphere air represents a mechanical mixture of gases in a constant relation to each other and the amount of impurities whose quantity is variable. Dry air denotes the air without water vapor and other impurities. Dry air is composed of 78% nitrogen, 21% oxygen, 0.9% argon, 0.037% carbon dioxide, etc. There are other ingredients in the air in the form of aerosols, diameter of 10^{-8} to 10^{-5} cm.

Out of particles in the air the most represented is water vapor which evaporates from water surface of the Earth, then vegetation cover, etc. Every minute, from the Earth evaporates about 10^9 tons of water. In the terrestrial atmosphere the presence of the water vapor ranges from 0.2% in the polar to 2.6% in the equatorial areas. In comparison to all the other gases in the atmosphere, water vapor has the greatest impact on the Earth's climate because it intensely absorbs infrared radiation from the Sun and Earth.

Up to the altitude of 11 km there is about 75% of the atmosphere mass whose density varies with the altitude following the barometric formula:

$$\rho = \rho_o e^{-mgh/(kT)} \tag{15}$$

where ρ_0—is air density at the sea level, h—is the altitude of the air, k—is Boltzmann constant, T—is air temperature at the altitude h [36].

2.1.1 Atmosphere Temperature

The temperature of the atmosphere is lower than the temperature of the Earth because the atmosphere absorbs less solar radiation in relation to Earth. Atmosphere absorbs better infrared radiation from the Earth in relation to the infrared radiation from the Sun, because the infrared radiation from the Earth has a greater wavelength.

The air temperature is highest at the surface of the Earth. With increasing altitude there is a decrease in air temperature. Reducing the temperature of air for each 100 m altitude is called a thermal gradient. The value of the thermal gradient of air depends on the latitude, the season, the location of the given place, humidity, etc., and is of the order of 0.6 °C/100 m. With increasing latitude, the value of the thermal gradient of air is reduced. The thermal gradient is lower in winter than in summer (0.1–0.2)°C/100 m. With increasing concentration of water vapor in the air, there is a decrease in air temperature and thermal gradient.

The heated air in the earth as lighter rises up thus forming so-called convective currents, and in its place colder, denser air (which is then heated and goes up) is lowered. In this way, heat is transferred from the Earth' surface into the air.

2.1.2 Significance of Atmosphere

The atmosphere protects the Earth from ultraviolet radiation harmful effects and excessive cooling. The atmosphere is also one of the tanks in the water cycle in nature and protects the Earth from space meteors.

Without Atmosphere the living world could not exist, sound could not be extended, there would be no solar radiation reflection, alternation of day and night would be instantaneous, the temperature difference between day and night would be around 200 °C, etc. [31, 32, 36].

2.2 *Absorption in Atmosphere*

Atoms and ions can absorb photons whose energy is equal to the difference of the energy of two discrete bound states in the atom:

$$h\nu_{vez} = E_i - E_j \tag{16}$$

Photoexcitation or absorption at the resonant frequency denotes the transition of an electron from a lower to a higher energy state under the influence of the incident electromagnetic radiation.

Photoionization means the transition of an electron from a bound to the free state under the influence of an incident electromagnetic radiation:

$$h\nu = E_{\text{jon}} + mv^2/2 \tag{17}$$

In this process, the absorbed photon energy is spent on ionization of atoms (E_{jon}) and kinetic energy of electron ($mv^2/2$).

During the emission of solar radiation through the Earth's atmosphere it can be absorbed, scattered, refracted, scintillated, etc.

The most important influence on the attenuation of solar radiation in the atmosphere and its own atmosphere radiation is exerted by water vapor and water in the atmosphere. Absorption is made on the water vapor in the atmosphere and refraction of the incident solar radiation is made on the water droplets. Water droplets appear in the form of stable mixtures-aerosols (mist) or unstable mixtures-hydrometeors (fog, clouds, dew, snow).

Up to the altitude of 70 km relative relations of gases in the atmosphere are almost constant, whereas the amount of water vapor varies.

Average composition of the dry atmosphere below 25 km is shown in Table 4.

The atmosphere absorbs most part of the ultraviolet and X radiation.

Due to the large number of different gases and particles, atmosphere differently absorbs and passes different wavelengths of solar radiation. Atmospheric window denotes the region of wavelengths where solar radiation without absorption, passes through the atmosphere (Fig. 27).

Table 4 Average composition of the dry atmosphere below 25 km [32]

Component	Symbol	Volume (%) (dry air)	Molecular weight
Nitrogen	N_2	78.08	28.02
Oxygen	O_2	20.98	32.00
Argon	Ar	0.93	39.88
Carbon dioxide	CO_2	0.035	44.00
Neon	Ne	0.0018	20.18
Helium	He	0.0005	4.0
Ozone	O_3	0.00006	48.0
Hydrogen	H	0.00005	2.02
Crypton	Kr	0.0011	
Xenon	Xe	0.00009	
Methane	CH_4	0.0017	

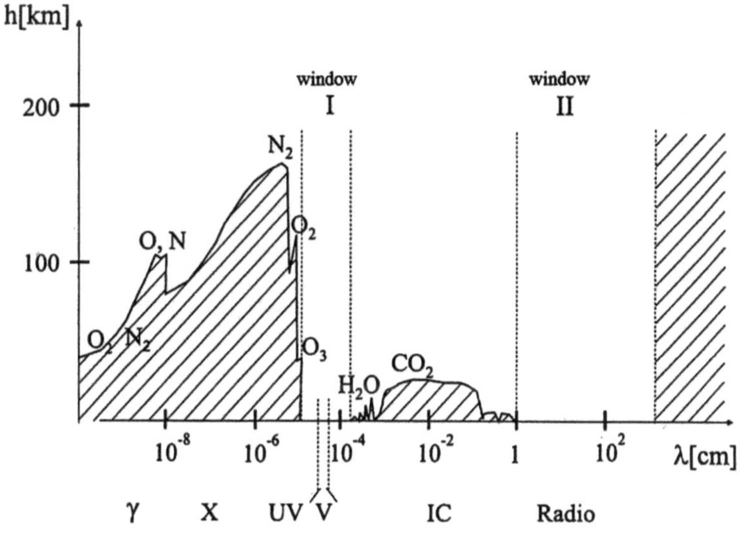

Fig. 27 Atmospheric windows [11]

Atmospheric absorption is significant in the infrared region grater than 3000 nm, with 100% absorption around 4200–4400 nm (mainly due to CO_2), 5500–7300 nm (mainly due to H_2O), and 14,000–16,000 nm (mainly due to CO_2), with atmospheric windows in between. Since water vapor is a good absorber of infrared radiation, most of the infrared radiation emitted by the Earth does not get through clouds [11, 31, 32].

2.3 *Atmospheric Scattering*

Solar radiation scattering is the process by which small particles suspended in a medium of a different index of refraction diffuse a portion of the incident radiation in all directions. In doing so, the incoming solar radiation on a small scale, can change its frequency and wavelength.

Scattering of solar radiation on small particles (aerosols) has a great influence on the propagation of radiation through the atmosphere.

The weakening of the radiation due to scattering depends on the radius of the aerosol particles r and their concentrations in the atmosphere $n(r)$. In scattering, particle size factor $\alpha = 2\pi r/\lambda$ is used, where r—is the radius of the particle, and λ—is the wavelength of the incident solar radiation.

There are four different types of scattering: Rayleigh, Mie, non-selective and Thompson scattering (Fig. 28).

2.3.1 Rayleigh Scattering

Rayleigh scattering is dominantly elastic scattering of solar radiation by atoms and molecules in atmosphere, much smaller than the wavelength of the incident radiation ($r < 0.1\ \lambda$). Rayleigh scattering results from the electric polarizability of the atoms and molecules. The oscilating electric field of a light wave acts on the charges within atoms and molecules, causing them to move at the same frequency. The atoms and molecules therefore become a small radiating dipole whose radiation we see as scattered light.

Rayleigh scattering of sunlight in the atmosphere causes diffuse sky radiation which is the reason for the blue color of the sky and the yellow tone of the Sun itself.

The amount of the scattering is inversely proportional to the fourth power of the incident wavelength ($I_\lambda \sim \lambda^{-4}$).

Intensity of the light scattered by any one of the small spheres of diameter (d) and refractive index (n) from a beam of unpolarized light of wavelength (λ) and intensity (I_0) is given by:

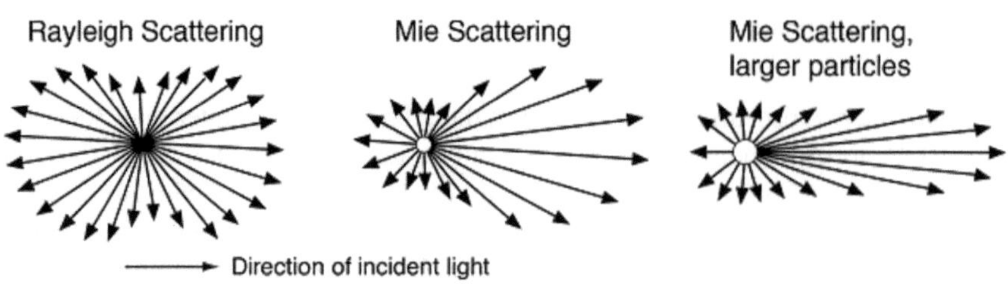

Fig. 28 Atmospheric scattering [37]

$$I = I_0 \frac{1 + \cos^2 \theta}{2R^2} \left(\frac{2\pi}{\lambda} \right)^4 \left(\frac{n^2 - 1}{n^2 + 2} \right)^2 \left(\frac{d}{2} \right)^6 \tag{18}$$

where R—is distance of the particles and θ—is scattering angle. Averaging this over all angles gives the Rayleigh scattering cross-section:

$$\sigma_s = \frac{2\pi^5}{3} \frac{d^6}{\lambda^4} \left(\frac{n^2 - 1}{n^2 + 2} \right)^2 \tag{19}$$

The strong wavelength dependence of the scattering ($\sim\lambda^{-4}$) means that shorter (blue) wavelengths are scattered more strongly than longer (red) wavelength. These results of the indirect blue light coming from all regions of the sky. The scattering at 400 nm is 9.4 times as great as that at 700 nm for equal incident intensity.

Rayleigh scattering occurs in the upper atmospheric layers, usually on tiny particles of dust. Rayleigh scattering is negligible at wavelengths greater than 1 μm [37].

2.3.2 Mie Scattering

Mie scattering occurs when the wavelength of the electromagnetic radiation is similar size to the atmospheric particles. Mie scattering generally influences radiation from the near UV through the mid-infrared parts of the spectrum. Mie scattering mostly occurs in the lower portions of the atmosphere where larger particles are more abundant, and dominates when cloud conditions are overcast. Pollen, dust and smog are major cause of Mie scattering. Mie scattering produces general haze in images.

2.3.3 Non-Selective Scattering

Non-selective scattering occurs when the diameter of the particles in the atmosphere are much larger than the wavelength of incident radiation ($d > \lambda$). Non-selective scattering is primarily caused by water droplets in the atmosphere. Non-selective scattering scatters all radiation evenly through infrared portions of the spectrum-hence the term non-selective.

2.3.4 Thompson Scattering

Thompson scattering is the elastic scattering of solar radiation by free electrons in atmosphere. In the low-energy limit, the electric field of the incident radiation accelerates free electrons, causing it, in turn, to emit radiation at the same frequency as the incident radiation.

Thompson scattering is an important phenomenon in plasma physics and was first explained by the physicist J. J. Thompson. The free electrons move in the direction of the oscilating electric field, resulting in electromagnetic dipole radiation. The moving particle radiates most strongly in a direction perpendicular to its acceleration and that radiation will be polarized along the direction of its motion. This scattering does not depend of wavelength of the incident solar radiation.

The solar K-corona is the result of the Thompson scattering of solar radiation from solar coronal electrons.

X-ray cristallography is based on Thompson scattering [36].

2.4 Atmospheric Refraction

Propagation of solar radiation through the atmosphere is shown in Fig. 29. The exponential change of the atmosphere density is given by using thin layers in which the density is constant. The light that falls on the border of the two layers with different densities is refracted, or is apparently bent upward towards the horizon.

For any two successive layers k, $k + 1$, the law of refraction of light applies:

$$n_k \cdot \sin i_k = n_{k+1} \cdot \sin i_{k+1} \tag{20}$$

where n_k and n_{k+1}—are indices of the refraction of adjacent layers, and i_k and i_{k+1}—are angles which incident and diffracted beam form with respect to the normal onto the layer. For all layers $k = 1, 2, \ldots, N$ in the atmosphere, the following expression can be written:

$$n_1 \cdot \sin i_1 = n_2 \cdot \sin i_2 = \ldots = n_N \cdot \sin i_N \tag{21}$$

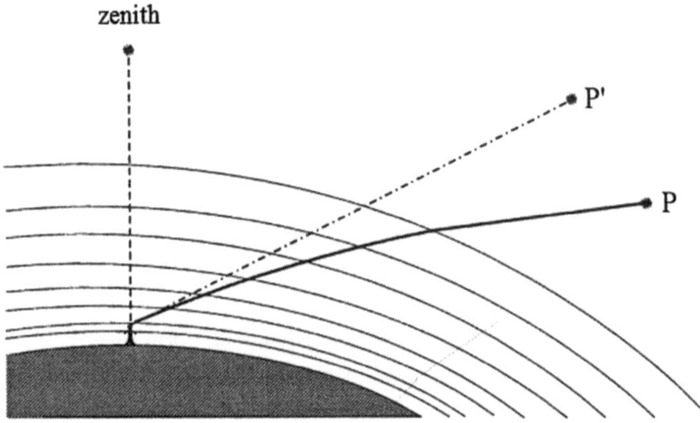

Fig. 29 Refraction in atmosphere [11]

Bearing in mind that the light beam on its way to the Earth passes through the more thickening layers where $n_1 < n_2 < \cdots < n_N$, it follows that refraction in the atmosphere apparently elevates the Sun above the horizon. Refraction in the atmosphere is increasing with the Sun approaching the horizon. Because of refraction in the atmosphere following occurs:

- extension of daylight, even though the Sun is below the horizon,
- discs of the Sun and the Moon near the horizon appear flattened, vertically deformed, because light rays in the lower points of the disk, closer to the horizon, suffer greater refraction.

The refractive index in the non-homogenous and non-stationary environment is in the function of location, time and the wavelength of the radiation:

$$n = n(x, y, z, t, \lambda) \tag{22}$$

Due to the dependence of the refractive index on the wavelength of the incident light, on the border of the two layers a dispersion of white (polychromatic) light occurs, that is a strong refraction of shorter wavelengths occurs [11, 12].

2.5 Scintillation in Atmosphere

Scintillation (flashing) of light means the chaotic change in its intensity during the propagation of light through the optically inhomogeneous and non-stationary (turbulent) Earth's atmosphere. In the turbulent environment there are many different size inhomogeneities of varied dimensions, which change their shape, size and position in space. In doing so, each non-homogeneity with respect to the incident solar radiation, behaves like a lens. Under the influence of temperature in the non-homogeneous environment there is a change in density, refractive index and deformation of the incident electromagnetic wave. Changes in the amplitudes of the light wave changes are manifested in shine changes while creating the impression of light source flickering.

2.6 Solar Radiation Extinction Coefficient

When solar radiation passes through the Earth's atmosphere it weakens due to its absorption and scattering.

The weakening of the solar radiation when passing through an environment can be described by the attenuation or extinction coefficient. In the following there will be discussed attenuation of radiation of wavelength λ and intensity I_λ through a layer thickness of s which absorbs and scatters radiation (Fig. 30).

If radiation is normally incident to the center of the density $\rho(x)$, reduction of the intensity of radiation—$\mathrm{d}I_x$ in a thin layer thickness $\mathrm{d}x$, can be calculated using the

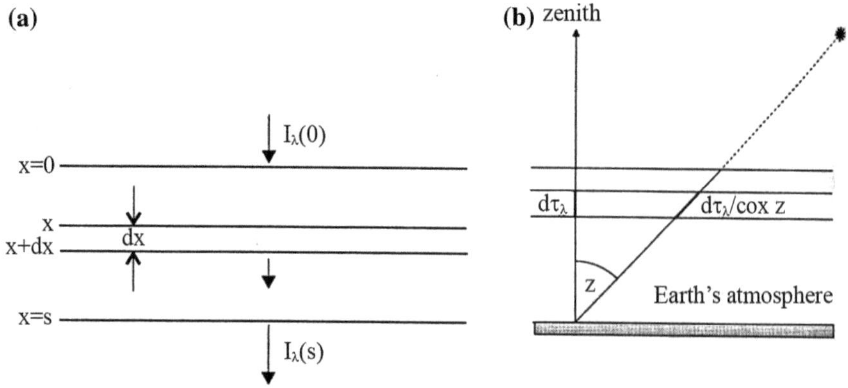

Fig. 30 **a** Solar radiation extinction, **b** atmospheric extinction [11]

following expression:

$$-\mathrm{d}I_x = k_\lambda(x)\rho(x)I_\lambda(x)\mathrm{d}x \tag{23}$$

where $k_\lambda(x)$—is the coefficient of extinction, which characterizes the layer opacity at a wavelength λ. Dividing the above expressions with I_λ and integration from $x = 0$ to $x = s$ following is obtained:

$$I_\lambda(s) = I_\lambda(0)\mathrm{e}^{-\int_0^s k_\lambda(x)\rho(x)\mathrm{d}x} = I_\lambda(0)\mathrm{e}^{-\tau_\lambda} \tag{24}$$

Integral in exponent:

$$\tau_\lambda = \int_0^s k_\lambda(x)\rho(x)\mathrm{d}x \tag{25}$$

is called the optical depth and is a measure of the environment opacity for the radiation of wavelength λ.

If the Sun is not at its zenith ($z = 0$), but at the zenith distance z, the weakening of the radiation is given by:

$$I_\lambda(\tau_\lambda) = I_\lambda(0)\mathrm{e}^{-\tau_\lambda/\cos z} \tag{26}$$

where τ_λ—is the optical depth of the atmosphere at a wavelength λ, measured normal to the surface, $I_\lambda(0)$—is the radiation intensity at the upper limit of the atmosphere, and $I_\lambda(\tau_\lambda)$—is the radiation intensity on the Earth's surface.

Transmittance of the atmosphere depends on the wavelength of the incident radiation. The atmosphere passes radiation through two atmospheric windows, as follows:

(1) The window I passes radiation in the wavelength range of 300–1200 nm, smaller part of the UV radiation (300–390 nm), visible light (390–750 nm) and part of the infrared radiation (760–1200 nm).
(2) Window II (radio window) passes radiation of 1 cm up to about 15–20 m.

In the visible region atmosphere transmittance is about 80%. The atmosphere is almost entirely non-transmittable for X and UV radiation. The radiation with $\lambda < 300$ nm is almost entirely absorbed by ozone and molecules and atoms of oxygen and nitrogen. In IR area most radiation is absorbed by water vapor and carbon dioxide. Observations from the Earth can be done through a few narrow windows between molecular absorption bands of 1200–2000 nm.

Bearing in mind that the IR radiation is absorbed in the lower parts of the atmosphere, this part of the spectrum can be obtained at low altitudes using probes on balloons or observatories in the high mountains.

Radio waves longer than 15–20 m are reflected in the ionosphere. Limit frequency of the waves passing through the ionized path is calculated using the following expression:

$$\nu_0 = 9 \times 10^3 (n_e)^{1/2} \tag{27}$$

where n_e—is the concentration of electrons in a given environment. Change in the intensity of UV radiation leads to the change of n_e, and thus to the limit frequency of the reflected radio waves.

Owing to the development of receivers for radio waves from 1 cm to 15–20 m, there occurred a development of radio astronomy that can be applied to gain better knowledge of the phenomena in the universe [11].

2.7 *The Greenhouse Effect*

When solar radiation reaches the Earth's atmosphere about 25% is reflected from clouds, about 20% is absorbed in the atmosphere and about 50% reaches the Earth. In the highest layers of the atmosphere gamma rays and X-rays are absorbed. Ultraviolet radiation is absorbed in the ozone layer at an altitude of 19-48 km above the Earth's surface.

About 85% of the solar radiation energy that reaches the Earth's surface is absorbed by soil, plants, water, etc. The rest of the solar radiation is reflected from the surface of the Earth, snow, ice, water areas, deserts, etc, and in the form of infrared radiation is returned to the atmosphere.

Greenhouse effect means heating of the Earth under the influence of infrared radiation from the atmosphere. The atmosphere is heated by absorbing radiation from the Sun and infrared radiation from the Earth on the molecules of water vapor, CO_2, N_2O, fluorine and other gases and compounds that are found in the atmosphere.

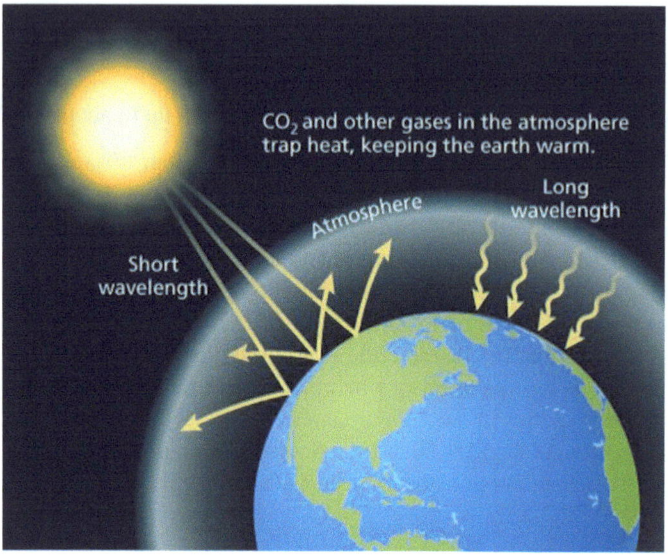

Fig. 31 Greenhouse effect [38]

In this way, the atmosphere prevents heat leaving the Earth to space and the natural cooling of the Earth (Fig. 31).

The greenhouse effect is the result of the interaction of solar radiation and infrared radiation from the Earth with the greenhouse gases in the Earth's atmosphere up to 100 km above the Earth's surface.

Maintenance of the energy balance between the Earth's surface, atmosphere and space is of great importance for maintaining the climate and life on Earth.

Gases that trap heat in the atmosphere are known as greenhouse gases. In the absence of the greenhouse gases an average temperature on the Earth would be—19 °C, as opposed to the current 15 °C.

Due to the greenhouse effect during the last century an average temperature on the Earth has increased by 0.5 °C. According to some predictions, an average temperature on the Earth will by 2100 increase by 2 °C. Due to the greenhouse effect, there occurred ice melting in the Antarctica, the disappearance of certain plant and animal species, etc. [31, 32, 38].

3 The Earth Revolution

3.1 Ecliptic

The Earth moves around the Sun in an elliptical orbit in a plane called the ecliptic. The trajectory of the Earth is an ellipse which is a little different from the circle.

Minimum distance from the Earth to the Sun (perihelion) is 147.3 × 10^6 km, and the largest (aphelion) 152.1 × 10^6 km. The average distance of the Earth from the Sun is 149.5 × 10^6 km. The volume of the Earth's orbit around the Sun is 940 × 10^6 km. The Earth orbits the full path for 365 days, 5 h, 48 min and 46 s. The average speed of the Earth orbiting along the ellipse is 29.8 km/s. In perihelion the speed of the Earth's orbiting is 30.3 km/s, and in aphelion it is 29.2 km/s. The slope of the plane of the ecliptic to the celestial equator is 23°27′ (Fig. 32).

In the context of the ecliptic, there are three major lines as follows: apsidal, equinox and solstice lines.

Apsidal line passes through the focus of the ecliptic and connects perihelion and aphelion. Apsidal line length is 299 × 10^6 km.

Equinox line is located at the intersection of the ecliptic and the celestial (Solar) level equator. Equinox line is perpendicular to the Solstice line. At the intersections of the ecliptic and the celestial equator levels are equinoctial points, when the lengths of day and night are equal.

Equinox line connects the positions of the Earth on the ecliptic when the day length in the northern hemisphere is the biggest (22 June) and the shortest (22 December). The equinox (solstice) line is at an angle of 11° in relation to the apsidal line. 700 years ago solstice line was matching with the apsidal line. This mismatch between the apsidal and solstice lines was due to the gravitational influence of the Sun, planets and the Moon on the movement of the Earth.

The Earth is moving along the ecliptic in the counterclockwise direction.

Due to the movement of the Earth around the Sun there follow:

- changes in elevation of the Sun above the horizon at noon,
- changes in the position of the Sun during the year,
- unequal duration of the day and night in any place during the year,

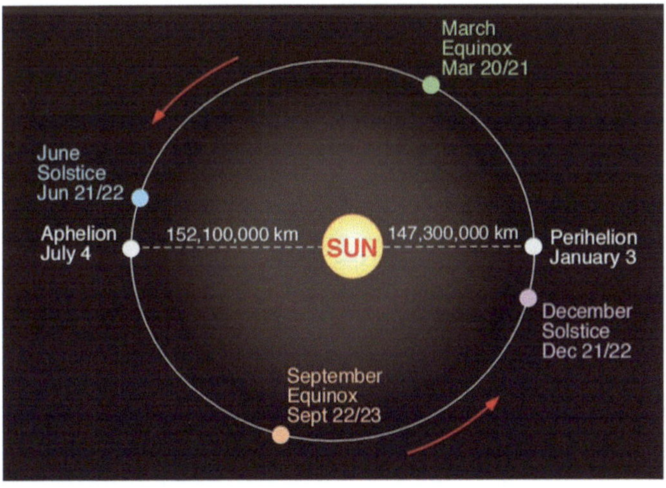

Fig. 32 The Earth movement around the Sun [39]

- unequal duration of the day and night in different places on the Earth on the same date,
- the change of the seasons,
- the existence of temperature zones on the Earth, etc. [11, 21–25, 30–36].

3.2 Zenith, Altitude and Azimuth Angle

Solar zenith angle is the angle between the zenith and the center of the Sun's disc.

The *altitude* (*elevation*) *angle* is the angle between horizon and the Sun's disc (Fig. 33).

The altitude angle (sometimes referred to as the "solar elevation angle") describes how high the Sun appears in the sky. The angle is measured between an imaginary line between the observer and the Sun and the horizontal plane the observer is standing on. The altitude angle is negative when the Sun drops below the horizon.

The *solar azimuth angle* is the angular distance between the south and the projection of the line of sight to the Sun on the ground. A positive solar azimuth angle indicates a position east of south, and a negative azimuth angle indicates west of south. Note that in this calculation, southern hemisphere observers will compute azimuth angles around $\pm 180°$ near noon (Fig. 34).

Altitude and azimuth are used to describe the location of an object in the sky as viewed from a particular location at a particular time [11, 12].

Fig. 33 Solar zenith and altitude (elevation) angles [40]

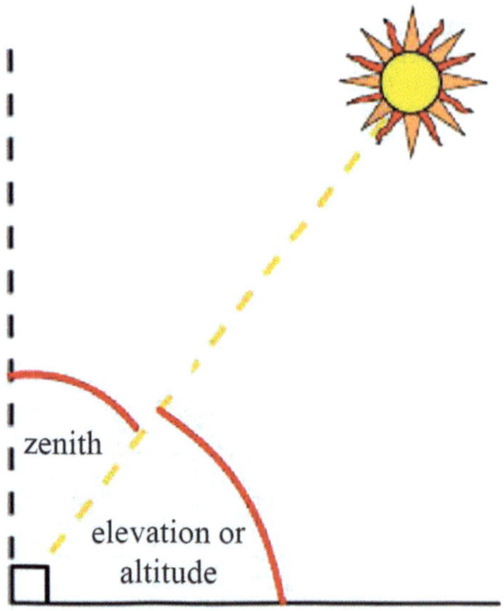

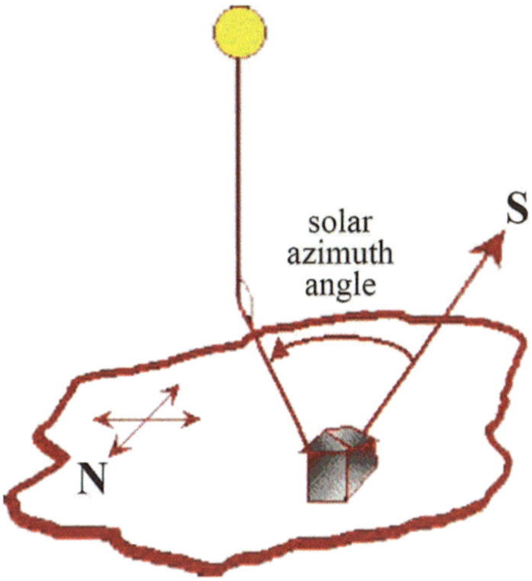

Fig. 34 Solar azimuth angle [41]

3.3 Precession and Nutation

Precession of the Earth means the gyro movement of the Earth around its axis of rotation for periods of 26,000 years. During precession motion of the Earth, the axes of its rotation permanently retains its inclination of 23°27′ to the plane of the ecliptic.

Precession of the Earth is caused by the gravitational effects of the Sun, Moon and planets on the Earth and a partial flattening of the Earth.

If Earth had a spherical shape and was completely a solid body of equal density throughout its volume, the direction and period of rotation of the Earth would remain unchanged over time. However, the shape of the Earth is similar to the spheroid shown in Fig. 35.

In Fig. 35, F stands for the force of the Moon attraction, and F_1 and F_2 mark forces acting on the protruding parts of the sphere. Since $F_1 > F_2$, this pair of forces tends to turn the axis of the Earth rotation, so that the equator plane is placed normal in relation to the Moon. Due to the simultaneous action of the attractive force of the Sun and the Moon, and the rotation of the Earth around its axis, this can lead to the gyroscopic effect, so that the axis of the Earth rotation $P_N P_S$ describes a cone around the axis of the ecliptic (Fig. 36).

In addition to the precession, the Earth's axis of rotation oscillates a little around its middle position, and this phenomenon is called nutation of the Earth's axis of rotation. This occurs because the precession forces of the Sun and the Moon constantly change its direction and intensity over time [11, 12].

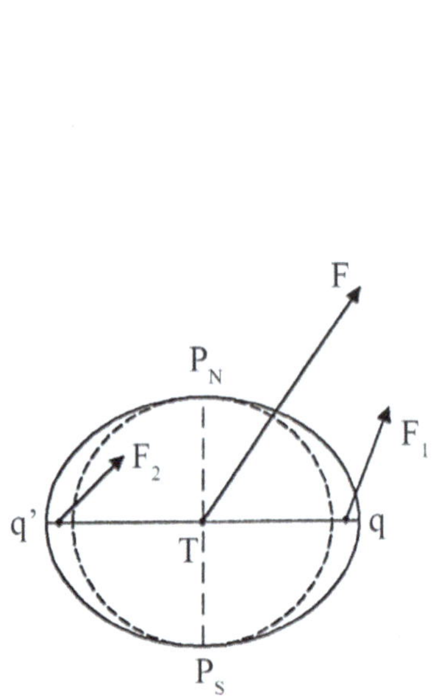

Fig. 35 The Earth shape [12]

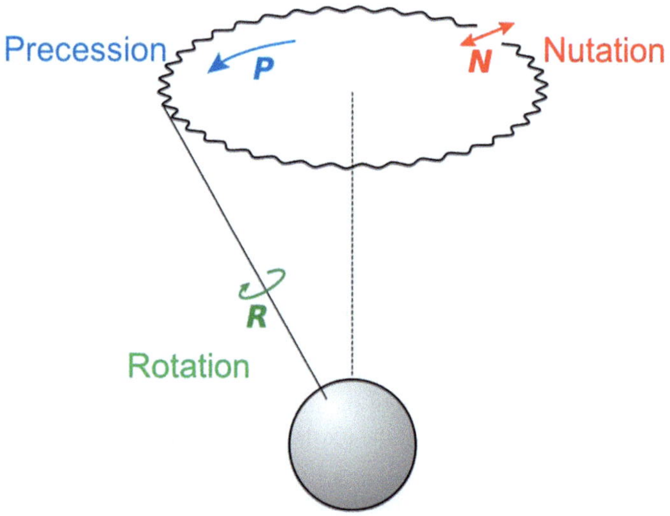

Fig. 36 The Earth precession and nutation [42]

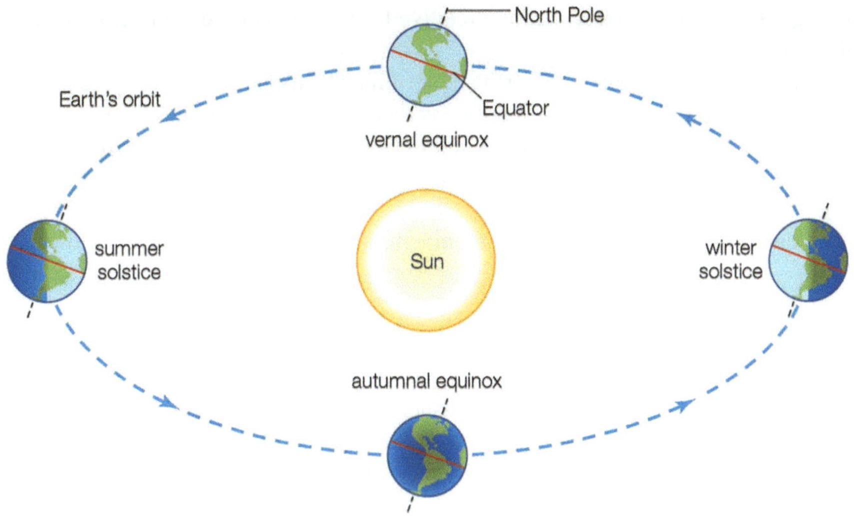

Fig. 37 Seasonal configuration of Earth and Sun [43]

3.4 Night and Day

Due to the inclination of the ecliptic plane towards the solar equator plane (23°27′) and the inclination of the Earth's axis to the plane of the ecliptic (66°33′), the Earth has five characteristic parallels, namely: the equator, two at 23°27′ northern and southern hemispheres and two at 66°33' northern and southern hemispheres (Fig. 37).

On 21.03 and 23.09 the country lies in the plane of the solar equator in one of the nodes of the equinox line, when the Sun‘s rays fall at midday, at a right angle to the equator. Then, the lengths of the days and nights on the northern and southern hemisphere are equivalent.

Summer Solstice occurs on 22.06 when the Sun‘s rays fall perpendicular to all the places in the northern hemisphere, with the latitude of 23°27′. Then, in the northern hemisphere the day is the longest and the shortest on the south hemisphere.

Winter Solstice occurs on 22.12 when the Sun‘s rays fall perpendicular to all the places on the southern latitude of 23°27′. That day, in the northern hemisphere the day is the shortest, and the night is the longest. At the same time, in the southern hemisphere the day is the longest and the night is the shortest.

The Sun is, in the summer and winter solstice, at the opposite ends of the apsidal line [11, 12, 35].

3.5 Temperature Zones of the Earth

The Earth rotates around its axis like the right screw that goes from the southern to the northern pole in a counterclockwise direction. The Earth rotation axis passes through its terminals and is inclined relative to the plane of the ecliptic at an angle of 66°33′.

The position of any place on Earth is determined by its latitude and longitude coordinates expressed in degrees. Longitude is determined in relation to the Greenwich meridian as the eastern and western ranging from 0° to 180°. Latitude is determined in relation to the equator, as north and south ranging from 0° to 90° (Figs. 38, and 39).

Generally speaking, there are five temperature zones on the Earth as follows:

1. One tropical zone	which extends from 0° to 23°27′ north and south latitude, where the average temperature during the year is higher than 20 °C
2. Two temperate zones	which extend from 23°27′ to 66°33′ north and south latitude, where the average temperature during the year is ranging from 10 to 20 °C
3. Two polar zones	which extend from 66°33′ to 90° north and south latitude, where the average temperature during the year is lower than 10 °C

Bearing in mind the temperatures and precipitation, 11 climate zones can be extracted on the Earth, namely: one equatorial, two tropical and two subtropical, two moderate, two sub-polar and two polar. Sharp boundaries between the mentioned zones do not exist [12, 21–25, 30–36, 40–45].

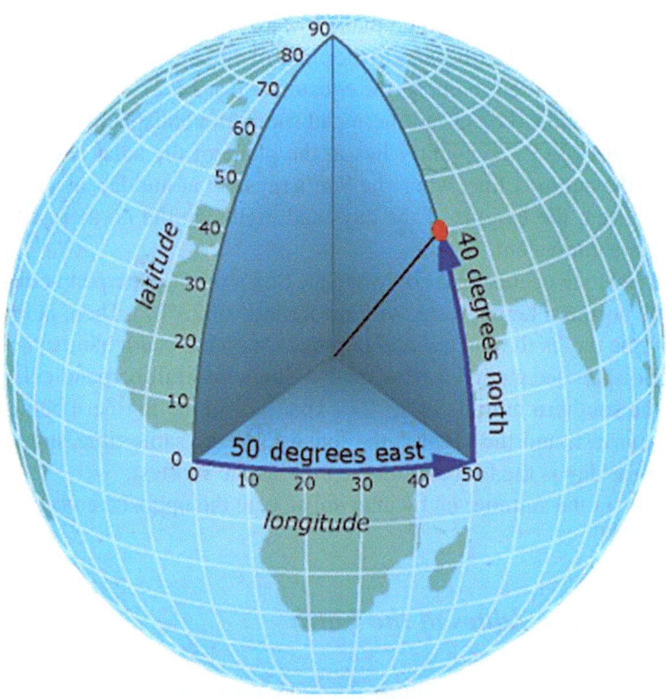

Fig. 38 Geographical latitude and longitude [44]

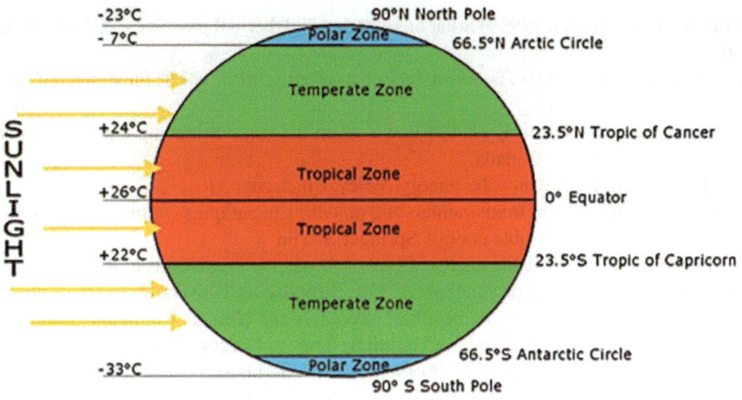

Fig. 39 Temperature zones of the Earth [45]

References

1. Babatunde BE (2012) Solar radiation. InTech
2. Unsold A, Baschek B (1983) The new cosmos. Springer, New York
3. Mayers DR (2013) Solar radiation: Practical Modeling for Renewable Energy Applications. CRC Press, Boca Raton
4. Foster R, Ghassemi M, Cota A (2010) Solar energy: renewable energy and the environment. CRC Press
5. https://en.wikipedia.org/
6. https://www.reddit.com/r/space/comments/3kf44p/a_sunspot_up_close/
7. http://astro.wsu.edu/worthey/astro/html/lec-sun.html
8. https://www.popularmechanics.com/space/deep-space/a11441/can-we-predict-solar-flares-andprotect-our-satellites-17341922/; https://www.nasa.gov/archive/content/goddard/soho-and-hinode-offer-insight-into-solar-eruptions/
9. Hsieh SJ (1986) Solar energy engineering. Prentice-Hall, Englewood Cliffs, New Jersey
10. McVeigh J (1980) Sun power. Pergamon Press, Oxford
11. Vukicevic-Karabin M, Atanackovic-Vukmanovic O (2004) General astrophysics. Zavod zaizdavanje udžbenika, Belgrade (in Serbian)
12. Vujicic B, Đurovic S (1995) Astrophysics with astronomy. Faculty of Sciences and Mathematics,Novi Sad (in Serbian)
13. https://hchem2017.blogspot.com/2017/05/nuclear-fusion-in-sun.html
14. https://www.pveducation.org/pvcdrom/properties-of-sunlight/atmospheric-effects
15. https://www.uwyo.edu/cpac/_files/docs/kasia_lectures/introduction.pdf
16. http://lib.convdocs.org/docs/index-232788.html
17. https://commons.wikimedia.org/wiki/File:Sun-path-polar-chart.svg
18. Pogosjan PC, Turketti LS (1975) Wolken, wind and weather. Verlag Mir, Moscow
19. http://synergyfiles.com/2015/10/what-is-the-optimal-angle-for-maximizing-solar-gain/
20. https://software.ecmwf.int/static/ERA40_Atlas/docs/section_B/parameter_ntotafosrpd.html#
21. Pavlović TM, Tripanagnostopoulos Y, Mirjanić LD, Milosavljevć DD (2015) Solar energy in Serbia, Greece and the Republic of Srpska, Academy of Sciences and Arts of the Republic of Srpska, Banja Luka (2015)
22. Furlan G (ed) (1981) Nonconventional energy. Plenum Press, New York
23. Harris N, Miller C, Thomas I (1985) Solar Energy Systems Design. John Willey, New York
24. Kalogirou AS (2014) Solar energy engineering - processes and systems, 2nd edn. Elsevier Academic Press

25. Duffie AJ, Beckman AW (1999) Solar engineering and thermal processes, 2nd edn. Wiley, New York
26. Scharmer K, Greif J (2000) The European solar radiation atlas, Vol. 1: Fundamentals and maps. L'Ecole des Mines, Paris
27. Chiras D, Aram R, Nelson K (2009) Power from the sun—achieving energy independence. New society Publishers, Canada
28. Garg HP (1982) Treatise on solar energy. Wiley, Chichester
29. Sen Z (2008) Solar energy fundamentals and modeling techniques—atmosphere, environment, climate change and renewable energy. Springer, Berlin
30. Solensen B (2004) Renewable energy. Elsevier Academic Press, Amsterdam
31. Cohen AR (2008) Physics of atmosphere. East Stroodsburg University
32. Barry GR, Chorley JR (1995) Atmosphere, weathers and climate. Routledge, London
33. Szokolay VS (1976) Solar energy and building. The Architectural Press, London
34. Gajic D (2005) Physics of the Sun. Prosveta, Nis (in Serbian)
35. Ducic V, Radovanovic M (2005) Climate in Serbia. Zavod za Udzbenike, Belgrade
36. Dukic D (1998) Climatology. Zavod za udzbenike, Belgrade
37. http://www.icrr.u-tokyo.ac.jp/YMAP/event/conf2018/slide/7-3_kiriki_icecube.pdf
38. http://www.oxfordpresents.com/ms/krohne/global-warming-and-ecological-communities/
39. http://earthfacts1.blogspot.com/2011/07/distance-from-earth-to-sun.html
40. https://ryanmccarthy.com.au/2010/01/04/sun-path-diagram-solar-chart/
41. https://susdesign.com/popups/sunangle/azimuth.php
42. http://freehostspace.firstcloudit.com/steveholmes/dwrntm/obliq/precnut/precnut2.htm
43. https://www.britannica.com/story/its-winter-its-summer
44. http://www.juntadeandalucia.es/educacion/descargasrecursos/plc/html/secundaria/locating_places.pdf
45. https://www.thegeographeronline.net/weather--climate.html

Photovoltaic Solar Energy Conversion

Tomislav Pavlovic, Plamen Ts. Tsankov, Nikola Dj. Cekić and Ivana S. Radonjić Mitić

Abstract In this chapter, general information about photovoltaic solar energy conversion, silicon and other solar cells, solar modules, solar batteries, charge controller, inverter, urban and rural application of solar cells, PV solar plants, solar module efficiency dependence on their orientation and tilt angle, solar modulessoiling, smart systems and mini-grids, economy of PV systems, and sustainability of the green economy is given.

1 General Information

1.1 Historical Overview

Photovoltaic solar radiation conversion is the process of converting solar radiation energy into the electrical energy.

The photovoltaic conversion of solar radiation takes place in solar cells made of semiconductor materials, which are of simple construction, have no mobile parts, are environmentally friendly, and have a long-life shelf.

T. Pavlovic (✉) · I. S. Radonjić Mitić
Faculty of Sciences and Mathematics, University of Niš, Niš, Serbia
e-mail: pavlovic@pmf.ni.ac.rs

I. S. Radonjić Mitić
e-mail: r_ivana_fizika@yahoo.com

P. Ts. Tsankov
Faculty of Electrical Engineering and Electronics,
Technical University of Gabrovo, Gabrovo, Bulgaria
e-mail: plamen@tugab.bg

N. Dj. Cekić
Faculty of Civil Engineering and Architecture, University of Niš, Niš, Serbia
e-mail: ncekic@yahoo.com

T. Pavlovic (ed.), *The Sun and Photovoltaic Technologies*,
Green Energy and Technology, https://doi.org/10.1007/978-3-030-22403-5_2

Development of solar cells originates in 1839 when Becquerel observed that the strength of current between two electrodes in the electrolyte increases upon exposure of electrodes to the light.

The same effect on the solid body (selenium) was first noticed by *W. G. Adams* and *R. E. Day* in 1877. Thanks to this, very soon a device for measuring the intensity of light was developed. Immediately afterward, the researchers turned to solving the problems of using solar cells as the commercial sources of electricity.

Rapid development of solar cell began in 1954 when *Pearson, Fuller*, and *Chapin* produced first solar cell made of monocrystalline silicon.

Starting with the launch of the first satellite in 1958, solar cells represent an irreplaceable source of electrical energy on satellites, space ships, and stations. On the earth, from the very beginning of their development solar cells have been applied in the isolated buildings, lighthouses, airports, research platforms at sea, residential, and industrial buildings, etc.

In respect of the primary material, solar cells can be divided into:

- Inorganic:
  - Silicon-based: single crystal silicon (c-Si), polycrystalline silicon (p-Si), thin-film-based amorphous silicon (a-Si), and micromorph (tandem combination of crystalline and amorphous silicon).
  - Non-silicon thin-film-based: cadmium telluride (CdTe), copper indium selenide (CIS), copper indium gallium selenide (CIGS).
- Organic: based on conductive polymers, conductive blends, or conductive polymer nanocomposites.
- Hybrid: dye-sensitized solar cells (DSSCs) which utilize organic liquid dyes in combination with titanium dioxide (TiO_2) immersed in an electrolyte solution (catalyst) and which utilize solid-state ABX_3 perovskite-type structure as the dye where AB is an organic–inorganic compound like methylammonium lead or tin while X is a halogen [1].

1.2 Principles of Solar Cell Operation

Solar cell is composed of *p* and *n* semiconductors in which due to the solar radiation absorption in the *p–n* junction, pairs of electron holes occur.

If the pair is formed far away from the *p–n* junction, it soon comes to its recombination. During the absorption of solar radiation in or near the *p–n* junction, the internal electric field separates the electrons and holes. Then, the electrons move toward the *n* side and the holes to the *p* side. As a result of such a movement of the electrons and holes, at the ends of the solar cell occurs a potential difference or voltage (Fig. 1).

Under the influence of contact potential on the *p–n* junction, the electrons move down the inclination of the conductive band and the holes up the inclination of the

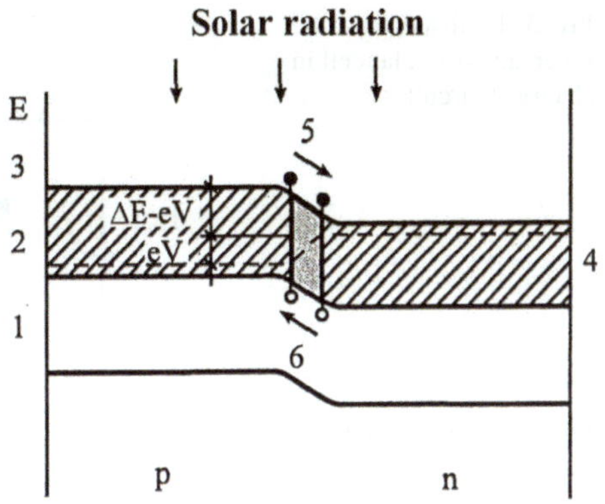

Fig. 1 Schematic representation of the processes that take place under the influence of solar radiation on the *p*–*n* junction: (1) valence zone, (2) restricted area, (3) conductive zone, (4) fermi level, (5) movement of electrons, (6) movement of holes

valence band, which leads to a reduction in the contact potential difference of the *p*–*n* junction. In this way, one achieves a new equilibrium state of the *p*–*n* junction with the potential difference of the open-circuit U_{oc}, which depends on the intensity of the incident solar radiation. If the solar cell is connected to an external consumer, a direct current will be generated in the electrical circuit.

1.3 Output Parameters of Solar Cells

Solar cell in electrical circuit is a source of direct current (Fig. 2).

Output parameters of solar cells are: short-circuit current I_{sc}, open-circuit voltage U_{oc}, nominal current I_{mpp}, nominal voltage U_{mpp}, maximum power P_{mpp}, fill factor F, and efficiency η.

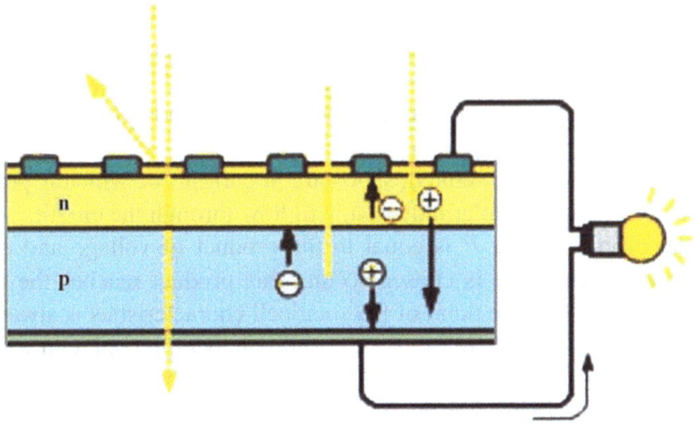

Fig. 2 Solar cell in electrical circuit

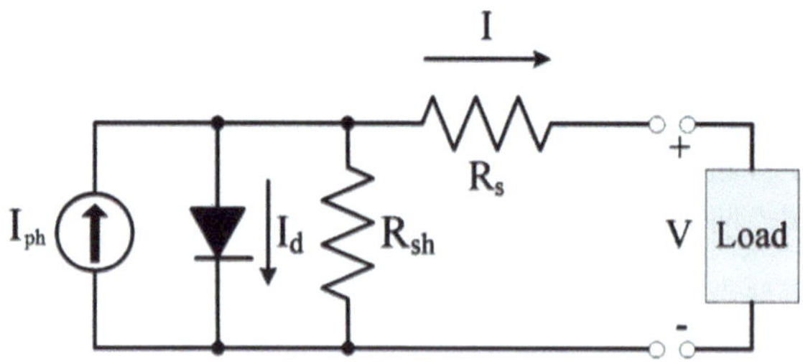

Fig. 3 Equivalent schematics of solar cell in electrical circuit

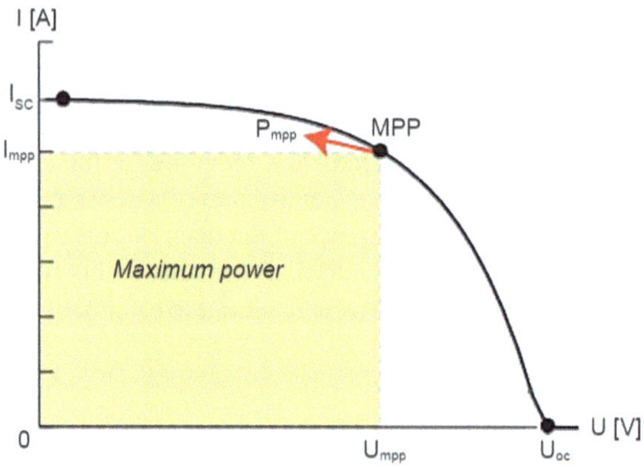

Fig. 4 Current–voltage characteristics of solar cell

Solar cell as a source of direct current possesses serial resistance R_s derived from its layer resistance and the resistance of the surface layers—electrodes. Occurrence of micro-defects within the solar cell causes the existence of its parallel resistance R_{sh} (Fig. 3).

Current–voltage characteristics of solar cell usually appear in the first quadrant, as shown in Fig. 4.

The important characteristics of solar cell are open-circuit voltage (U_{oc}) and short-circuit current (I_{sc}).

Open-circuit voltage is the maximum voltage at the ends of the solar cell in the open circuit.

If the solar cell is short-circuited, short-circuit current I_{sc}, which is proportional to the intensity of incident solar radiation, will flow through the circuit.

Since the electric power P is equal to the product of voltage and current, in practice working resistance is chosen so that this product reaches the maximum value. Product $U \cdot I$ at some point of the solar cell characteristics is always smaller than the product $U_{oc} \cdot I_{sc}$. Therefore, for the optimal operating point, in which useful power is maximum power, that is $P_{mpp} = U_{mpp} \cdot I_{mp} = F \cdot U_{oc} \cdot I_{sc}$ is smaller than one. This relationship represents the fill factor of solar cells:

$$P_{mpp} = U_{mpp} \cdot I_{mp} = F \cdot U_{oc} \cdot I_{sc} \tag{1}$$

$$F = \frac{U_{\text{mpp}} \cdot I_{\text{mpp}}}{U_{\text{oc}} \cdot I_{\text{sc}}}. \tag{2}$$

The fill factor is a measure of how close the given solar cell is to the ideal one, that is, how big is the influence of the serial resistance on the solar cell efficiency. In good solar cells, the fill factor ranges from 0.7 to 0.9.

Efficiency (rate of useful activity) of the solar cell is expressed by the relation of the used energy and the total energy of solar radiation energy incidence on the solar cell.

Efficiency of the solar cell can be calculated in the following way:

$$\eta = \frac{I_{\text{mpp}} \cdot U_{\text{mpp}}}{I_{\text{s}} \cdot S} = \frac{F \cdot I_{\text{sc}} \cdot U_{\text{oc}}}{I_{\text{s}} \cdot S}, \tag{3}$$

where U_{oc} and I_{sc} are voltage and current of the short circuit, I_{s} is the intensity of solar radiation, and S is the surface of the solar cell.

To increase the efficiency of the solar cell, it is necessary to increase U_{oc} and I_{sc} and that F is around one.

1.4 *Materials for Solar Cells*

To produce monocrystalline solar cells, most frequently used are silicon (Si), indium phosphide (InP), gallium arsenide (GaAs), cadmium sulfide (CdS), cadmium telluride (CdTe), aluminum antimonide (AlSb), gallium phosphide (GaP), cadmium selenide (CdSe), etc. One uses a *p–n* junction on the basis of one or more semiconductor materials from which heterogeneous junctions are formed such as a junction of Cu_2S/CdS. In addition to this, a combination of semiconductor *p*- or *n*-type with the metal is also used, the so-called Schottky barrier, for example Au/Si. The literature also contains information about the properties of several tens of different solar cells. However, today solar cells are mostly produced of monocrystalline, polycrystalline and amorphous silicon, gallium arsenide (GaAs), and copper sulfide/cadmium sulfide (Cu_2S/CdS).

1.5 *Coefficient of Absorption*

Dependence of the absorption coefficient of Si and GaAs for solar cells on the energy of incident radiation is given in Fig. 5.

Figure 5 shows that the absorption coefficients of materials from which one makes solar cells decrease with increasing wavelength of the incident radiation.

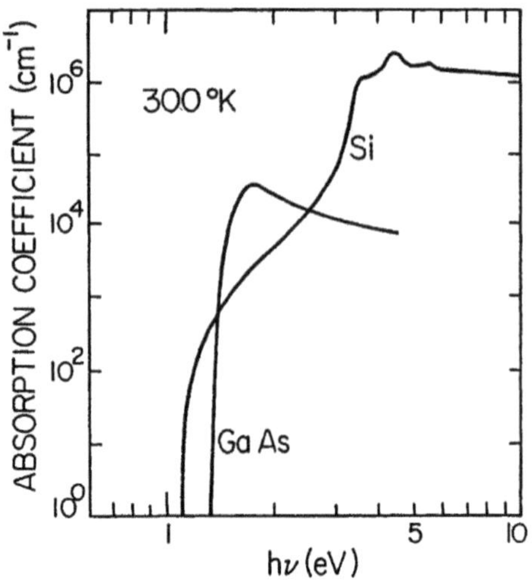

Fig. 5 Dependence of the absorption coefficient of Si and GaAs for solar cells on the energy of incident radiation [30]. *Courtesy* Elsevier

During the absorption of solar radiation, an electron–hole pair is formed in the solar cell, whose energy depends on the energy of the incident photons. The absorption of incident radiation occurs if the condition is met:

$$h\nu \geq E_g \tag{4}$$

or

$$\lambda \leq \frac{hc}{E_g} \tag{5}$$

where h is Planck's constant, ν is frequency of the incident photons, E_g is width of the forbidden zone of semiconductors, and c is the speed of light.

Photons with wavelengths $\lambda > hc/E_g$ are not absorbed in the solar cell. During absorption of photon with energy $h > E_g$, the excess energy $h - E_g$ is surrendered to a semiconductor, thereby increasing its internal energy. Based on this, it could be concluded that in order to produce a solar cell more suitable is a semiconductor with the smallest width of the forbidden zone, because it can absorb a wider range of wavelengths of solar radiation. However, it has been experimentally established that with increasing width of the forbidden zone of semiconductors there is a decrease of the inverse saturation current and increase of the open-circuit voltage of the solar cell. Based on the above, it ensures that the efficiency of a solar cell is a complex function of the width of the forbidden zone, and that for solar cell production semiconductor materials with small width of the forbidden zone are not suitable.

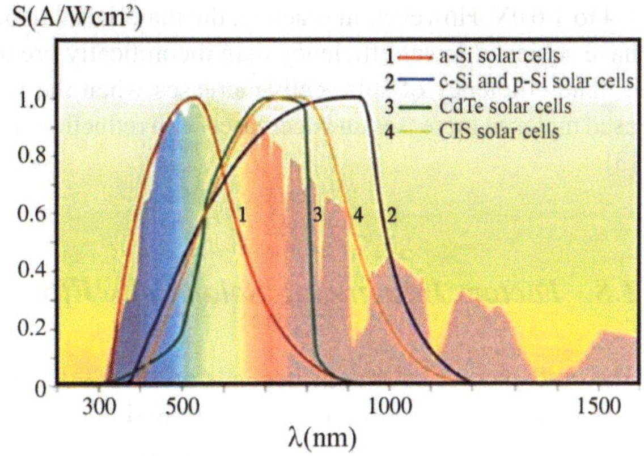

Fig. 6 Spectral sensitivities of different types of solar cells [31]

1.6 Solar Cells'Spectral Sensitivity

Solar cells are not equally sensitive to all wavelengths of the solar radiation spectrum. The spectral sensitivity of solar cells depends on the nature of semiconductors, present ingredients, technologies of cells forming, etc.

Photons with short wavelengths are absorbed at the entrance of the solar cell, away from the *p–n* junction, and do not significantly affect the solar cells' current. Photons with large wavelengths (infrared region) pass through the *p–n* junction, are absorbed at the bottom of the solar cell, or are reflected from the rear electrode, and do not contribute much to solar cells' current increase. Only photons with wavelengths in the area of maximum sensitivity of solar cells contribute significantly to solar cells' current [2].

Spectral sensitivities of different types of solar cells are shown in Fig. 6.

1.7 Efficiency of Solar Cells

Efficiency of solar cells initially increases with the increase of E_g, it achieves maximum, and with further increase of E_g, it decreases. Maximum efficiency of 20% is achieved for $E_g = 1.5$ eV, that is, when the photons have a wavelength $\lambda = 0.83$ μm [3].

Today, a larger number of semiconductors which can be used for solar cell production are known. In practice, many of them are not used due to the complicated production technology, high prices, or low efficiency.

In case of monocrystalline silicon, forbidden zone width is 1.1 eV, which corresponds to the efficiency of about 20%. Silicon theoretically is not the most suitable material for the solar cell production. Theoretically, it would be best to use for solar cell production semiconductor materials with the forbidden zone width ranging from

1.4 to 1.6 eV. However, in practice, the materials that have this forbidden zone width have achieved lower efficiency than theoretically predicted efficiency.

The efficiency of solar cells increases when the width of the forbidden zone of used material increases and decreases with reduction in the intensity of solar radiation [3].

1.8 Factors Influencing Solar Cells'Efficiency

The efficiency of solar cells depends on several factors including: reflection on the surface of the solar cell, losses in infrared and ultraviolet area, losses due to the thickness of the solar cell, losses due to voltage factor, losses due to recombination, and losses in the serial resistance.

Reflection on the surface of the solar cell. Optical reflection on the solar cell depends on its surface micro-roughness. With increased micro-roughness of the solar cell's front surface, reflection from it reduces. In order to reduce the reflection, appropriate anti-reflection coatings are applied to the solar cell. In solar cells with anti-reflection coatings, optical reflection can be reduced to 3%.

Losses in the infrared area. In the solar spectrum, photons with wavelengths $\lambda > hc/E_g$, where E_g is the energy gap of the semiconductor material the solar cell is made of, do not generate photocurrent but lead to a rise in solar cell temperature. In monocrystalline and polycrystalline silicon solar cells, in this way efficiency losses amount to about 23%.

Losses in the ultraviolet area. In monocrystalline silicon (Si) solar cell, photons with energies above 1.1 eV generate photocurrent and surrender the excess energy to a monocrystal which is then heated. In this way, solar cell efficiency loss is around 33%.

Losses due to the thickness of the solar cell. In solar cells, sensitive part is not thick enough to absorb all the incident photons. Namely, one part of the flux passes through the solar cell and is absorbed on the rear electrode. Losses due to the thickness of the solar cell can be reduced below 1% by using reflective rear electrode which returns to the solar cell photons that reach it.

Losses due to voltage factor. During the solar radiation absorption, electrons do not receive the full amount of energy absorbed in the material of solar cell. Consequently, the voltage at the ends of the solar cell is less than expected. In this way, about 17% of solar cell efficiency is lost.

Losses due to the filling factor. Product $I_n \cdot U_n$ in the I–U curve because of its shape can never be equal to the area under the curve. In the best case, fill factor can reach a value of 0.9. Because of this, about 5% of solar cell efficiency is lost.

Losses due to recombination. The generated electrons and holes in the solar cell during the absorption of solar radiation have a certain life span after which they recombine, leading to a 4% loss in its effectiveness.

Losses in the series resistance. On the series resistance of the solar cell as a diode, about 1% of its efficiency is lost.

Some of these loss factors of the solar cell efficiency are determined by fundamental physical laws so that they cannot be reduced. Efficiency losses which depend on the technology of solar cell formation can be reduced. If the losses which depend on the applied technology were reduced to the minimum, maximum theoretical efficiency of the monocrystalline solar cells would amount to 22%.

Finally, it should be noted that the efficiency of solar cells essentially depends on the band gap in the semiconductor material which they are made of.

2 Silicon Solar Cells

2.1 Silicon

Silicon ($_{14}Si^{28,086}$) is next to oxygen most abundant element in the earth's crust (27.6%). Most of it is found in the form of oxide SiO_2 which occurs as quartz, sapphire, chalcedony, agate, opal, etc. Natural silicon is composed of three stable isotopes $_{14}Si^{28}$ (92.28%), $_{14}Si^{29}$ (4.67%), and $_{14}Si^{30}$ (3.05%).

Silicon belongs to the group IV of the periodic system of elements and has an electronic configuration 3s 3p. Silicon can be easily obtained and processed, and it is not toxic and does not form compounds which are harmful to the environment. Silicon builds SiO and SiO_2 with oxygen, which fall within the dielectric materials. In terms of structure, silicon may be amorphous, polycrystalline, and monocrystalline.

2.2 Polycrystalline Silicon

In the modern electronics industry, silicon is the main semiconductor element. The electronic components of silicon are stable at temperatures up to 200 °C.

For industrial purposes, silicon of metallurgical (98%) and semiconductor (99.999999%) purity is used.

2.2.1 Metallurgical Silicon

Metallurgical silicon is industrially prepared by the reduction of SiO_2 with carbon at the temperature of 1500–1750 °C:

$$SiO_2 + 2Cl \rightarrow Si + 2CO \tag{6}$$

Composition of permitted impurities in metallurgical silicon is given in Table 1.

Table 1 Permitted impurities in metallurgical silicon

Element	Weight proportion (%)	Element	Weight proportion (%)
Fe	0.35–1.5	P	0.02
Al	0.2–0.5	S	0.01
Ca	0.03–0.07	B	0.005
Mn	0.2	As	0.005
Cr	0.1	C	0.05

Metallurgical silicon has a polycrystalline structure. In order to obtain one kilogram of metallurgical silicon, three kilograms of SiO_2 and 14 kWh of electricity are required.

2.2.2 Semiconductor Purity Silicon

Purification of metallurgical silicon to silicon semiconductor purity (one atom of impurity on 10^6 atoms of silicon) is performed at 300 °C according to the Siemens procedure:

$$Si + 3HC1 \leftrightarrow SiHCl_3 + H_2 \tag{7}$$

Boiling point of trichlorosilane is 31 °C.

Semiconductor silicon of high purity in the form of granule or rod (ingot) diameter of 20–200 mm and a length of 0.5–1 m is obtained by multiple fractional distillation and thermal decomposition of $SiHCl_3$ at 1200 °C.

Permitted impurities in semiconductor silicon are shown in Table 2.

2.3 *Monocrystalline Silicon*

Semiconductor silicon is polycrystalline. In order to be converted into monocrystalline state, it is necessary to melt it at 1400 °C and to use one of the known methods

Table 2 Permitted impurities in semiconductor silicon

Impurities	ppm
Elements of group III	0.3
Elements of group V	1.5
Heavy metals	0.1
Carbon	300
Oxygen	50
Other	0.001

to convert it into the monocrystalline state. Industrially, monocrystalline silicon is obtained by *Czochralski pulling method* and float zone method.

2.3.1 Czochralski Pulling Method

To obtain monocrystalline silicon by Czochralski pulling method, the installation shown in Fig. 7 is used.

The installation for the preparation of single crystal silicon by Czochralski pulling method comprises a chamber with the container with molten semiconductor purity silicon and single crystal silicon seed holder which can rotate and move along the vertical axis of the chamber. Single crystal silicon by Czochralski pulling method is obtained by immersing monocrystalline silicon seed into the melt of silicon and lightly drawing it from the melt while rotating it. Owing to a somewhat lower temperature at the contact point of the seed and melt and to the effect of surface charge forces, on the seed from the melt single crystal silicon of the same orientation as

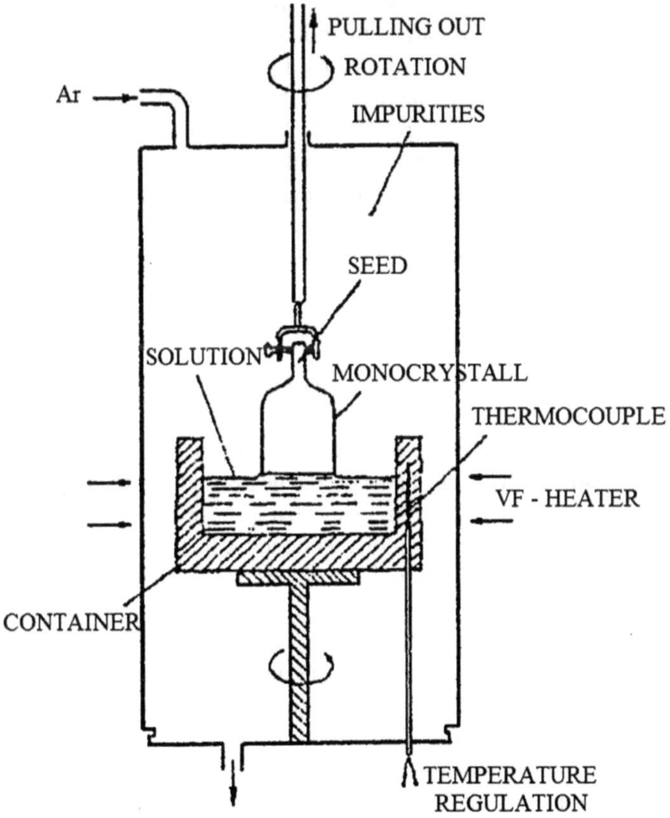

Fig. 7 Installation for obtaining monocrystalline silicon by Czochralski pulling method

the seed is formed. Doping of single crystal silicon is carried out by inserting the appropriate impurities into the melt. The rate of single crystal silicon pulling is 1 mm to 2 cm per hour, and the diameter is several centimeters.

The oxygen concentration in single crystal silicon obtained by Czochralski pulling method is of the 10^{-6} order, and specific resistance is ~80 Ω cm. Crystallographic structure is very good, and the number of dislocations is less than 100 per cm^2. Ingots or parts of the crystal obtained by Czochralski pulling method are often used as seeds for the preparation of the single crystal silicon by float zone method.

2.3.2 Float Zone Method

To obtain monocrystalline silicon by float zone method, the installation shown in Fig. 8 is used.

The installation consists of a chamber in which there is a holder which can rotate and move along the vertical axis of the chamber. Polycrystalline silicon rod is attached to the lower part of the holder, and it heats and melts at its upper end. Single crystal silicon seed is fixed to the underside of the upper part of the holder which rotates and immerses into the melted part of the polycrystalline rod. Pure silica is binding to the seed while impurities, due to the zonal purification, move toward the lower parts of the polycrystalline rod. The rotation rate of the seed is several revolutions per minute, and the speed of crystal growth is 1–2 mm/min. In the float zone method, a melt is not in contact with the walls of the container which allows for growth of very pure silicon crystal without dislocations, length up to 1 m and diameter 12–15 cm. Monocrystalline silicon ingots are cut to disks of certain thickness and are used in the semiconductor industry for the production of integrated circuits, microprocessors, solar cells, etc.

2.3.3 Monocrystalline Silicon Properties

Monocrystalline silicon atoms are interconnected by covalent bonds to the surface-centered cubic lattice. Distances between the nearest atoms of the monocrystalline silicon lattice are 0.543 and 0.235 nm, respectively. Monocrystalline silicon is black, opaque, very glossy, tough, and poorly conductive for electricity. With adequate impurities, a monocrystalline silicon becomes a good conductor of electricity (Fig. 9).

Basic physical and chemical properties of monocrystalline silicon and SiO_2 are given in Table 3.

Experiments have found that monocrystalline silicon forbidden zone width varies with temperature. Dependence of the monocrystalline silicon forbidden zone on temperature is given in Fig. 10.

Figure 10 shows that monocrystalline silicon forbidden zone width decreases with increasing temperature. It is also experimentally established that different elements are differently soluble in monocrystalline silicon.

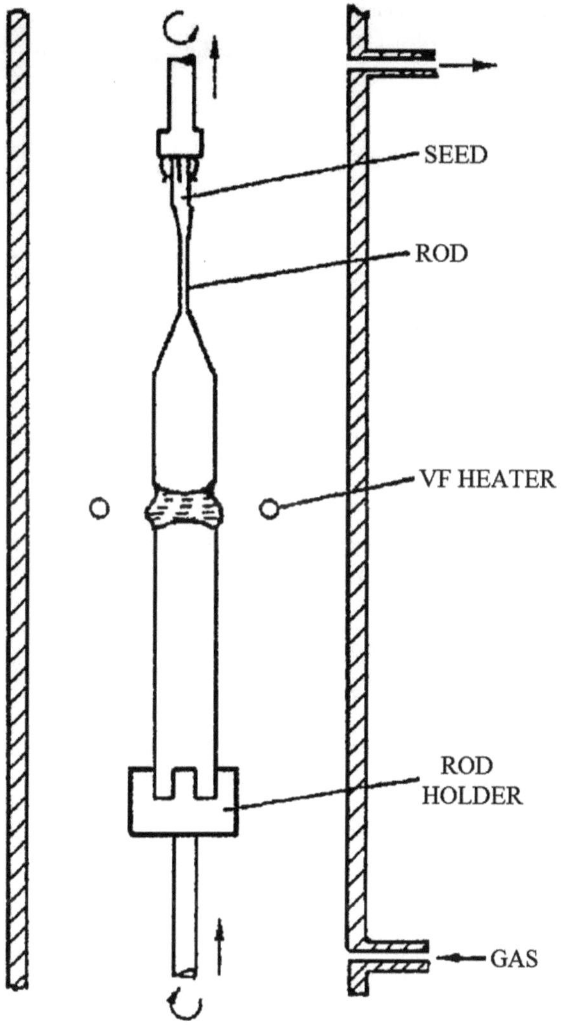

Fig. 8 Installation for obtaining monocrystalline silicon by float zone method

Energy levels of impurity atoms in silicon are located in silicon forbidden zone and represent recombination centers for free electrons.

In the crystalline state, silicon is very inert, is insoluble in acids and highly soluble in bases, and is a mixture of $HN0_3 + NF$. To eliminate macro-defects on monocrystalline silicon solutions, KOH and NaOH are used. Chemical polishing of silicon is carried out by nitric acid solution.

Fig. 9 Monocrystalline silicon cubic lattice [32]

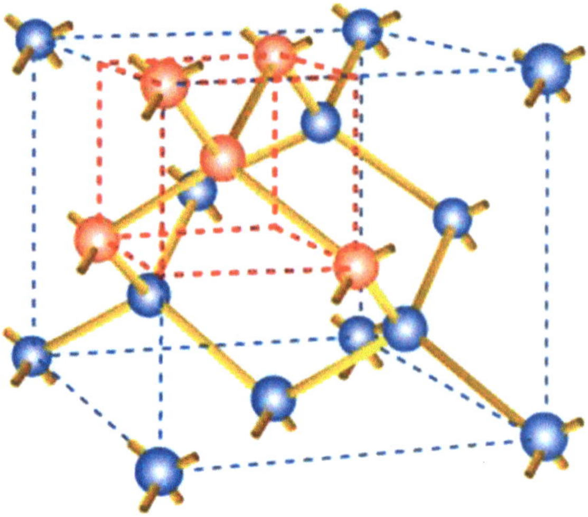

Table 3 Basic physical and chemical properties of monocrystalline silicon and SiO_2 at $T = 27$ °C

Property	Si	SiO_2
Group	14	
Relative atomic or molecul. mass	28.08	60.08
Atomic or molecular density (cm^{-3})	5.0×10^{22}	2.3×10^{22}
Crystal structure	Diamant	Tetrahedron bond 50% covalent 50% ion
No of atoms in a single cell	8	
Lattice constant (nm)	0.543	
Density (g cm^{-3})	2.33	2.27
Forbidden zone (eV)	1.11	8
Effective state density (cm^{-3}) Conductive zone Valent zone	2.8×10^{19} 1.04×10^{19}	
Carrier concentration (cm^{-3})	1.45×10^{10}	
Mobility (cm^{-3} V^{-1} s^{-1}) Electrons Holes	1350 480	Insulator $p = 10^{16}$ Ωcm at $T = 300$ K
Dielectric constant	11.7	3.9
Melting point (°C)	1415	1700
Specific heat capacity (Jg^{-1} K T^{-1})	0.7	1.0
Heat conductivity (W cm^{-1} K^{-1})	1.5	0.014
Linear coef. of expansion heat (K T^{-1})	2.5×10^{-6}	0.5×10^{-6}

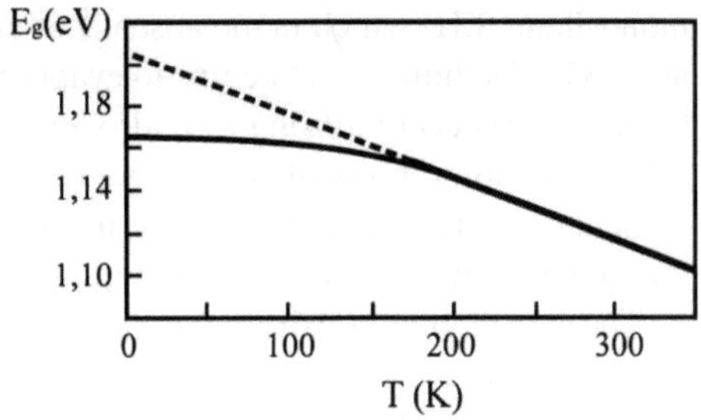

Fig. 10 Dependence of monocrystalline silicon forbidden zone on temperature [33]

2.4 Amorphous Silicon

Due to high cost and time-consuming process of monocrystalline silicon manufacturing, as well as the great losses when monocrystalline Si ingot cutting into the plates for solar cells, attempts were made to directly obtain monocrystalline silicon plates or monocrystalline silicon strips. The success has not been achieved in obtaining thin monocrystalline or polycrystalline plates, but in obtaining the amorphous silicon (a-Si) from the gas phase.

One of the reasons to explore the possibility of using amorphous silicon rather than a crystalline for solar cells lies in the fact that the thickness of amorphous siliconsolar cells is 300 times less than the thickness of the monocrystalline siliconsolar cells.

The main difference between the amorphous and the crystalline silicon is reflected in the arrangement of their structure. In crystalline silicon structure, arrangement is correct and periodical, and in amorphous one it is irregular and statistical. In pure state, amorphous silicone is not interesting for the semiconductor industry because it does not have good photoconductive features. For this very reason over years, amorphous silicon has not had a wider practical application. Sudden changes in the application of amorphous silicon occurred in 1975 when *Spear* and *Le Comber* published the results of their research related to obtaining a-Si by thermal decomposition of silane in a glow discharge with a substrate temperature of 300 °C.

From the point of application, amorphous layers are required to have as much as possible non-arranged structure and the energy zone as close to the energy zones of the crystalline materials. State density in a forbidden zone of undoped amorphous materials should be as small as possible in order to reach the effect of doping on the change of their semiconductor properties.

2.4.1 Amorphous Silicon Growing

Amorphous silicon can be grown by thermal decomposition of silane (SiH_4) in glow discharge by reactive sputtering, chemical layering from the vapor phase, photolytic methods, etc.

Silicon does not react directly with hydrogen. By special procedure, from silicon and hydrogen, silane (SiH) can be obtained. Among silanes, the most important is monosilane SiH_4 which in the absence of air is stable and colorless gas. Monosilane at −185 °C solidifies and begins to evaporate at −111.9 °C. It is stable up to 400 °C when it starts decomposing to hydrogen and silicon. By thermal decomposition of SiH_4, pure amorphous silicon is obtained.

Schematic representation of the installation for obtaining amorphous silicon by thermal decomposition of silane in glow discharge is shown in Fig. 11.

The installation consists of a vacuum chamber, electrodes, and electronics for the thermal decomposition of silane in glow discharge and the auxiliary containers for containing and supplying diborane (B_2H_6) and phosphine (PH_4) for amorphous silicon doping.

Experiments have shown that substrate heat is not sufficient to clear down the bond between the silicon and the hydrogen in silane. Silane decomposition occurs in collisions of electrons with energies of a few eVs with silane molecules. Thereby in silane plasma, molecule fragments are formed which in contact with the heated substrate form layers of amorphous silicon.

An important intermediate step in silane decomposition is formation of silane (SiH_2). Increment of a-Si layers is proportional to the extent of silane formation. Using higher silanes (Si_2H_6), degree of formation of SiH_2 and the yield of a-Si layers are greatly increased.

Quality of a-Si layers depends on the pressure and the flow rate of gas, plasma properties, substrate temperature, etc.

In addition to non-uniformness of the amorphous silicon structure, there are unsaturated (hanging) bonds that lead to an increase in the allowed energy levels in the forbidden zone and make amorphous silicon doping difficult. For the purpose of amorphous silicon doping, it should be first alloyed with hydrogen (a-Si: H) or fluorine (a-Si: F). Alloying with hydrogen is carried out by silane discharging, wherein

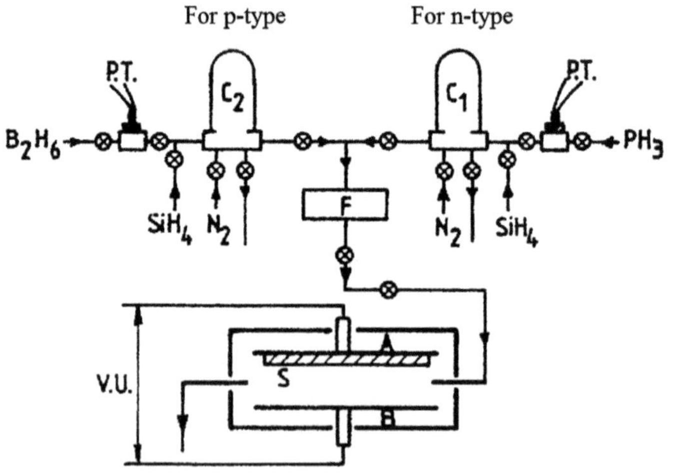

Fig. 11 Schematic representation of the installation for obtaining amorphous silicon by thermal decomposition of silane in glow discharge [34]

a-Si: H contains 5–10% H. Hydrogen atoms complete unsaturated bonds and other micro-defects in amorphous silicon structure. Alloyed amorphous silicon, there is no difficulty, can dope phosphorus or boron from the gas phase. Alloyed amorphous silicon can be doped by phosphorus or boron from gas phase. First, amorphous silicon alloying and doping were carried out in 1975.

The content of hydrogen in a-Si layers depends on the substrate temperature. Hydrogen significantly improves the electrical properties of amorphous silicon. This improvement only goes up to a certain concentration of hydrogen in silicon (up to 6 at.% H). When further increasing the concentration of hydrogen, it is binding in the form of clusters or as a thin layer on the surface of amorphous silicon, causing the weakening of the electric characteristics of the amorphous silicon. Chemically, hydrogen represents an impurity in silicon, while electrically it contributes to improving the electric characteristics of the amorphous silicon. The bonds between the silicon and the hydrogen are ion and infrared active. During heating of a-Si: H, there occurs dissolution of Si: H bonds and the reconstruction of Si: Si bonds.

Development of technology has led to a-Si alloying with F and H, to give a-Si: F: H structure which has proved suitable for the formation of a-Si solar cells.

Doping of a-Si: H is carried out in the gas phase which contains a well-controlled concentration of impurities. For amorphous silicon doping, phosphine (PH_3) and diborane (B_2H_6) are used. By introducing phosphine, a-Si layers become *n*-conductive, and introducing diborane *p*-conductive.

Amorphous silicon obtained by various methods shows different properties. Amorphous silicon obtained by thermal decomposition of silane in glow discharge has better transport properties than the amorphous silicon obtained by sputtering or vacuum vapor deposition.

2.4.2 Properties

Each atom of amorphous silicon is linked to four neighboring silicon atoms by covalent bonds similar to those of the crystalline silicon. Due to the structural disorder, amorphous silicon shows deviation from the mutual spacing of the silicon atoms and the angles between them, when compared to crystalline silicon.

To test the structural characteristics of a-Si layers, scanning electron microscopy, X-ray diffraction method, method of Raman scattering, and others are used. Structural model of monocrystalline, polycrystalline, and amorphous silicon is shown in Fig. 12.

Schematic representation of unsaturated bonds in amorphous silicon is shown in Fig. 13.

Amorphous silicon has a fine fibrous structure with almost regular half sphere endings on the surface. Alloy of a-Si: H has a larger width of the forbidden zone $E_g = (1.55–1.87)$ and higher solar radiation absorption coefficient than crystalline silicon.

Amorphous siliconabsorbs UV radiation better than crystalline silicon. In the visible part of solar radiation spectrum, amorphous silicon absorption is ten times greater

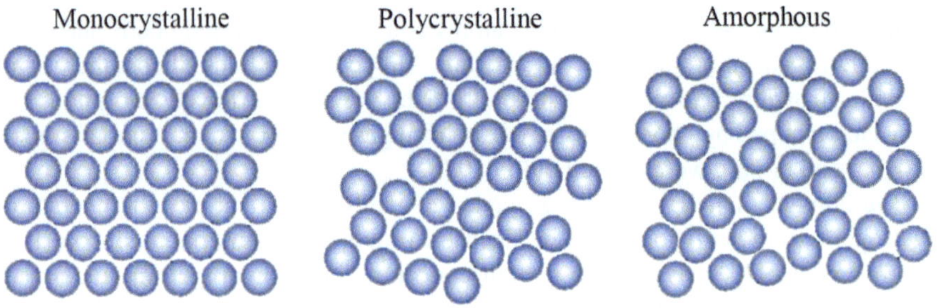

Fig. 12 Schematic of allotropic forms of silicon: monocrystalline, polycrystalline, and amorphous silicon [35]

Fig. 13 Schematic representation of unsaturated bonds in amorphous silicon [32]

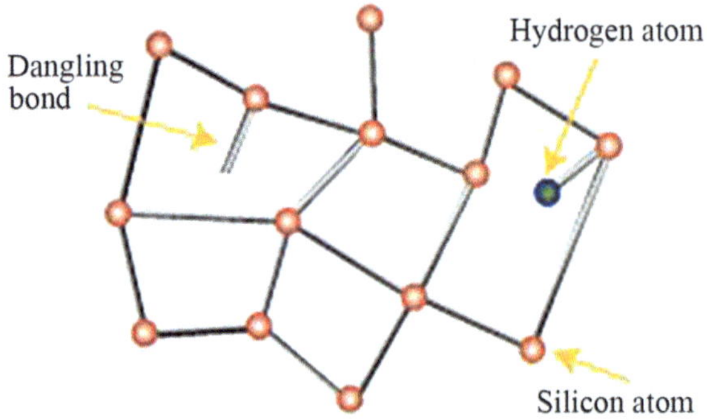

than the absorption of crystalline silicon. Amorphous silicon thickness of 0.2 μm completely absorbs green color containing maximum of solar radiation spectrum.

In the near-infrared solar radiation spectrum (0.75–2.5 μm), optical absorption of doped a-Si is ten times higher than the optical absorption of the undoped a-Si: H. This is due to higher density of states in doped a-Si with respect to a-Si: H. Photoconductivity of the illuminated a-Si is 10^5 times higher than the photoconductivity of non-illuminated a-Si.

2.5 Polycrystalline Silicon Solar Cells

Polycrystalline siliconsolar cells are made from polycrystalline silicon of semiconductor purity in the form of a strip, which can be obtained in several different ways: *method of strip with deformed edge growth, method of dendrite networking, method of horizontal, vertical, and oblique strip pulling, silicon growth on ceramic, method of rotating mold*, etc.

Method of strip pulling with defined edge is based on the effect of melt surface tension which raises the molten silicon in the tool of the spinner bar (Fig. 14).

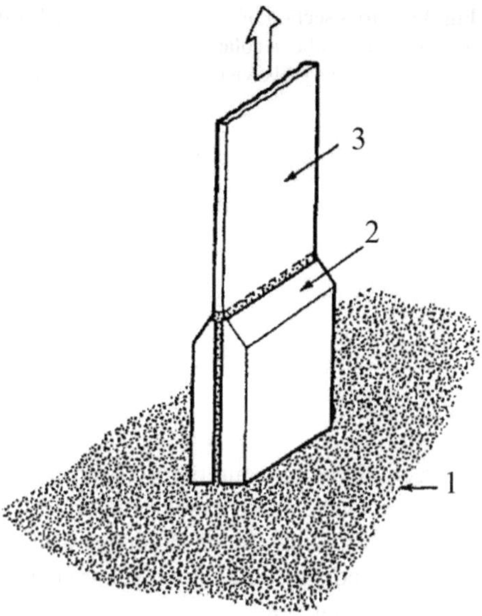

Fig. 14 Method of polycrystalline strip with deformed edge pulling from molt: (1) molten silicon, (2) graphite holder, (3) polycrystalline silicon strip [4]. *Courtesy* M. A. Green

When entering monocrystalline silicon seed into the melt, on contact surface of the seed and melt a meniscus is formed with clearly pronounced upper and lower edges. The strip of polycrystalline silicon is formed by moving the crystal seed upward at a constant rate of 10 cm/min and cooling the strips' thickness of 0.02 cm and width of 10 cm. The resulting strip is cut into the appropriate dimensions and is used to make polycrystalline Si solar cells.

Preparation of polycrystalline strip for solar cell production is performed by chemical etching of its surface, wherein on the surface of the strip a pyramidal structure is formed with the height of the pyramid of 10 μm.

Thanks to a pyramidal structure of the polycrystalline silicon strip surface, incident light is multiply reflected and absorbed on it. Using the appropriate anti-reflection coating, efficiency losses of the polycrystalline Si solar cell due to the reflection from its surface have been reduced to a minimum.

Cross section of the solar cell on the basis of polycrystalline silicon is shown in Fig. 15.

Figure 15 shows that solar cell consists of the polycrystalline *p*–*n* junction, front and rear electrode.

The grain size of polycrystalline silicon ranges from several microns to several millimeters. In polycrystalline solar cells, very significant are boundary areas between polycrystalline material grains. The boundary between two grains acts as a series resistance which opposes the movement of electrons. At the boundary between the grains, there is an electric field similar to the field that exists on the boundary of the metal–semiconductor junction. Grain boundaries can be seen as defects in the crystalline silicon with the energy levels in the forbidden zone. These levels represent

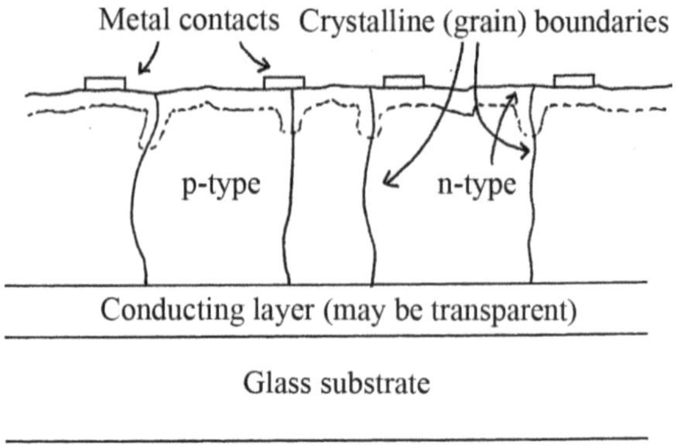

Fig. 15 Cross section of polycrystalline silicon solar cell [30]. *Courtesy* Elsevier

the recombination centers for electrons that are removed from the atoms under the influence of the incident solar radiation.

Polycrystalline silicon doping, formation of compound, deposition of electrical contacts, and anti-reflection layer are performed similarly as in the case of the single crystal Si solar cells.

Minority charge carriers (holes) which are formed near the *p–n* junction recombine on it, and those formed near the grain boundaries recombine at the border grain boundaries. Minority charge carriers do not contribute to the solar cell current. In order to reduce current losses, grain boundary length from the front to the back side of the solar cell has to be greater than the length of the diffusion barrier of the minority charge carriers. Due to diffusion of impurities in the grain boundaries in the course of *p–n* junction formation, the grain boundaries represent alternative routes for the movement of electric charges through the *p–n* junction. In order to achieve significant reduction in power losses that occur due to charging recombination at grain boundaries, the grains should be a few millimeters long. Regardless of this, the thickness of the commercial polycrystalline Si solar cells is less than 1 mm (Fig. 16).

Polycrystalline Si solar cells are produced in various shapes and sizes. Commercial polycrystalline Si solar cell has the efficiency of about 15% and laboratory ones about 22.3%.

2.6 *Monocrystalline Silicon Solar Cells*

A method of fabricating monocrystalline silicon solar cell is as follows.

2.6.1 Preparation of Wafers

From single crystal silicon, the wafers' thickness of 200–300 μm is cut (Fig. 17).

Fig. 16 Photovoltaic polycrystalline Si module

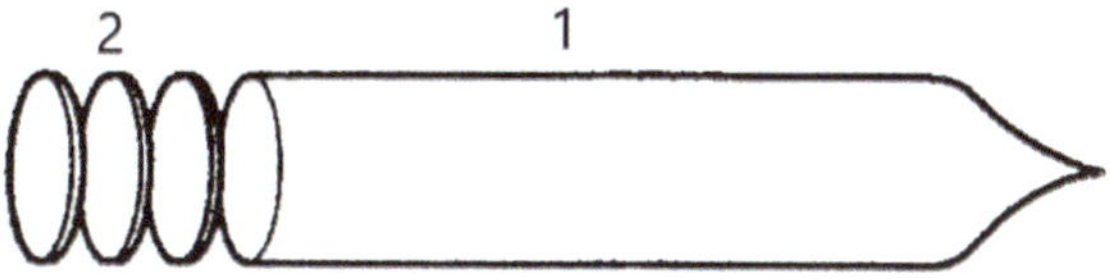

Fig. 17 Cutting of monocrystalline silicon ingots into wafers: (1) ingot and (2) cutted wafers [4]. *Courtesy* M. A. Green

After cutting, wafers are polished and cleaned in a diluted solution of hydrochloric and nitric acid.

2.6.2 Doping

For the preparation of *n*-type semiconductor silicon, doping is carried out with phosphorus, and for obtaining *n*-type semiconductor doping is carried out with boron. For silicon doping and formation of *p*–*n* junction, the following methods are applied: *gas-phase diffusion*, *solid-state diffusion*, *epitaxial growth of doping layer*, *ion implantation*, etc. For silicon doping with phosphorus, applying gas-phase diffusion installation whose schematic representation is given in Fig. 18 is used.

During the formation of monocrystalline silicon, boron is added to the silicon melt, so that the inner part of the Si wafer is *p*-type semiconductor. Silicon wafers are placed in a quartz tube at the temperature of 800–900 °C. Under the influence of

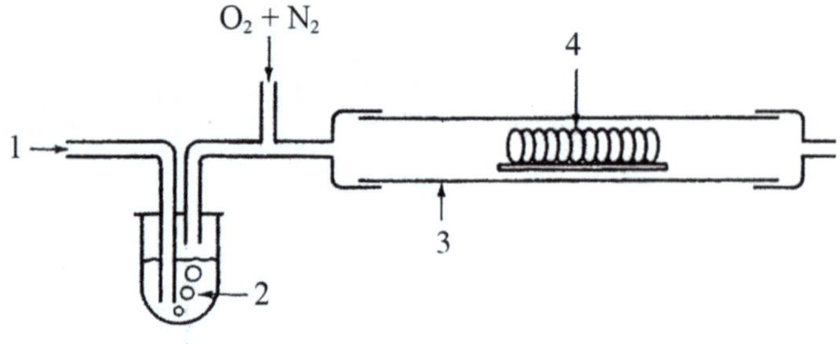

Fig. 18 Installation for silicon doping with phosphorus applying gas-phase diffusion: (1) gas, (2) liquid $POCl_3$, (3) quartz tube, (4) Si wafers [4]. *Courtesy* M. A. Green

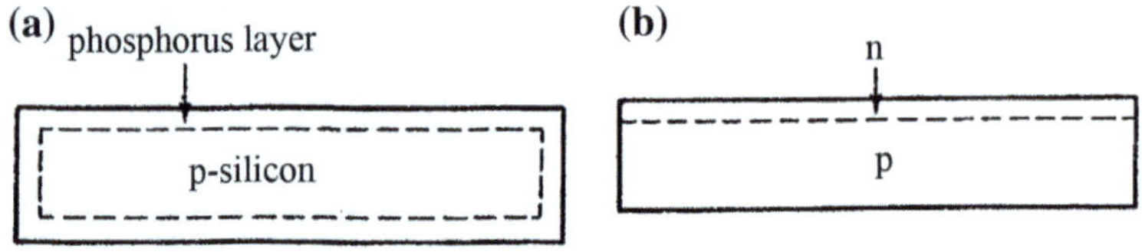

Fig. 19 Phosphorus distribution: **a** immediately after diffusion and **b** after removing *n*-layer from the back and the sides of the Si wafers [4]. *Courtesy* M. A. Green

gas which is fed to a solution of $POCl_3$, it starts evaporating and phosphorus transits into the quartz tube in which the phosphorus, by diffusion, is installed into the surface parts of the Si wafers. Twenty minutes later, the concentration of phosphorus in the Si wafers' surface areas is much greater than the concentration of boron, so that on the Si wafers' surface *n*-type semiconductor is formed. Removing of *n*-layer from the back and the sides of the Si wafers is performed by chemical and mechanical means (Fig. 19).

2.6.3 Rear Side Preparation

Since the rear side of the Si wafer is away from the *p–n* junction and thereby away from the influence of the internal electric field, there is a problem of collecting the charge on it. This is solved by doping the rear side of the Si wafer stronger than the front side. If the base material is *p*-type, rear side will be richer doped (p^+), so as to obtain the n^+pp^+ cell type, which is known as BSF cell, or cells with the field along the rear side.

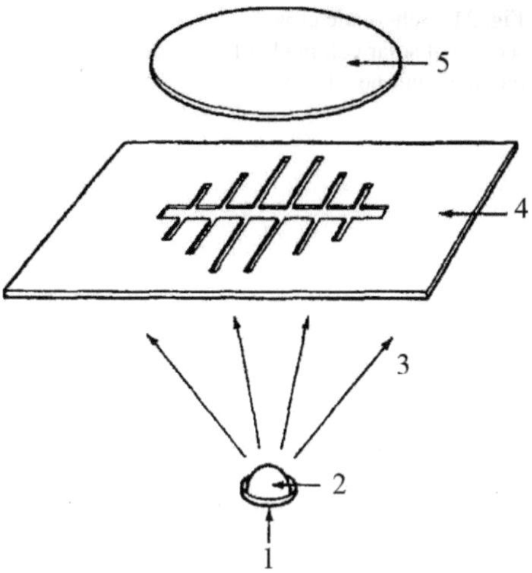

Fig. 20 Vacuum vaporing of the upper metal contact: (1) vapor source, (2) vaporing material, (3) vapored material, (4) mask, (5) Si wafer [4]. *Courtesy* M. A. Green

2.6.4 Metallization Contacts

Metal contacts are formed by vacuum vaporing of the respective metals on the Si wafer, as shown in Fig. 20.

2.6.5 Anti-reflection Layer

For this purpose, commonly used are Ti/Pd/Ag coating. On the silicon layer, first Ti layer having a good adhesion to the silicon is deposited, then Pd layer is applied on it, and finally Ag layer is applied above it. In order to improve the contact between the wafer and Ti/Pd/Ag, and for a metal contact to have the smallest possible contact resistance, Si wafer with metal contact is for a certain period of time subjected to the temperature of 500–600 °C.

Anti-reflection coating is used to reduce reflectance and charge surface recombination velocity. Due to a high refractive index of silicon (3–6), reflection of solar radiation from the Si solar cell ranges from 30 to 60%. For an anti-reflection layer, materials with a refractive index of 1.5–2 can be used. Such materials include: SiO, SiO_2, TiO, TiO_2, and Ta_2O_3. Depending on the material from which an anti-reflection layer is made, monocrystalline silicon solar cells in different colors can be produced [4].

A schematic cross section of monocrystalline silicon solar cell is given in Fig. 21.

2.6.6 Characteristics

Monocrystalline silicon solar cell is sensitive in the wavelength ranging from 0.4 to 1.1 μm and that its maximum sensitivity ranges in the wavelengths between 0.8 and

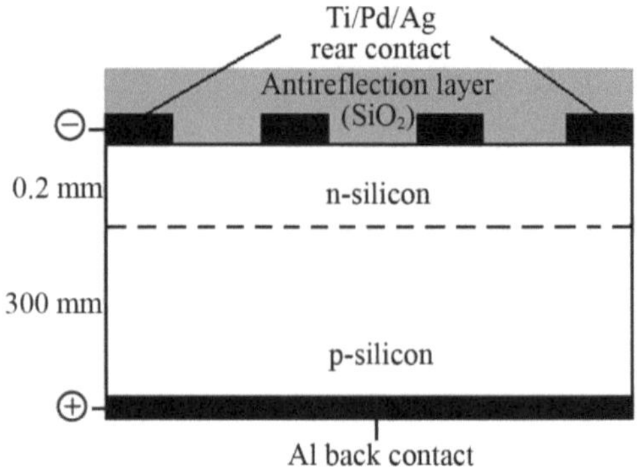

Fig. 21 Schematic cross section of solar cell made of monocrystalline silicon

0.9 μm. Maximum spectral sensitivity of the monocrystalline silicon solar cell does not coincide with the maximum of the spectral distribution of solar radiation. Based on this, it can be concluded that the monocrystalline silicon is not an ideal material for solar cell manufacturing [2] (Fig. 22).

In 2017, the laboratory record of monocrystalline silicon solar cell efficiency was 26.7% and commercially about 17%.

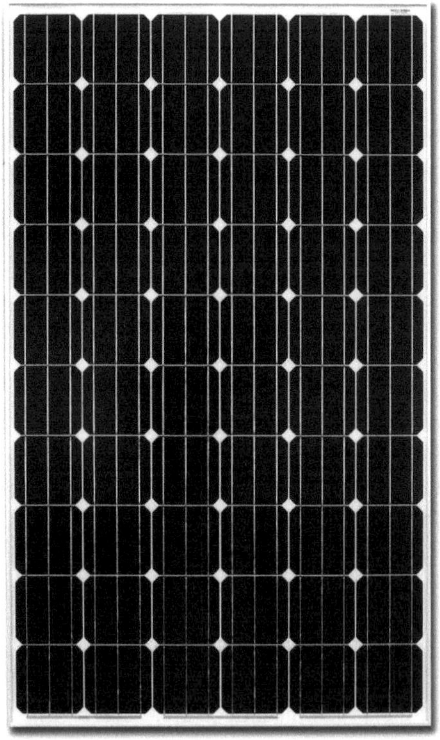

Fig. 22 Photovoltaic monocrystalline silicon module

2.7 *Amorphous Silicon Solar Cells on Glass Substrate*

The first amorphous silicon solar cell was formed in 1974 by *D. E. Carlson* in RCA Laboratory in the USA. The first commercial a-Si solar cell appeared in 1980 and had an efficiency of 3%.

2.7.1 Non-transparent

Amorphous siliconsolar cells are formed by a capacitive radio frequency discharge of silane (SiH_4) with diborane (B_2H_6) as a source of *p*-donor and a phosphine (PH_4) as a source of *n*-donor.

The formation of a-Si solar cell on glass occurs in the following way:

- Glass is thoroughly cleaned.
- On the preheated glass (200–300 °C) by vacuum vaporing a transparent SnO_2 electrode or an electrode which is a mixture of In_2O_3 and SnO_2, thickness of ~0.4 μm is applied.
- Using laser, a transparent electrode is cut into narrow parallel strips.
- Over the cut electrode by capacitive radio frequency discharge of silane, diborane, and phosphine, a-Si layers are applied, *n* and *p*, where *i* labels undoped a-Si layer,
- Using laser, *n*- and *p*-layers are cut into narrow parallel strips; thickness of the *n*-layer is 0.02 μm, *i*-layer 0.5 μm, and the *p*-layer about 0.08 μm.
- Over the cut *n*- and *p*-layers by vacuum vaporing the Al electrode, thickness of ~0.4 μm is applied.
- Al electrode is cut into narrow parallel strips using laser.
- On SnO_2 and Al electrodes, back contacts are placed.
- Protective transparent plastic film is applied over solar cells [5].

Schematic representation of a-Si solar cell on glass formation is given in Fig. 23.

Schematic representation of the cross section of a-Si solar cell on glass is given in Fig. 24.

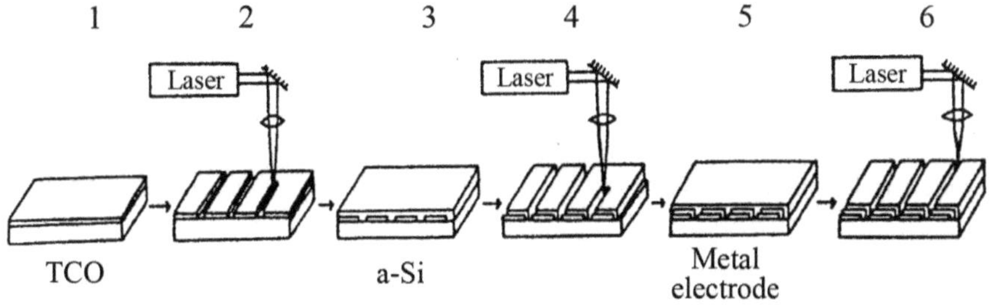

Fig. 23 Schematic representation of a-Si solar cell on glass formation: (1) vacuum vaporing of transparent SnO_2 electrode (TCO), (2) laser cutting of SnO_2 electrode, (3) deposition of *n*- and *p*-layers, (4) laser cutting of *n*- and *p*-layers, (5) vacuum vaporing of Al electrode, (6) laser cutting of Al electrode

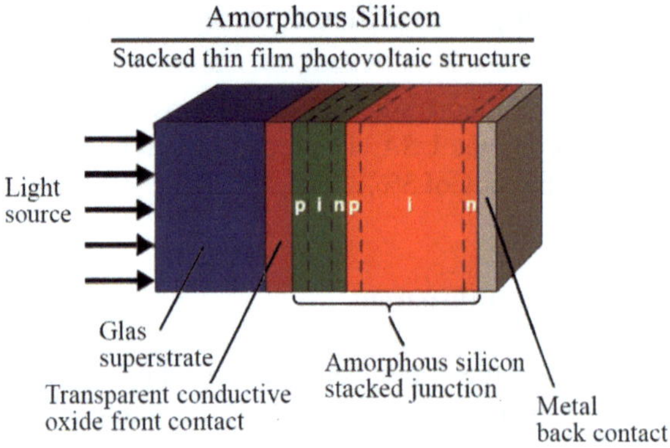

Fig. 24 Schematic representation of cross section of a-Si solar cell on glass

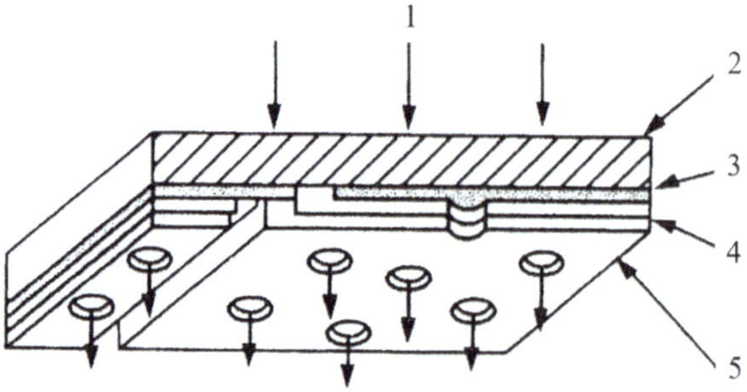

Fig. 25 Schematic representation of semitransparent a-Si solar cells on glass: (1) solar radiation, (2) glass, (3) transparent electrode, (4) a-Si layers, (5) metal electrode

2.7.2 Semitransparent

The Japanese company Sanyo manufactures semitransparent a-Si solar cells on glass that absorb 30% of the incident solar radiation and generate electricity (Fig. 25).

2.7.3 Spectral Sensitivity

Modern development of solar cells takes place in the direction of semitransparent solar cell improvement in order to facilitate their incorporation into the existing and new architectural settings.

Amorphous silicon solar cell on glass has a maximum sensitivity in the wavelength ranging from 400 to 600 nm, with the greatest intensity of solar radiation. With increasing wavelengths above 600 nm spectral sensitivity of the a-Si solar cell on glass drops abruptly [6].

Solar module dimensions of $91.5 \times 30.5\ \text{cm}^2$ with serial-connected a-Si solar cells on glass at $t = 25\ °\text{C}$, with radiation of $1000\ \text{W/m}^2$, have the following characteristics: short-circuit current 1.08 A, open-circuit voltage $U_{\text{oc}} = 20$ V, nominal current $I_{\text{n}} = 0.88$ A, nominal voltage $U_{\text{n}} = 13.5$ V, and nominal power $P_{\text{n}} = 12$ W.

2.8 *Amorphous Silicon Solar Cells on Plastic Substrate*

The first a-Si solar cell on plastic foil was formed in 1987. A schematic representation of the installation for the production of a-Si solar cells on a plastic polyethylene film is shown in Fig. 26.

The installation consists of a vacuum chamber in which there are two rollers with polyethylene foil thickness of 100 μm and a width of 250 mm and a system for depositing *n*- and *p*-amorphous silicon layers from a gaseous phase on movable foil [7].

2.8.1 Amorphous Silicon Solar Cells on Plastic Films

Amorphous siliconsolar cells on plastic films consist of plastic film substrate with metal electrode on it above which there are amorphous silicon layers, transparent electrode, metal electrode, and upper protective layer (Fig. 27).

Amorphous siliconsolar cells on plastic film are flexible and light and can be circuit bent at least in diameter of 5 cm (Fig. 28).

Advantages of amorphous silicon for the production of solar cells as compared to the monocrystalline silicon are the following:

- Amorphous silicon has higher coefficient of solar radiation absorption than monocrystalline silicon.
- To manufacture siliconamorphous solar cells, remarkably smaller quantity of material is needed as compared to the monocrystalline siliconsolar cells.
- Amorphous siliconsolar cells can be formed on glass substrate, plastic films, or metal substrate on significantly larger surfaces than monocrystalline siliconsolar cells surfaces.
- Semitransparent amorphous siliconsolar cells on glass substrate are being increasingly used in practice.

Disadvantages of siliconamorphous solar cells as compared to the monocrystalline and polycrystalline siliconcellssolar are the following:

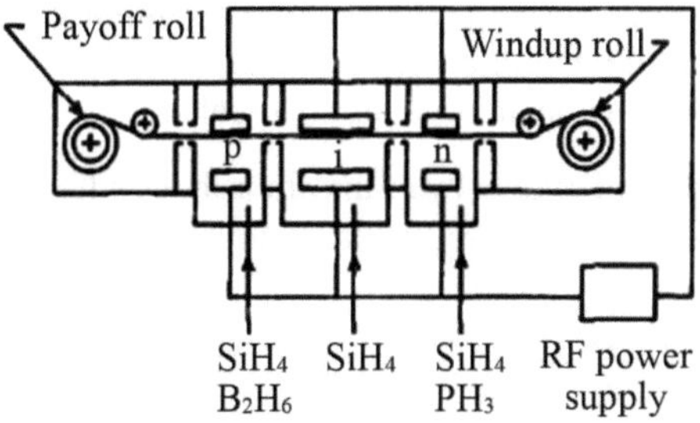

Fig. 26 A schematic representation of the installation for the production of a-Si solar cells on a plastic polyethylene film [7]. *Courtesy* Elsevier

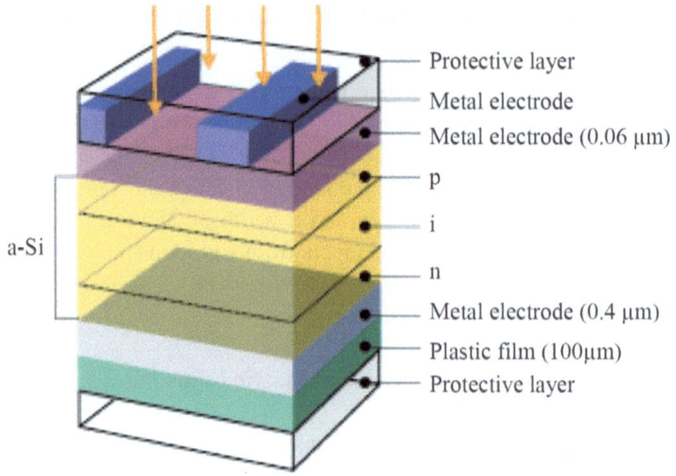

Fig. 27 Cross section of amorphous silicon solar cell on plastic film

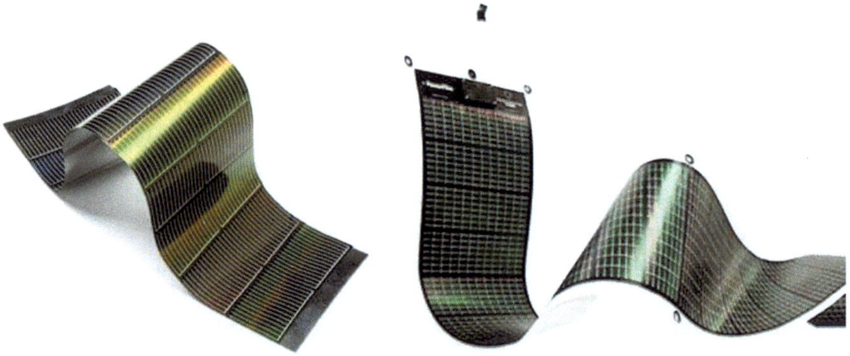

Fig. 28 Amorphous silicon solar cell on plastic foil

- Amorphous siliconsolar cells have smaller efficiency (5–7%) as compared to the efficiency of polycrystalline (14%) and monocrystalline Si solar cells (15%).
- Longer period of illumination causes some degradation of the optic and electric characteristics of siliconthe amorphous solar cells.

Since the upper limit of efficiency of the amorphous siliconsolar cells is 16%, the researchers have a wide scope of research ahead to increase their efficiency.

3 Other Solar Cells

3.1 GaAs Solar Cells

Thanks to the width of the forbidden zone of 1.45 eV, the absorption coefficient of ~10^5 cm^{-1}, and the melting point of 1238 °C, GaAs is an ideal material for the formation of the solar cell.

GaAs solar cells are made of monocrystalline and polycrystalline GaAs. Monocrystalline GaAs is made from polycrystalline GaAs applying *Bridgeman* or *Czochralski* method. The crystal structure of GaAs is similar to the crystal structure of silicon (Fig. 29).

Today, the commercial GaAs solar cells are formed in two ways: *by doping GaAs and heteroepitaxy deposition of AlAs or* $Al_xGa_{1-x}As$ *from liquid or gas phase on the monocrystalline GaAs.* Efficiency of doped GaAs cells derived by heteroepitaxy deposition of AlAs or $Al_xGa_{1-x}As$ is about 28%.

Polycrystalline GaAs cells are formed of polycrystalline GaAs thickness of 2 μm.

Since the GaAs solar cells are thermostable, they are often used in photovoltaic systems with solar radiation concentrators. The efficiency of GaAs solar cells with concentrators is about 30–35%.

GaAs cells are about one hundred times more expensive than monocrystalline Si solar cells. Since GaAs solar cells have already reached maximum efficiency, further investigations are aimed at reducing their prices.

3.2 CdTe Solar Cells

Cross section of a cadmium telluride solar cell is shown in Fig. 30. A layer of cadmium sulfide is deposited from solution onto a glass sheet coated with a transparent

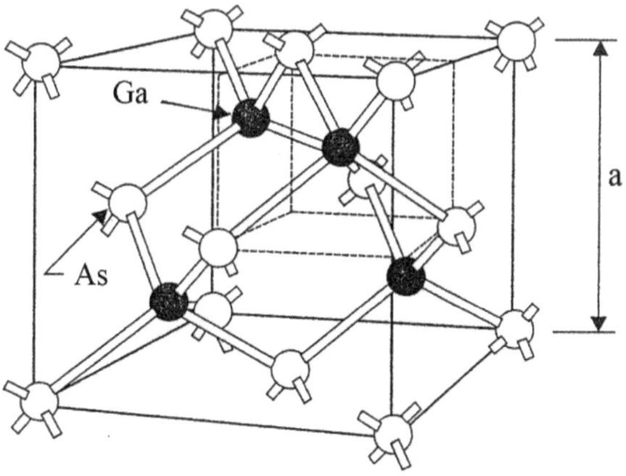

Fig. 29 Crystal structure of GaAs

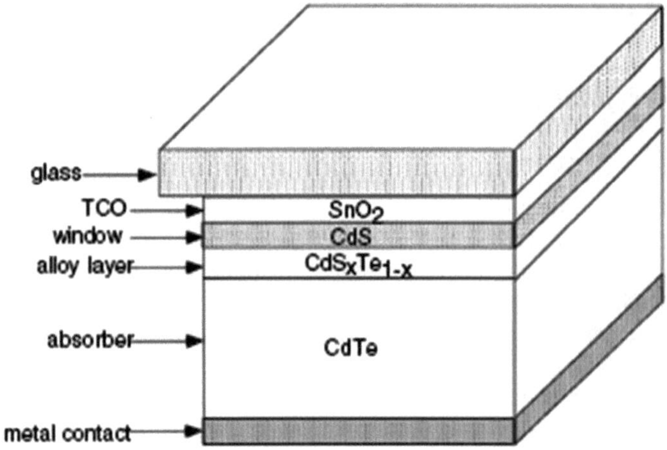

Fig. 30 Schematic cross section of cadmium telluride solar cell

conducting layer of thin oxide. This is followed by the deposition of the main cadmium telluride cell by variety of techniques including close-spaced sublimation, vapor transport, chemical spraying, or electroplating.

CdTe solar cells have been used as low cost, high efficiency, thin-film photovoltaic applications since 1970. With the forbidden zone width of ~1.5 eV and the coefficient of absorptionabsorption ~10^5 cm^{-1}, which means that a layer thickness of a few micrometers is sufficient to absorb ~90% of the incident photons, CdTe is almost an ideal material for solar cell manufacturing. CdTe solar cell is sensitive in the wavelength of 0.3–0.95 μm, and maximum of its sensitivity is in the wavelength range of 0.7–0.8 μm. Laboratory CdTe cells have the efficiency of 16% and commercial ones around 8%. Great toxicity of tellure and its limited natural reserves diminish the prospective development and application of these cells.

Laboratory maximum CdTe solar cells efficiency is 21% and commercially module efficiency is 16%.

3.3 CIS Solar Cells

The materials based on $CuInSe_2$ that are of interest for photovoltaic applications include several elements from groups I, III, and VI in the periodic table. CIS is an abbreviation for general chalcopyrite films of copper indium selenide ($CuInSe_2$).

CIS technology is a star performer in the laboratory with 19.5% efficiency demonstrated for small cells but has proved difficult to commercialize. Unlike other thin-film technologies, which are deposited onto a glass substrate, CIS technology generally involves deposition onto a glass substrate as shown in Fig. 31. An additional glass top cover is then laminated to the cell/substrate combination. Present designs require

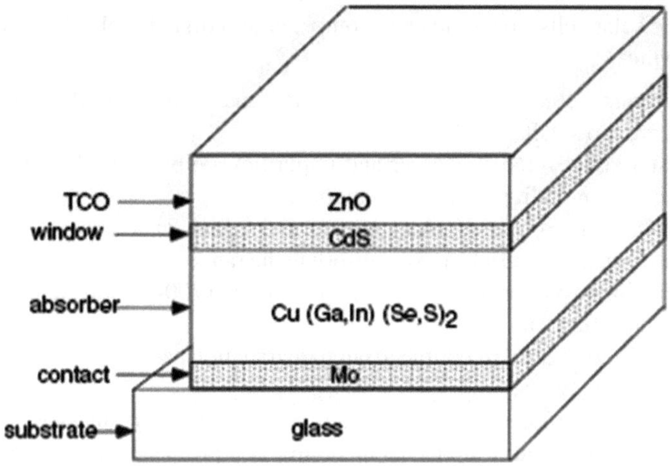

Fig. 31 Schematic cross section of solar cell made of copper indium diselenide (CIS)

a thin layer of CdS deposited from solution. Considerable effort is being directed to replacing this layer due to the issues associated with the use of cadmium, as previously noted. However, a long-term issue with CIS technology is the lack of materials for their production. All known reserves of the indium would only produce enough solar cells to provide a capacity equal to all present wind generators.

$CuInSe_2$ with its optical absorption coefficient exceeding 3×10^4 cm^{-1} at wavelengths below 1000 nm, and its direct band gap being between 0.95 and 1.2 eV, is a good material for solar cells. CIS solar cell is sensitive in the wavelength of 0.4–1.3 μm, and maximum of its sensitivity is within the wavelength range of 0.7–0.8 μm.

However, manufacturing costs of CIS solar cells at present are high when compared to siliconsolar cells but continuing work is leading to more cost-effective production processes. Laboratory maximum CIS solar cells' efficiency is 21.7%, and commercially module efficiency is 9%.

3.4 *Organic Solar Cells*

Organic solar cells (OSCs) are composed of a blend active layer of a *p*-type organic semiconductor (*p*-OS) donor and an *n*-type organic semiconductor (*n*-OS) acceptor sandwiched between a transparent bottom electrode and a top metal electrode. The commonly used *p*-OS donor materials include *p*-type conjugated polymers or solution-processable conjugated organic small materials, or *n*-type inorganic semiconductor nanocrystals. Indium-doped tin oxide (ITO) glass is usually used as the transparent electrode, and low work function metals such as Ca, Mg, and Al are used as the top electrode.

Organic solar cells' low efficiency compared to inorganic solar cells arises from the following:

- Low dielectric constant of organic materials means that the electron diffusion lengths are too small.
- No crystal lattice—local disorder and impurities are present which leads to low charge carrier mobility of organic materials.
- Spectral mismatch between solar spectrum and organic materials which absorb the sunlight. Organic materials mostly absorb in the range of visible light in spectrum, while maximum photon density of sunlight is around 700-nm wavelength (between visible and infrared).
- Organic materials are susceptible to photodegradation particularly when in contact with water or oxygen.
- In the polymer-fullerene-based OSCs, the growth of the crystal size of cell components with time, at a higher temperature, is a problem for their morphology stability.
- Flexible device requires encapsulation to be performed by applying a flexible plastic layer with barrier layers applied to reduce the permeability to ambient degrading agents.

3.4.1 Advantages and Disadvantages of Organic Solar Cell Technology

Conducting organic materials, in particular π-conjugated polymers, have emerged as a new class of semiconductors since high conductivity was observed in doped polyacetylene in 1977. Numerous advantages of this type of photovoltaic include:

- Environmental impact—energy consumed during manufacturing is fairly low while utilized organic compounds, including solvents, are not too hazardous to the environment.
- Inexpensive and versatile materials and processing—polymers are relatively inexpensive to synthesize, compared to methods of silicon production, and can be processed from solutions into thin films by various simple and inexpensive methods like coating, printing, and roll-to-roll technologies all of which are highly compatible with various substrates and have low energy and temperature requirements.
- Their intrinsic properties like low weight, flexibility, and transparency due to the molecular nature of the materials, which opens up new market opportunities in applications where these properties are explicitly demanded.
- Tailoring properties—functionality of polymers can be tuned through careful molecular design and synthesis in order to fit an application, as organic molecules are easier to handle than inorganic atoms like silicon. Molecular engineering can be used to modify molecular mass, band gap of polymers, length, and functional groups of polymers, thereby changing charge generation ability, etc. This gives rise to a wide array of multifunctional polymers with specific optical and electrical properties for application in a particular photovoltaic market.

Even with all of these extraordinary possibilities like their intrinsic properties, successful large-scale commercialization organic photovoltaics is still not feasible since, out of the three criteria: cost, efficiency, and lifetime, only cost is met, which limits them to a niche market.

Laboratory maximum organic solar cells' efficiency is 11.2% [1, 8].

3.5 *Multiple Solar Cells*

Higher efficiency of photovoltaic conversion of solar radiation can be achieved if instead of one single semiconductor with *p–n* junction one uses structure with dual or multiple *p–n* junctions of different semiconductor materials. The first semiconductor needs to have a greater forbidden zone width, and it needs to absorb shortwave part and to pass long-wave part of the spectrum of solar radiation. The second semiconductor should absorb long-wave part of the spectrum of solar radiation.

Such structures can be realized in two ways: (a) using dichroic mirrors and (b) superposition of layers of different semiconductor materials to produce the tandem (cascade) cells.

3.5.1 Solar Cells with Dichroic Mirrors

In the first way, semipermeable dichroic mirror separates the spectrum of solar radiation into the shortwave and the long-wave part by focusing it on two separate *p–n* junctions, the first with a higher and the other with a smaller width of the forbidden zone. On this principle, a combined solar cell of the Si cells (1.1 eV) and AlGaAs/GaAs cells (1.65 eV) was produced, which in the concentrator photovoltaic system has an efficiency of 27%.

3.5.2 Tandem Solar Cells

To make solar cells more competitive with other sources of electricity, their cost per peak watt should drop below two dollars, with a solar radiation conversion efficiency of above 15%. Monocrystalline solar cells and their modules do not meet the first, but they meet the second condition.

In solar cells based on amorphous silicon, the price condition is almost met, while their efficiency is still far below 15%.

The fastest and most direct solution to this problem is now seen in the formation of the so-called tandem solar cells that would have two or four electrodes. In terms of construction, there are two types of tandem solar cells, mechanical and monolithic tandem solar cells.

Monolithic tandem solar cell is composed of two solar cells that are in direct contact one above the other. Upper cell is made of a material with a higher and lower with a smaller width of the forbidden zone. Open-circuit voltage of this solar cell is equal to the sum of voltages at each of the cells. Short-circuit current is equal to the lesser current of the component cells. This condition determines the design of tandem solar cells. The thicknesses of the component cells must be selected in such a way that the absorbed solar radiation in each of the cells generates equivalent photocurrent. In addition, solar radiation absorption coefficient of the upper cell material must be smaller than the absorption coefficient of the lower cell.

In *mechanical tandem solar cells*, two solar cells made of materials with different widths of forbidden zones are located one above the other. The component cells are separated by a layer of transparent insulators, and each has two electrical electrodes through which they can be connected serial, parallel or combined.

The existing tandem solar cell technologies can be classified into three groups in which a-Si solar cell is one component. In the first group, only thin amorphous materials total thickness of up to 1 μm and in a monolithic configuration are used. For the first cell, one uses a-Si: H with $E_g \sim 1.7$ eV, and for the other cell with a smaller E_g one uses alloy Si–Ge or Si–Sn. In laboratory conditions, the efficiency of these cells reaches 13% and on larger areas 10%, which is not a huge improvement in relation to the single a-Si: H solar cells. This group includes also a-Si hetero-junctions with amorphous alloys Si–C or Si–N with $E_g > 2.2$ eV.

The second group comprises mechanically linked hetero-junctions with four connections of amorphous silicon with CdS/CdTe or $CuInSe_2$ or polycrystalline structures with an efficiency of 15.6% in the laboratory cells and 12.3% on 30×30 cm^2 solar modules.

The third group comprises hybrid tandem solar cells of amorphous–polycrystalline silicon. This tandem with the energy band gap of 1.7 eV (a-Si) and 1.1 eV (c-Si) almost optimally covers the spectral distribution of solar radiation energy. The component cells are interconnected by a transparent insulating layer. Upper solar cell could reach an efficiency of 9% and lower 6%, which would contribute to the total efficiency of a-Si/c-Si tandem solar cell of 15%. Research on this type of tandem solar cells is in progress.

Since the tandem solar cells are significantly more expensive than Si solar cells, currently they are used to power satellites and in photovoltaic systems with solar radiation concentrators.

3.6 Solar Cells with Concentrators

In order to increase the efficiency of photovoltaic conversion of solar radiation, one uses photovoltaic conversion systems with concentrators as Fresnel lens or suitably inclined flat mirrors.

3.6.1 Fresnel Lenses

Fresnel lenses, which, instead of continuously curved surfaces, have curved segments on a flat surface. The segments have the same curvature and are placed under such an inclination that the paraxial beams reach out the lens focus after refraction. The dimensions of the segments are limited by the thickness of the lens. With the increase in the thickness of the lens, the number of its segments decreases (Fig. 32).

Fresnel lenses are made of plastic acrylic materials by casting and pressing in appropriate molds. For their production, considerably less material is needed than for the conventional classic lenses.

Fresnel lenses can have a dotted or linear focus, similar to spherical and cylindrical lenses. Due to the effects of transition regions between the segments, focus of the Fresnel lenses has smaller sharpness than the focus of the classical collecting and cylindrical lenses.

The disadvantages of the spherical and cylindrical lenses consist in the fact that the concentration factor of one lens is limited and it is difficult to produce a conventional lens with a small focal length. The concentration factor of the spherical lens is proportional to the ratio of the lens diameter and its focus distance. In Fresnel lenses, the advantages of the system with several lenses and individual lenses' features are combined, whereby each part of the Fresnel lenses concentrates sunlight on the receiver. Additional advantage of Fresnel lenses is reflected in their low thickness.

The concentrator with Fresnel lenses uses only direct sunlight. The concentration factor is 100–1000, and the working temperature ranges from 300 to 1000 °C. Fresnel lenses are used in thermal conversion of solar radiation and in the photovoltaic generators with sun radiation concentrators (Fig. 33).

In photovoltaic systems with concentrator, cells that are stable at higher temperatures and that have efficiency greater than 20% (GaAs) are used.

Fig. 32 Fresnel lens [36]

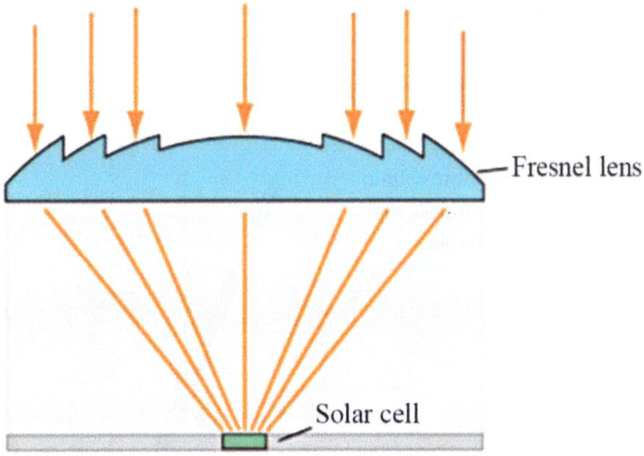

Fig. 33 Solar photovoltaic system with Fresnel lens as concentrators of solar radiation [37]

3.6.2 Flat Mirrors

By using lenses and mirrors on solar cells, only direct solar radiation can be focused. The application of photovoltaic systems with concentrator in areas with a lot of cloudy days is not recommended (Fig. 34).

The efficiency of solar photovoltaic systems with concentrator is between 30 and 35%.

3.7 Development of Solar Cells' Efficiency

Development of laboratory solar cells' efficiency is given in Fig. 35.

Efficiency comparison of technologies: best laboratory cells versus best laboratory modules, is given in Fig. 36.

Solar cells and solar modules' efficiency in 2017 is given in Table 4.

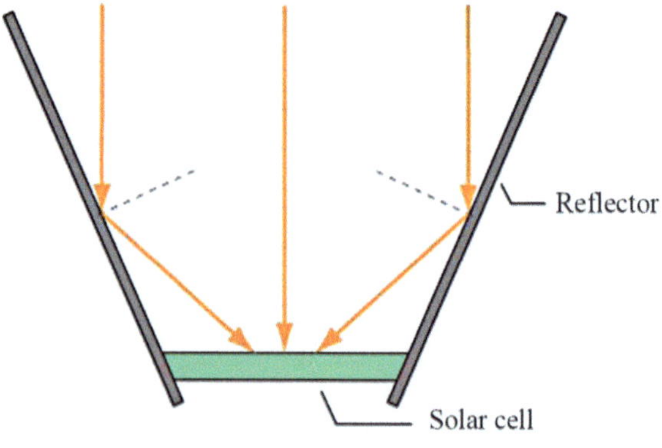

Fig. 34 Solar photovoltaic system with flat mirrors as concentrators of solar radiation [36]

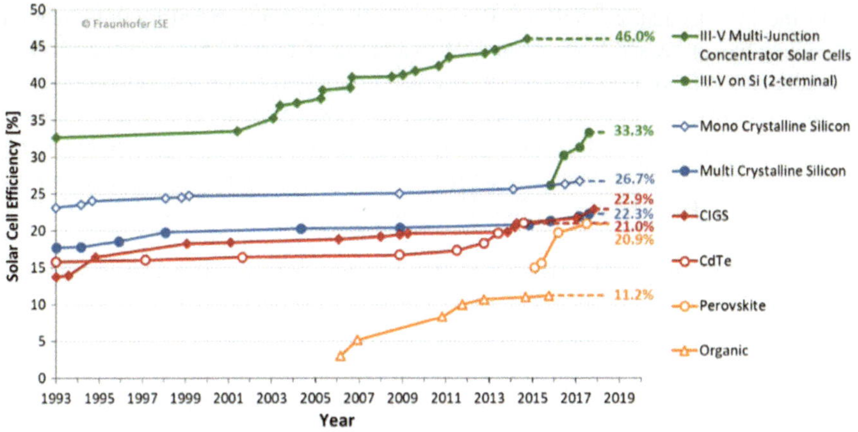

Fig. 35 Development of laboratory solar cell efficiency [38]. *Courtesy* Fraunhofer Institute

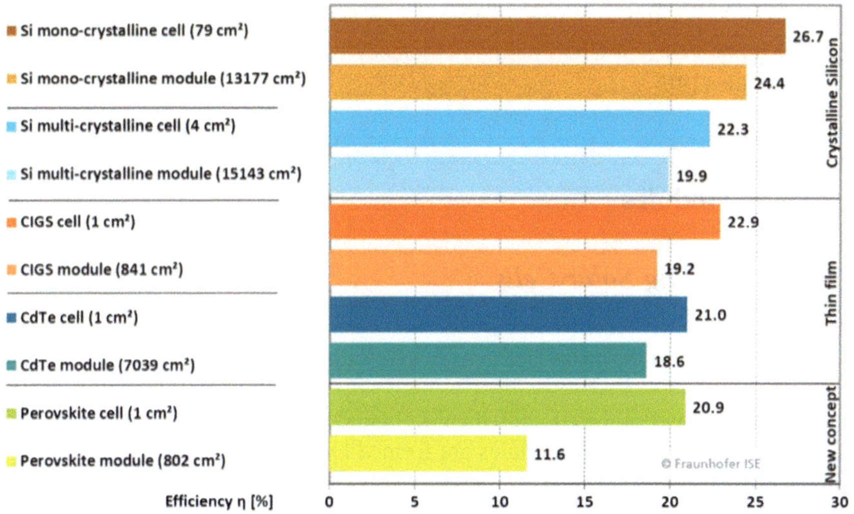

Fig. 36 Efficiency comparison of technologies: best laboratory cells versus best laboratory modules [38]. *Courtesy* Fraunhofer Institute

Table 4 Solar cell and solar module efficiency in 2017 [38]

Solar cells	Conc.	c-Si	p-Si	CIS	CdTe	Organic	Multi-junction with conc.
Efficiency	(%)						
Record cell efficiency	46.8	26.7	22.3	21.7	21.0	11.2	46.8
Commercial module efficiency		17	15	9	16		

In the last decade, crystalline silicon wafer-based commercial module average efficiency has increased from about 12% to a range of 17 to 17.5%. Best performing modules in the laboratory can currently reach up to 24.4% efficiency. Current crystalline module efficiency is typically at least 2% lower than the efficiency at the cell level due to losses caused by various factors such as: module border, cell spacing, cover reflection, and cell interconnection. However, cell and module efficiencies are intrinsically linked and current developments in best cell efficiency levels suggest that continued improvements in the average efficiency of modules will continue in the foreseeable future. For instance, by 2024, industry expectations place the range of stabilized cell efficiency for mass production of crystalline silicon-based cells at 19.8–25% depending on cell type and architecture up from a current range of 18.8–23.5%.

The two most deployed thin-film technologies are cadmium–telluride (CdTe) and copper indium gallium selenide (CIGS). First Solar (the largest CdTe manufacturer) reported fleet average efficiencies increasing from 12.9% in 2012 to 16.6% in 2016 for their CdTe modules. For CdTe cells, module efficiency record for the moment is 18.6%. The best CIGS-reported efficiencies so far were 17.5% for modules. Solar Frontier reports current CIGS module efficiency between 12.2 and 13.8% for their CIGS modules [9].

4 Solar Modules

4.1 Connecting Solar Cells

4.1.1 Connection in Series

Solar module is part of the circuit which is obtained by mutual electrical connecting of several solar cells. Solar modules are formed by serial or parallel connection of solar cells.

The serial connection of solar cells in solar module is shown in Fig. 37.

In serial connection of solar cells, the voltage of solar module is equal to the sum of the voltages on each individual solar cell, while the total current is equal to the

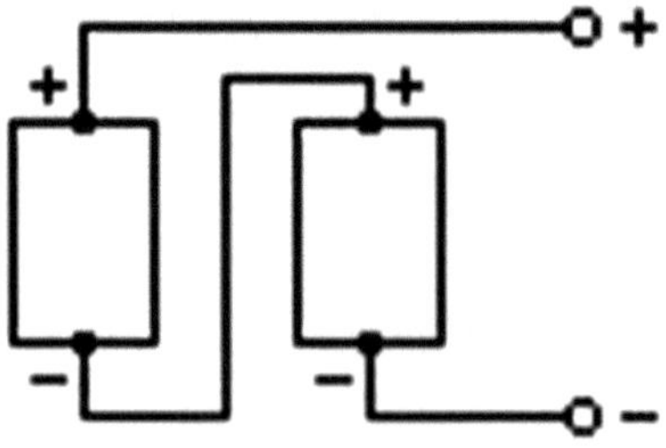

Fig. 37 Serial connection of solar cells in solar module

current provided by the individual solar cell in series:

$$U = \sum_{i}^{n} U_i, \ I = I_i \tag{8}$$

where U is the voltage of solar module, U_i is the voltage of a solar cell, I is the current of solar module, and I_i is the current of an individual solar cell.

4.1.2 Connection in Parallel

The parallel connection of solar cells in solar module is shown in Fig. 38.

In parallel connection of solar cells, the voltage of solar module is equal to the voltage on each individual solar cell in it, while the current is equal to the sum of currents provided by solar cells in module:

$$U = U_i, \ I = \sum_{i}^{n} I_i \tag{9}$$

where U is the voltage of solar module, U_i is the voltage of each single solar cell, I is the current of solar module, and I_i is the current of each individual solar cell.

In the serial and parallel connection of solar cells in solar modules, all solar cells should be of the same type and produced by the same manufacturer.

4.2 Basic Characteristics of Solar Modules

Solar cells are connected in solar modules to obtain higher output power compared to the power of each cell separately.

Solar module should have a good mechanical stability, resistance to the effects of weathering, stability in the temperature range from −50 to +90 °C, resistance to ultraviolet radiation, the safety of electric shock and other hazards related to the use of photovoltaic systems, etc.

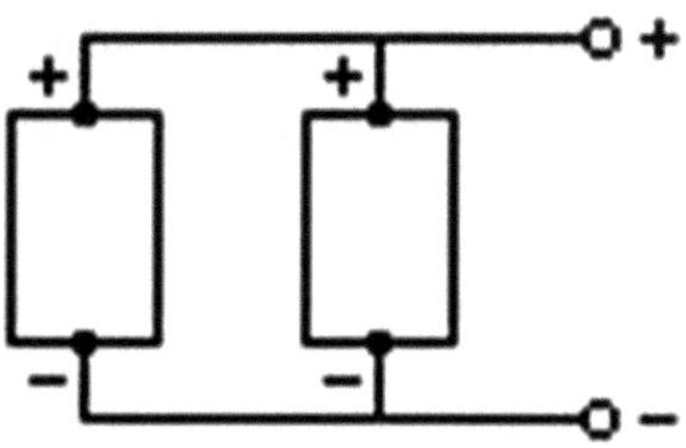

Fig. 38 Parallel connection of solar cells in solar module

Solar modules can be divided according to the type of solar cells used, the housing, the front protection, etc.

Energy efficiency of solar modules decreases with the increasing temperature, reduction of solar radiation intensity, increased soiling of the front cover, etc.

Commercial solar modules typically have a power of 50–250 W.

4.3 Standard Test Conditions for Solar Modules

Standard test conditions (STC) are conditions under which the solar modules are tested in a laboratory. Module testing is carried out in the following conditions: solar radiation intensity of 1000 W/m^2, optical air mass of AM 1.5, temperature of solar module of 25 °C, and wind speed of 1 m/s. The prospects of solar module manufacturers contain the following data: the value of maximum power P_{mpp} (W_P), open-circuit voltage U_{oc} (V), short-circuit current I_{ks} (A), voltage at the maximum power U_{mpp} (V), current at maximum power I_{mpp} (A), and the value of the nominal operating temperature of the solar module T_{NOCT} (°C).

The real working conditions of solar modules are different from the conditions under which they are tested. Climate conditions in which solar modules operate are changing during the day and the year, leading to losses due to angular and spectral distribution and low values of solar radiation intensity, solar cell temperature, etc.

1. ***Angular distribution of solar radiation***

Due to changes in the position of the sun during the day and the presence of a diffuse component, solar radiation does not reach at the right angle the surface of solar module, as is the case during module testing in a laboratory. Low values of solar radiation reach the surface of the solar module at high incident angle of solar radiation or when solar radiation is largely diffused. Reflection and transmission of solar radiation on the material solar cells are made of depends on the angle of solar radiation incidence. When the angle of solar radiation incidence reaches the surface of the module greater than 50°, there is a significant increase in reflection from the solar cells. It should be noted that the angular dependence of solar radiation reflectance is often left out in simulations and calculations of solar system efficiency. However, in practice, with the horizontal and vertical mounted solar modules as facade elements, this effect cannot be ignored.

2. ***Spectral distribution of solar radiation***

At the same intensity, different parts of solar radiation spectrum give different photocurrents in accordance with the spectral response of the solar cell. Spectral distribution of solar radiation changes with the change of the sun's position, weather, and air pollution, so it rarely reaches the optical air mass AM 1.5.

3. ***Solar radiation intensity***

At a constant temperature of the solar cell, the efficiency of the solar module is reduced with a decrease in solar radiation intensity, due to the logarithmic dependence of the open-circuit voltage on the photocurrent. At low values of solar radiation intensity, decrease in the efficiency of the solar module is larger and less predictable.

4. ***Solar cell temperature***

Ambient temperature and solar radiation intensity in real climatic conditions change during the day and year, and affect the change of solar module temperature. Under the influence of solar radiation, heating of solar cells occurs, i.e., solar modules. Higher temperature of solar cells leads to a reduction in solar modules' efficiency.

4.4 Mounting Solar Modules

Solar modules are mounted on the respective metal structures on the ground, roofs, and walls of houses and other buildings, balconies, canopies over parking spaces, and similar places at a certain angle relative to the horizontal plane (Fig. 39).

In order to maximize solar radiation incidence in the Northern Hemisphere, solar modules should be oriented to the south at an angle equal to the latitude of a given location $+15°$ for the winter period and $-15°$ for the summer period. In the Southern Hemisphere, solar modules should be oriented to the north at an angle equal to the latitude of a given location $-15°$ in winter and $+15°$ in summer.

Approximate efficiency loss in the Northern Hemisphere by not facing solar modules directly south is given in Table 5.

For the Southern Hemisphere countries, the previous chart should be revised.

Fig. 39 Installation angle of solar modules depending on the latitude of a given place in Northern Hemisphere: (1) maximum summer insolation, (2) maximum annual insolation, (3) maximum winter insolation

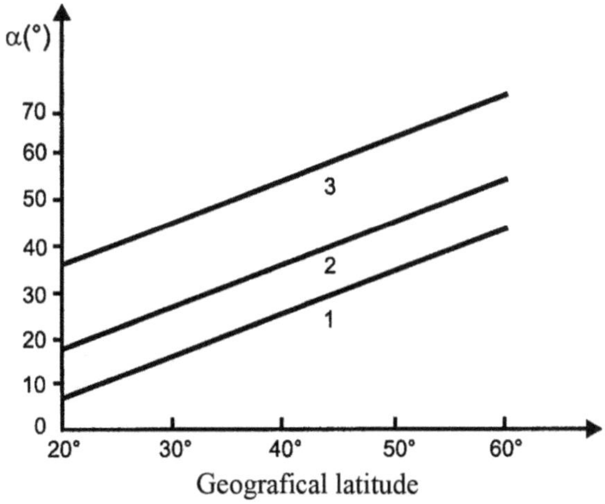

Table 5 Approximate efficiency loss in the Northern Hemisphere by not facing solar modules directly south

Solar module position	Angle from south (°)	Efficiency loss (%)
South	0	0
South–west	22.5	5
South–east	22.5	5
South–west	45	10
South–east	45	10
South–west	67.5	15
South–east	67.5	15
West	90	20
East	90	20

In the Northern Hemisphere, solar modules will always work best if they are south-facing. In the Southern Hemisphere, solar modules will always work best if they are north-facing.

It is not always possible to position solar modules so that they are facing exactly the right way. It was found that the average efficiency drops of a solar module mounted away from the south in the Northern Hemisphere are around 1.1% for every five degrees. This means, if solar module faces east or west, it can be expected around 20% loss of the efficiency compared to modules facing in optimal position [10].

4.5 Shading Solar Modules

Shaded solar cell does not produce electricity and is an obstacle in the open circuit. When current passes through the overshaded solar cell, it is then heated and burned out. To protect the solar cells, one uses bypass diodes which are connected in parallel with each solar cell individually. To protect the sub-modules, one uses blocking diodes that are connected in series with sub-modules (Fig. 40).

Protection of solar cell by by-pass diodes is used when connecting solar modules to the solar generator.

4.6 Solar Generator

Solar generator denotes several interconnected solar modules which in the circuit are a source of electricity. Several interconnected solar modules comprise a series or a string. Solar generator with battery, battery charge controller, inverter, monitoring system, etc., makes up a photovoltaic system.

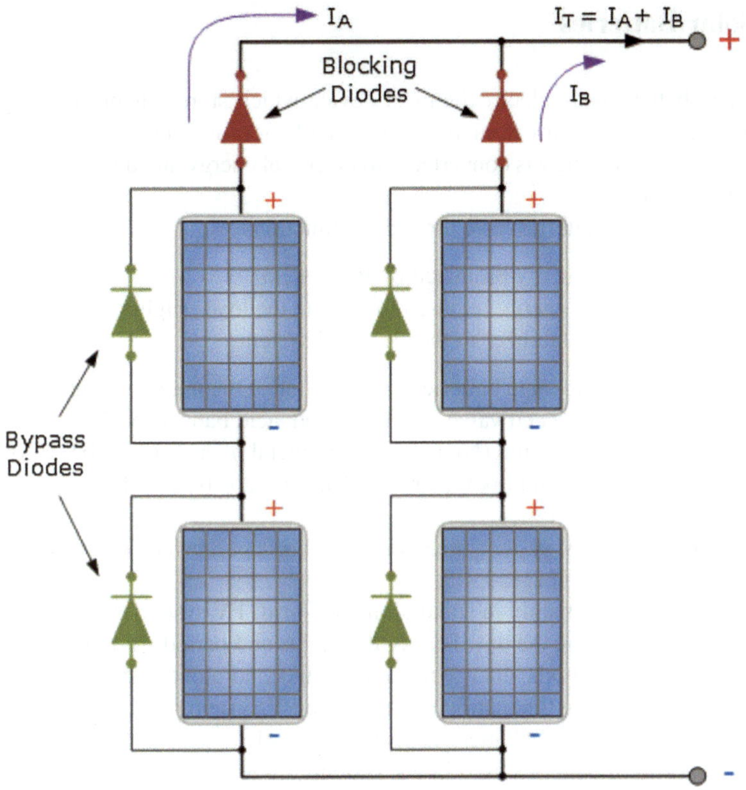

Fig. 40 Protection of solar module from the shading of solar cells using bypass and blocking diodes [39]

Output voltage, current, and power of the solar generator depend on the mode of solar module connection, solar radiation intensity, temperature, etc.

In the series connection of the same solar modules, output current is equal to the current of each individual module, and the output voltage is equal to the sum of all module voltages.

In the parallel connection of the same solar module, output current is equal to the sum of all currents of each individual module, and the output voltage is equal to the voltage of an individual module.

In solar generator operation, there can occur short-circuiting, ground break-throughs, breaks in the circuit, shading of modules, etc.

5 Solar Batteries

A storage battery is an electrochemical device which stores chemical energy that can be released as electrical energy. When the battery is connected to an external load, the chemical energy is converted into electrical energy and direct current flows through the circuit.

The PV system batteries have three main functions:

- To store electrical energy produced by PV system,
- To supply electrical energy required to operate the loads (lighting, pumps, etc.),
- To act as a voltage stabilizer in the electrical system.

Lead–acid and solid-state batteries can be found at the market. Lead–acid batteries can be wet (flooded) and valve-regulated lead–acid battery (VLRA). Solid-state batteries are nickel–cadmium (Ni–Cd) , nickel–metal hydride (Ni–MH or Ni–MH), lithium-ion (Li-ion), lithium polymer, etc. They are costly, and they need different types of controllers.

The batteries for PV systems must be able to accept repeated deep charging and discharging without damage.

Batteries are used mainly in stand-alone (off-grid) PV systems to store electrical energy produced during the day. During the nights and periods of low solar irradiation, battery can supply the energy to the load. In stand-alone systems, batteries are required because of the fluctuating nature of the PV system output.

Schematic diagram of a battery is given in Fig. 41.

The majority of batteries used in PV systems are lead–acid batteries. The chemistry of the lead–acid batteries avoids problems, such as stratification, freezing, and sulfation. Relatively low cost and general availability of lead–acid batteries mean that they are used in all but mostly in demanding PV environments [8, 10–14].

5.1 *Basic Characteristic of Batteries*

Main parameters of storage batteries are:

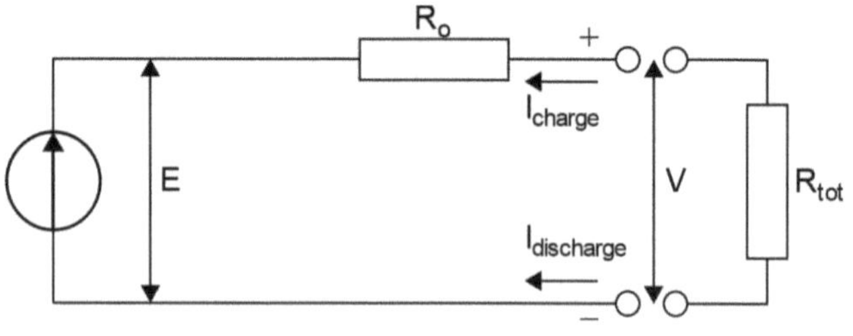

Fig. 41 Schematic diagram of a battery

- ***Nominal voltage***. The voltage at the battery as rated is the nominal voltage at which the battery is supposed to operate. The voltage is measured in volt (V). The so-called solar batteries or lead–acid batteries for PV applications are usually rated at 12 V, 24 V, or 48 V. The actual voltage of PV systems may differ from the nominal voltage.
- ***Capacity***. A term capacity refers to the amount of charge that the battery can deliver at the rated voltage. The capacity is directly proportional to the amount of electrode material in the battery. This explains why a small cell has a lower capacity than a large cell based on the same chemistry, even though the open-circuit voltage across the cell will be the same for both cells. Thus, the voltage of the cell is more chemistry-based, while the capacity is more based on the quantity of the active materials used. The capacity is measured in ampere hours (Ah). For batteries, Ah is the more convenient unit because in the field of electricity the amount of energy is usually measured in watt-hours (Wh). The energy capacity of a battery is simply given by multiplying the rated battery voltage measured in volt by the battery capacity measured in ampere hours, which results in the battery energy capacity in watt-hours.
- ***Battery efficiency***. For storage systems, usually the round-trip efficiency $h_{bat} = E_{out}/E_{in}$ is used, which is given as the ratio of the total storage output E_{out} to the total storage input E_{in}. For example, if 10 kWh is pumped into the storage system during charging, but only 8 kWh can be retrieved during discharging, the round-trip efficiency of the storage system is 80%. The round-trip efficiency of batteries can be broken down into two efficiencies: the *voltaic efficiency*, which is the ratio of the average discharging voltage to the average, and the *coulombic efficiency* (*or Faraday efficiency*), which is defined as the ratio of the total charge extracted from the battery to the total charge put into the battery over a full-charge cycle charging voltage. When comparing different storage devices, usually this round-trip efficiency is considered. It includes all the effects of different chemical and electrical non-idealities occurring in the battery.
- ***State of charge*** (***SOC***) is defined as the percentage of the battery capacity available for discharge. For example, a 10 Ah rated battery that has been drained by 2 Ah is said to have a SOC of 80%.
- ***Depth of discharge*** (***DOD***) is defined as the percentage of the battery capacity that has been discharged. For example, a 10 Ah battery that has been drained by 2 Ah has a DOD of 20%. The SOC and the DOD are complimentary to each other.
- ***Cycle lifetime*** is defined as the number of charging and discharging cycles after which the battery capacity drops below 80% of the nominal value. Usually, the cycle lifetime is specified by the battery manufacturer as an absolute number. However, stating the battery lifetime as a single number is an oversimplification because different battery parameters discussed so far are not only related to each other but also dependent on the temperature. Figure 42a shows cycle lifetime as a function of the DOD for different temperatures. Colder operating temperatures mean longer cycle lifetimes.

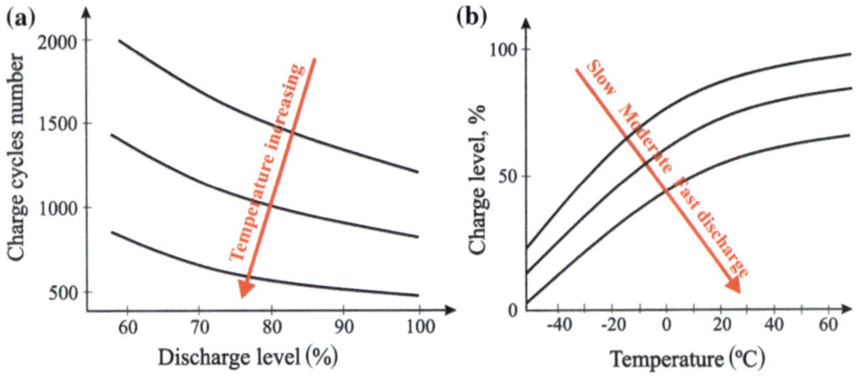

Fig. 42 Qualitative illustration of **a** the cycle lifetime of a Pb–acid battery in dependence of the DOD and temperature and **b** effect of the temperature on the battery capacity

Cycle lifetime depends strongly on the DOD. The smaller the DOD, the higher the cycle lifetime. The battery will last longer if the average DOD can be reduced during the lifetime of the battery.

- ***Temperature effects***. As seen in Fig. 42b, the lower the temperature, the lower the battery capacity. At higher temperatures, chemicals in the battery are more active, leading to an increased battery capacity. At high temperatures, it is even possible to reach an above-rated battery capacity. However, such high temperatures are severely detrimental to the battery health. Overheating can cause overcharging and subsequent overvoltage of the lead–acid battery. To prevent this, charge controllers are used. In Fig. 42b, it can be seen that battery capacity increases when the discharge current is lower. This is because the discharge process in the battery is diffusion limited: If more time is allowed, better exchange of chemical species between the pores in the plate and the electrolyte can take place.
- ***Aging***. The major cause for aging of the battery is sulfation. If the battery is insufficiently recharged after being discharged, sulfate crystals start to grow, which cannot be completely transformed back into lead or lead oxide. Thus, the battery slowly loses its active material mass and hence its discharge capacity. Corrosion of the lead grid at the electrode is another common aging mechanism. In case of lead–acid batteries, antimony poisoning is a major cause for accelerated aging. Corrosion leads to increased grid resistance due to high positive potentials. Further, the electrolyte can dry out. At high charging voltages, gassing can occur, which results in the loss of water. Thus, demineralized water should be used to refill the battery from time to time [15–18].

5.2 *Lead–Acid Batteries*

There are three main categories of lead–acid batteries: wet (flooded), gel, and absorbed glass mat (AGM) batteries. Most common at the market are wet-cell lead–acid batteries. They can be sealed or non-sealed. Sealed lead–acid batteries require no maintenance (Fig. 43).

Lead–acid battery consists of the galvanic elements that each has two Pb electrodes (anode and cathode) which are located in the electrolyte. By serial connection of galvanic elements, one obtains rechargeable battery of a nominal voltage corresponding to the number of galvanic elements multiplied by the voltage of one element.

In the lead–acid solar battery when charging, the following chemical reaction occurs:

$$2 \cdot P_bSO_4 + 2 \cdot H_2O \rightarrow P_bO_2 + 2H_2SO_4 + P_b \quad (10)$$

In the lead–acid solar batterybattery when discharging, the following chemical reaction occurs:

$$P_bO_2 + 2H_2SO_4 + P_bSO \rightarrow 2 \cdot P_bSO_4 + 2H_2O \quad (11)$$

Under aging of batteries, one implies irreversible physical and chemical processes on the electrodes and the electrolytes which over time reduce its capacity. These processes involve chemical disintegration of electrolytes, formation of sulfate crystals, corrosion, sludge, drying, etc.

These batteries require the addition of water to maintain electrolyte levels. Loss of electrolyte occurs from evaporation and gassing.

Fig. 43 Lead–acid solar battery

The service life of solar batteries with grating electrodes and liquid electrolyte is about 3–8 years.

Lead–acid batteries have following characteristics: low cost, strongly built, capable of high currents, no memory effect, good life span when correctly used, recyclability, low energy density (ratio capacity/weight), auto-discharge rate, temperature sensitive, risk of sulfation when stored for a long period in discharged state, etc.

In stand-alone PV systems, deep discharge lead–acid batteries are commonly used due to their deep-cycle capabilities and long life. Starting, lighting, and ignition batteries, commonly used in automobiles, are not recommended for PV applications due to their limited deep-cycle capabilities. Valve-regulated lead–acid (VRLA) sealed batteries are popular but have some requirements that are hard to meet in PV systems.

The life of a lead–acid battery is proportional to the average state of charge (SOC) of the battery if the battery is not overcharged, over-discharged, or operated at temperatures exceeding manufacturers' recommended specifications. A typical flooded, deep-cycle, lead–acid battery that is mentioned above 90% SOC can provide two or three times more full charge/discharge cycles than a battery allowed to reach 50% SOC before recharging. Similar but more dramatic results are found with sealed VRLA and lead-calcium alloyed grid batteries.

A lead–acid battery must not be run completely flat. Minimum of 20% state of charge should be maintained in battery at all times to ensure it is not damaged. It is better to design system in such a way that the battery charge rarely goes below 50%.

5.3 VRLA Batteries

A valve-regulated lead–acid battery (VRLA), more commonly known as a sealed battery, or maintenance-free battery, is a type of lead–acid rechargeable battery. Due to their construction, they can be mounted in any orientation and do not require constant maintenance (Fig. 44).

There are two primary types of VRLA batteries, gel cells and absorbed glass mat (AGM). Lead–acid cells consist of two plates of lead, which serve as electrodes, suspended in an electrolyte consisting of dilute sulfuric acid.

In AGM and gel-type VRLAs, the electrolyte is immobilized. In AGM, this is accomplished with a fiberglass mat; in gel batteries or gel cells, the electrolyte is in the form of a paste-like gel created by adding silica and other gelling agents to the electrolyte.

VRLA cells may be made of plates similar to a flooded lead–acid battery or may be made in a spiral roll form to make cylindrical cells.

VRLA batteries have a pressure relief valve which will activate when the battery starts building pressure of hydrogen gas, generally a result of being recharged.

Fig. 44 A 12 V VRLA battery, typically used in small uninterruptable power supplies and home security systems

AGM and gel batteries have the following characteristics:

- Shorter recharge time than flooded lead–acid,
- Cannot tolerate overcharging: overcharging leads to premature failure,
- Shorter useful life, compared to properly maintained wet-cell battery,
- Discharge significantly less hydrogen gas,
- AGM batteries are, by nature, safer for the environment and safer to use,
- Can be used or positioned in any orientation.

Water loss is less of a problem in gel batteries than in AGM batteries, because there is more acid to start with. For this reason, gel batteries are preferred to AGM batteries in PV systems where high operating temperatures are expected.

5.4 *Gel Batteries*

A modern gel battery (also known as a gel cell) is VRLA battery with a gelified sulfuric acid electrolyte mixed with fumed silica, which makes immobile gel-like mass. Gel battery does not need to be kept upright (Fig. 45).

In gel solar batteries, there are no unpleasant fumes. These batteries have twice the service life of solar batteries with grating electrodes. At the depth of discharge of 50%, their lifetime is around 1000 cycles. The main drawback of these batteries is somewhat higher price.

5.5 *AGM Batteries*

In an AGM battery, the electrolyte is a combination of micro-glass fibers and sulfuric acid. Glass fibers completely absorb acid. In these batteries, it is not necessary to add

Fig. 45 Gel solar battery

water, and they can be installed in any position, are resistant to low temperatures, have a longer shelf life, exhibit higher efficiency, and are more expensive than wet lead–acid batteries (Fig. 46).

AGM batteries are sealed, or more correctly, valve-regulated lead–acid batteries.

This type of battery uses an AGM between tightly packed flat plates. All the acid is absorbed in glass mat separator, but the pores of the glass mat are not completely filled. Oxygen formed at the positive electrode during charging has a pathway through empty (or part-empty) pores to move to the negative electrode or recombination.

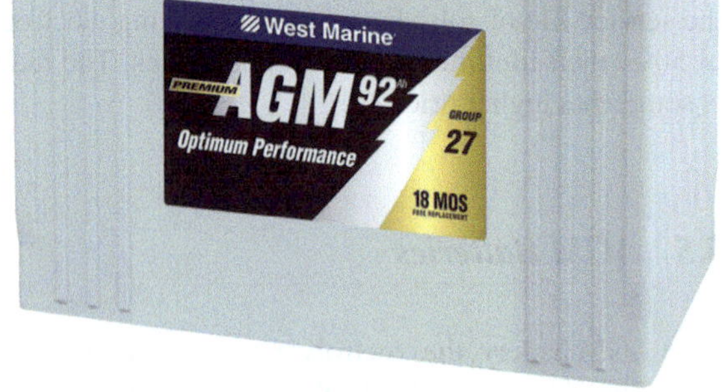

Fig. 46 AGM battery

AGM batteries have a good high current (short discharge time) performance. AGM batteries contain very little acid, and they are very susceptible to water losses that occur at high temperatures. They have a good resistance to being frozen solid.

AGM batteries can be made with either flat or tubular positive plates. Attention should be paid to the manufacturer's instructions regarding ventilation requirement. AGM batteries present better discharging characteristics than conventional batteries in different ranges of temperatures.

AGM batteries are commonly used in off-grid solar power and wind power installations as energy storage banks. AGM batteries are routinely chosen for remote sensors such as ice monitoring stations in the Arctic. AGM batteries, due to their lack of free electrolyte, will not crack and leak in this cold environment.

AGM and gel batteries are used in sailplanes, nuclear submarine fleet, for recreational marine purposes, in telecommunication, measurement equipment, etc.

5.6 Li-Ion Batteries

A lithium-ion or Li-ion battery (LiB) is a type of rechargeable battery in which lithium ions move from the negative electrode to the positive electrode during discharge and back when charging. Li-ion batteries use an intercalated lithium compound as one electrode material, component to the metallic lithium used in a non-rechargeable lithium battery (Fig. 47).

Li-ion batteries are common rechargeable batteries for portable electronics with a high density, tiny memory effect, and low self-discharge. Li-ion batteries are also growing in popularity for military, battery electronic, vehicle and aerospace applications, in telecommunication applications, etc.

Li-ion batteries pose a safety hazard since they contain a flammable electrolyte and may be kept pressurized. A battery cell charged too quickly could cause a short

Fig. 47 Li-ion battery

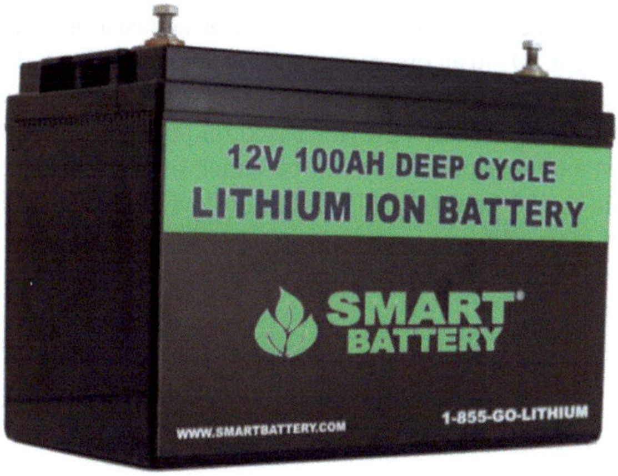

circuit, leading to explosions and fires. Also, they become crushed if a battery without overcharge protection is subjected to a higher load that it can safely handle. An external short circuit can trigger the batteries to explode.

Li-ion batteries were proposed by the British Chemist M. Stanley Whittinham in the 1970s.

5.6.1 Construction

The three primary functional components of a Li-ion battery are positive and negative electrodes and electrolyte. Generally, negative electrode of a conventional Li-ion cell is made from carbon. Positive electrode is a metal oxide. Electrolyte is a lithium salt in an organic solvent.

Pure lithium is highly reactive. It reacts vigorously with water to form lithium hydroxide (LiOH) and hydrogen gas.

Li-ion batteries are more expensive than Ni–Cd batteries but operate over wider temperature range with higher energy densities. They require a protective circuit to limit pick voltage.

Li-ion batteries are available in various shapes which can generally be divided into four groups:

- Small cylindrical (solid body without terminals, such as those used in older laptop batteries),
- Large cylindrical (solid body with large threaded terminals),
- Pouch (soft flat body, such as those used in cell phones and never laptops; also referred to as Li-ion polymer or lithium polymer),
- Rigid plastic case with large treated terminals (such a vehicles' fraction packs).

5.6.2 Liquid Electrolytes

Liquid electrolytes in Li-ion batteries consist of lithium salts such as $LiPF_6$, $LiBF_4$, or $LiClO_4$ in an organic solvent such as ethylene carbonate, dimethyl carbonate, and diethyl carbonate.

Composite electrolytes based on poly (oxyethylene) (POE) provide a relatively stable interface. They can be either solid (high molecular weight) and be applied in dry Li-polymer cells, or liquid (low molecular weight) and be applied in regular Li-ion cells.

Electrolyte degradation mechanisms include hydrolysis and thermal decomposition.

5.6.3 Solid Electrolytes

Solid ceramic electrolytes are mostly lithium metal oxides which allow lithium ion transport through the solid more readily due to the intrinsic lithium. The main benefit of solid electrolytes is that there is no risk of leaks, which is serious safety issue for batteries with liquid electrolytes. Solid ceramic electrolytes can be broken down into two main categories: ceramic and glossy. Glossy solid electrolytes are amorphous atomic structures made up of similar elements to ceramic solid electrolytes, but have higher conductivities due to higher conductivity at grain boundaries.

5.6.4 Performance

- Li-ion batteries offer good charging performance at cooler temperatures and may even allow "fast charging" within a temperature range 5–45 °C (41–113 °F). High temperature during charging may lead to battery degradation, and charging at temperatures above 45 °C will degrade battery performance. At lower temperatures, the internal resistance of the battery may increase, resulting in slower charging and prolonged charging times.
- Specific energy density 100–250 Wh/kg,
- Volumetric energy density 250–620 Wh/L,
- Specific power density 300–1500 W/kg.

Since Li-ion batteries contain less of toxic metals than other types of batteries which may contain lead or cadmium, they are generally categorized as non-hazardous waste.

5.6.5 Battery Life

Rechargeable battery life is typically defined as the number of full charge–discharge cycles before significant capacity loss. Inactive storage may also reduce capacity. On average, Li-ion batteries' lifetime consists of 100 cycles. Batteries for mobile phones, or other handhold devices in daily use, are not expected to last larger than 3 years.

5.7 *Nickel–Cadmium Batteries*

Nickel–cadmium (Ni–Cd) batteries in the charged state have positive plates with nickel oxy-hydroxide (NiOOH) as active material. Negative plates are cadmium metal as active material. Electrolyte is potassium hydroxide (KOH) in water (20–35% by weight). On this charge, the NiOOH of the positive plate is converted to $Ni(OH)_2$ and the cadmium metal of the negative plate is converted to $Cd(OH)_2$ (Fig. 48).

Fig. 48 Nickel–cadmium battery

Ni–Cd batteries are used in larger PV systems where there are extremely high (>40 °C) or extremely low (−10 °C) operating temperatures. Ni–Cd batteries can operate down to −50 °C and up to at least +50 °C.

Ni–Cd batteries are 3–4 times more expensive than lead–acid batteries.

Ni–Cd batteries in a PV system have a maximum DOD of 90%. Industrial Ni–Cd batteries used in PV systems are normally of the open-type design for standby use at low discharge rates. Most of standby Ni–Cd batteries are supplied with 20% KOH electrolyte as a standard and have freezing point of −25 °C. If Ni–Cd batteries have 30% KOH electrolyte, their freezing point is −58 °C.

There is a worldwide pressure to ban Ni–Cd batteries because of a toxic waste, and this has already happened in the EU for small consumer-type sealed batteries.

Ni–Cd batteries' advantages are: long life, reduced maintenance, deep discharge without damage, performance much less affected by temperature, voltage regulation not as important, excellent charge retention, and high capacity at low temperatures.

Ni–Cd batteries' disadvantages are: Cost per ampere hour is high, display a "memory" of a battery discharge history, etc.

5.8 Nickel–Metal Hydride Batteries

A nickel–metal hydride battery's (NiMH or Ni–MH) positive electrodes are similar to that of the Ni–Cd battery, electrolyte is nickel oxide hydroxide (NiOOH), and the negative electrode is hydrogen-absorbing alloy instead of cadmium. A Ni–MH battery can have 2–3 times higher capacity of an equivalent size Ni–Cd battery, and its energy density can approach that of a Li-ion battery (Fig. 49).

Fig. 49 Nickel–metal hydride battery

The operating range of Ni–MH batteries is between −20 and +40 °C. The price of Ni–MH batteries is similar to that of Ni–Cd batteries and is much higher than that of lead–acid batteries.

Nickel–metal hydride batteries (Ni–MH) are characterized by: good density of energy, no memory effect, important currents support, easy storage and transport, recyclability, fragile because do not support overcharge, requires specific automatic chargers, more expensive than Ni–Cd, life span less than for Ni–Cd, important auto discharge etc.

Ni–MH batteries have replaced Ni–Cd batteries for many roles, notably small rechargeable batteries. Applications of Ni–MH electric vehicle batteries include all electric plug-in vehicles.

5.9 Solar Armor Plate Battery

Solar armor plate batteries are most complex and most robust batteries, and have a shelf life of 15–20 years. Due to their large mass, they are not transferable. At present, there are two basic versions of these batteries at the market: liquid electrolyte and special insulators, and gel (Fig. 50).

In the depth of discharge of 50%, their lifetime is around 3500 cycles. Their drawbacks are great mass, higher price, and periodic maintenance of electrolyte.

5.10 Solar Block Batteries

Solar block batteries represent a combination of batteries with lattice and armor plates. They are designed only for the stationary use. In the depth of discharge of 50%,

Fig. 50 *Solar* armor plate battery

their lifetime is around 2100 cycles. They require maintenance every 0.5–3 years. Their drawback is a slightly higher price (Fig. 51).

Fig. 51 Solar block battery

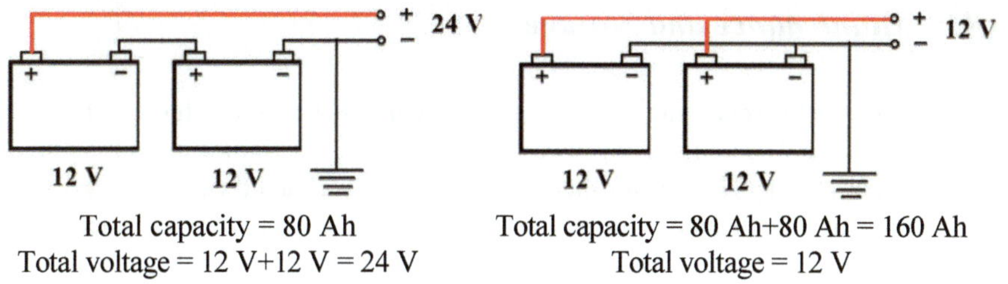

Fig. 52 In-series (left) and in-parallel (right) battery wiring

5.11 Wiring Solar Batteries

Most lead–acid batteries, like solar modules, can be connected together to form a longer battery bank (Fig. 52).

In serial connected batteries with same voltages and capacity, total capacity is equal to one battery capacity and voltage is a sum of batteries voltages.

In parallel-connected same batteries, total capacity is sum of battery capacity and voltage is equal to voltage of one battery.

5.12 Battery Requirements

Battery size is designed on the autonomy period (3 days), depth of discharge (e.g., 50%), and derating for round-trip efficiency (e.g., 75%).

Total daily load (Ah) requirement = daily energy (Wh)/system nominal voltage (V).

For lighting system, this is 12 V.

$$720\,\text{Wh}/12\,\text{V} = 60\,\text{Ah/day}$$

Required battery bank capacity = (days of autonomy × daily load (Ah)/total battery bank capacity).

$$(3\times 60\,\text{Ah})/(0.75\times 0.5) = 480\,\text{Ah}$$

Average daily depth of discharge = total daily load (Ah)/total bank capacity (Ah).

$$60\text{Ah/day}/480\,\text{Ah} = 12.5\%\ \text{daily}$$

For lead–acid batteries, generally one wants to design for 10–15% daily DOD.

5.13 Maintenance and Storage

Batteries are self-discharging devices which means that they discharge when not connected with PV systems. They have self-discharging rate indicated as the rate of discharge per month. This means that they need to be recharged periodically.

Batteries must be located in an area without extreme temperatures and adequately ventilated.

5.13.1 Overcharging

Overcharge is the excess Ah delivered to recharge the battery. Some overcharge is necessary to achieve full charge and to prevent sulfation. In PV systems, 1–4% overcharge is common. In conventional non-PV cycling systems, at least 10% overcharge is common.

5.13.2 Ventilation

Ventilation is needed to lose heat, especially internal heat produced on overcharge, and to get rid of gases. Sealed batteries in tightly enclosed surroundings will overheat often disastrously.

A charged open battery will produce hydrogen and oxygen gases. If the concentration of hydrogen in air exceeds 4%, there is an explosion hazard and the appropriate ventilation is necessary.

Sealed lead–acid batteries produce small amount of hydrogen due to internal corrosion. Sealed batteries required same ventilation more on cooling needs than on the removal of oxygen.

5.13.3 Temperature Control

In PV systems, batteries need protection from low and high temperatures. In a very hot climate, it is necessary to avoid direct sunlight on the battery enclosure, use light enclosure, allow of plenty air circulation by producing sufficient air space and ventilation.

5.13.4 Water Loss

For open batteries, the main maintenance requirement is to add distilled or demineralized water periodically [8, 10, 11, 13, 14].

5.14 Batteries for Electric Cars

Today, in the world of electric cars, batteries are used whose characteristics are given in Table 6.

In practice, the following batteries are most commonly used for EA: lithium-ion (Li-ion), lithium polymer (Li-poly), Na/NiC12, nickel–metal hydride (Ni–MH), nickel–cadmium (Ni–Cd) and lead.

EA batteries need to have high capacity, to quickly fill, to be emptied to the end, to have a high degree of useful effect, to have long lifetime, to have low cost, etc.

The EA autonomy, or its radius of movement between two battery recharges, depends on the capacity of the battery, efficiency of EA, driving mode, additional consumer in the vehicle, etc. Depending on driving conditions, the typical radius of the modern EA is 120–200 km.

Basic information about the battery capacity and the radius of electric cars moving are given in Table 7.

The world is working intensively to increase the charge speed and the radius of EA movement between two chargings. The charging speed and the EA motion range depend on charger characteristics and battery capacity.

Batteries need to be handled carefully because some types of batteries contain very toxic substances [19].

6 Charge Controllers

The solar controller (charge controller) is a device that controls the currents flowing between the battery, the PV array, and the load of the PV-LED systems and that

Table 6 Characteristics of several types of batteries used in electric cars

Type	Cell-rated voltage (V)	Specific energy (Wh/kg)	Energy density (Wh/l)	Specific power (W/kg)	Degree of efficiency (%)	Shelf life (cycle)	Price (USD $/kW)
Lead–acid battery	2.1	35–40	70	100–150	70–92	500–800	100–150
Ni–Fe	1.2	51	–	99	65	920	–
Ni–Cd	1.2	40–60	50–150	260	80	2000	–
Ni–MH	1.2	50–60	140–300	200–1000	76	600–1000	300–400
Zn–Br	–	75–85	–	40	75	350	–
Na–S	–	81	–	150	91	600	–
Li-ion	3.6	80–90	250	<1000	80	1200	–
Li-polymer	3.7	100	300	2000+	85	500–1000	–

Table 7 Information about battery capacity and radius of electric cars moving

Charge power	Connection type	Charge power (kW)	Charge current (A)	Vehicle autonomy/h (km)
Normal power	Alternating single phase	≤37	10–16	<20
Middle power	Alternating single phase or three phases	37–22	16–32	20–110
High power	Alternating three phases	>22	>32	>100
High power	Single phase	>22	>32	>100

ensures that the electrical parameters present at the battery are kept within the manufacturer specifications. It also protects the PV-LED elements against surges and short-circuit currents.

During the day sunlight time, the generated PV power is sent to the battery. When the battery is fully charged, and the PV panel is still connected to the battery, the battery might overcharge, which can cause several problems like gas formation, capacity loss, or overheating. This is avoided by decoupling the PV panel from the battery, which is one of the main functions of the solar controllers.

Another basic function of solar controllers is recognizing the start of night according to the voltage of the photovoltaic modules, in order to switch the supply of the load by the charged batteries. In some PV-LED systems supplying power to consumers during the day, the solar controller transfers power from the PV modules directly to the load using only the excess power for charging the battery.

During the cloudy days at low irradiance, the load energy needed may exceed the energy stored in the batteries, which can cause the heavily battery discharge. Over-discharging the battery has a detrimental effect on the cycle lifetime, as discussed above. The charge controller prevents the battery from being over-discharged by disconnecting the battery from the load.

For optimal performance, the battery voltage has to be within specified limits. The charge controller has to maintain an allowed voltage range in order to ensure a healthy operation. Further, the PV panel will have its U_{mpp} at different levels, based on the temperature and irradiance conditions. The best charge controllers perform an appropriate voltage regulation to ensure the battery operates in the specified voltage range, while the PV panel is operating at the MPP.

Solar controllers can be distinguished between *series* and *shunt* controllers—Fig. 53. In a series controller, overcharging is prevented by disconnecting the PV array until a particular voltage drop is detected, at which point the array is connected to the battery again. In a parallel or shunt controller, overcharging is prevented by short-circuiting the PV array. This means that the PV modules work under short-circuit mode, and that no current flows into the battery. These topologies also ensure over-discharge protection using power switches for the load connection, which are

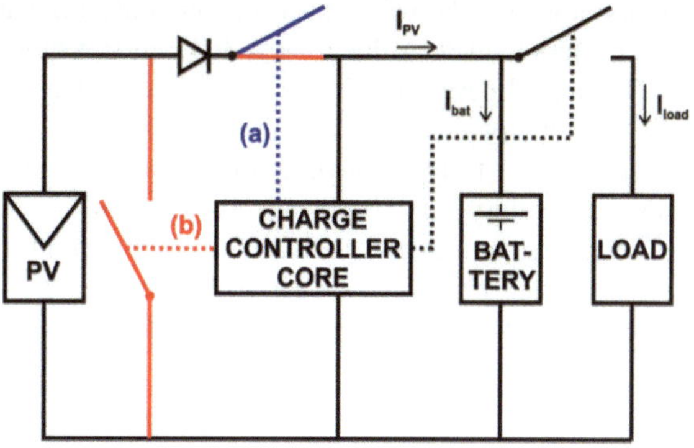

Fig. 53 Basic wiring scheme of **a** a series and **b** a shunt solar controller

appropriately controlled by the algorithms implemented into the charge controller algorithm [18].

6.1 Shunt Controller Designs

Since photovoltaic cells are current-limited by design (unlike batteries), PV modules and arrays can be short-circuited without any harm. The ability to short-circuit modules or an array is the basis of operation for shunt controllers.

The shunt controller regulates the charging of a battery from the PV array by short-circuiting the array internal to the controller. All shunt controllers must have a blocking diode in series between the battery and the shunt element to prevent the battery from short-circuiting when the array is regulating. Because there is some voltage drop between the array and the controller and due to wiring and resistance of the shunt element, the array is never entirely short-circuited, resulting in some power dissipation within the controller. For this reason, most shunt controllers require a heat sink to dissipate power and are generally limited to use in PV systems with array currents less than 20 A. The regulation element in shunt controllers is typically a power transistor or metal–oxide–semiconductor field-effect transistor (MOSFET), depending on the specific design.

There are a couple of variations of the shunt controller design. The first is a simple interrupting or on–off-type controller design. The second type limits the array current in a gradual manner, by increasing the resistance of the shunt element as the battery reaches full state of charge.

The *shunt-interrupting controller* completely disconnects the array current in an interrupting or on–off fashion when the battery reaches the voltage regulation set point. When the battery decreases to the array reconnect voltage, the controller connects the array to resume charging the battery. This cycling between the regulation voltage and the array reconnect voltage is why these controllers are often called "on–off" or "pulsing" controllers. Shunt-interrupting controllers are widely available and are low cost; however, they are generally limited to use in systems with array currents less than 20 A due to heat dissipation requirements.

The shunt-linear controller maintains the battery at near a fixed voltage by gradually shunting the array through a semiconductor regulation element, when a battery becomes nearly fully charged. In some designs, a comparator circuit in the controller senses the battery voltage and makes corresponding adjustments to the impedance of the shunt element, thus regulating the array current. In other designs, simple Zener power diodes are used, which are the limiting factor in the cost and power ratings for these controllers. There is generally more heat dissipation in shunt-linear controllers than in shunt-interrupting types [16].

6.2 Series Controller Designs

As the name implies, this type of controller works in series between the array and the battery, rather than in parallel as for the shunt controller. There are several variations to the series-type controller, all of which use some type of control or regulation element in series between the array and the battery. While this type of controller is commonly used in small PV systems, it is also the practical choice for larger systems due to the current limitations of shunt controllers.

In a series controller design, a relay or solid-state switch either opens the circuit between the array and the battery to discontinuing charging or limits the current in a series linear manner to hold the battery voltage at a high value. In the simpler series-interrupting design, the controller reconnects the array to the battery once the battery falls to the array reconnect voltage set point. As these on–off charge cycles continue, the "on" time becomes shorter and shorter as the battery becomes fully charged.

Because the series controller open-circuits rather than short-circuits the array as in shunt controllers, no blocking diode is needed to prevent the battery from short-circuiting when the controller regulates.

The most simple series controller is the series-interrupting type, involving a one-step control, turning the array charging current either on or off. The charge controller constantly monitors battery voltage, and disconnects or open-circuits the array in series once the battery reaches the regulation voltage set point. After a pre-set period of time, or when battery voltage drops to the array reconnect voltage set point, the array and battery are reconnected, and the cycle repeats. As the battery becomes more fully charged, the time for the battery voltage to reach the regulation voltage becomes shorter each cycle, so the amount of array current passed through to the battery becomes less each time. In this way, full charge is approached gradually in small steps or pulses, similar in operation to the shunt-interrupting-type controller. The principle difference is the series or shunt mode by which the array is regulated.

Similar to the shunt-interrupting-type controller, the series-interrupting-type designs are best suited for use with flooded batteries rather than the sealed VRLA types due to the way power is applied to the battery.

6.3 *Series-Interrupting, Two-Step Constant Current Design Type of Controller*

Series-interrupting, two-step constant current design type of controller is similar to the series-interrupting type; however, when the voltage regulation set point is reached, instead of totally interrupting the array current, a limited constant current remains applied to the battery. This "trickle charging" continues either for a pre-set period of time or until the voltage drops to the array reconnect voltage due to load demand. Then, full array current is once again allowed to flow, and the cycle repeats. Full charge is approached in a continuous fashion, instead of smaller steps as described above for the on–off-type controllers.

In a *series linear, constant voltage controller* design, the controller maintains the battery voltage at the voltage regulation set point. The series regulation element acts like a variable resistor, controlled by the controller battery voltage sensing circuit of the controller. The series element dissipates the balance of the power that is not used to charge the battery, and generally requires heat sinking. The current is inherently controlled by the series element and the voltage drop across it.

Series linear, constant voltage controllers can be used on all types of batteries. Because they apply power to the battery in a controlled manner, they are generally more effective at fully charging batteries than on–off-type controllers [16].

6.4 *Series-Interrupting, Pulse-Width Modulated (PWM) Design*

This algorithm uses a semiconductor-switching element between the array and the battery, which is switched on/off at a variable frequency with a variable duty cycle to maintain the battery at or very close to the voltage regulation set point. Although a series-type PWM design is discussed here, shunt-type PWM designs are also popular and perform battery charging in similar ways. Similar to the series linear, constant voltage algorithm in performance, power dissipation within the controller is considerably lower in the series-interrupting PWM design (Fig. 54).

By electronically controlling the high-speed switching or regulation element, the PWM controller breaks the array current into pulses at some constant frequency and varies the width and time of the pulses to regulate the amount of charge flowing into the battery. When the battery is discharged, the current pulse width is practically fully on all the time. As the battery voltage rises, the pulse width is decreased, effectively reducing the magnitude of the charge current—Fig. 55.

PWM controllers will operate close to the maximum power point but often slightly "above" it. An example of operating range is shown in Fig. 56.

The PWM design allows greater control over exactly how a battery approaches full charge and generates less heat. PWM-type controllers can be used with all battery types; however, the controlled manner in which power is applied to the battery makes

Fig. 54 PMW battery charge controller

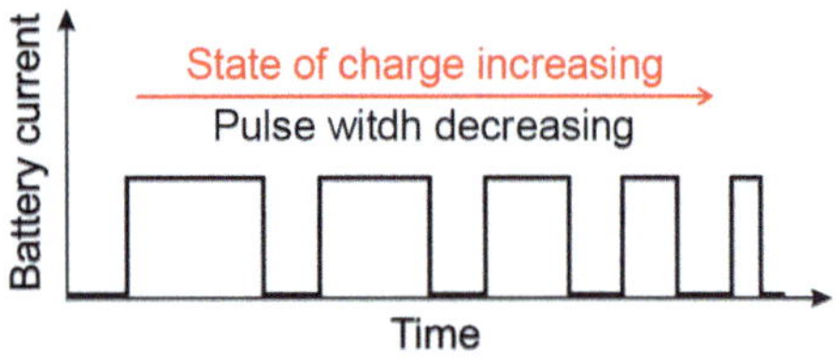

Fig. 55 Amendment of pulses in PWM charge controllers

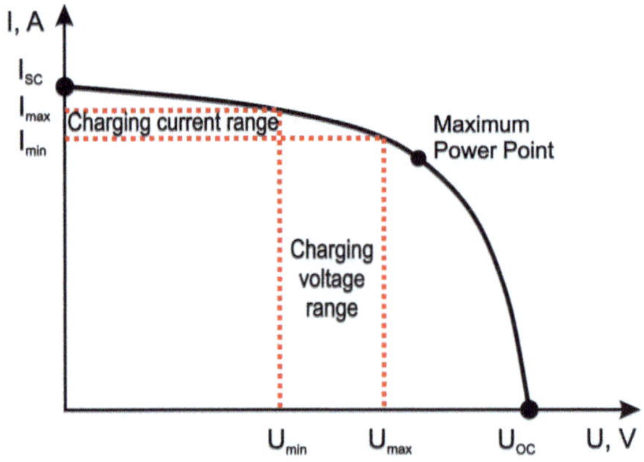

Fig. 56 Current and voltage charging ranges of PWM solar controllers

them preferential for use with sealed VRLA-type batteries over on–off-type controls. To limit overcharge and gassing, the voltage regulation set points for PWM and constant voltage controllers are generally specified lower than those for on–off-type controllers.

Fig. 57 MPPT battery charge controller

6.5 *Maximum Power Point Tracking (MPPT) Solar Controllers*

Unlike all of the controllers discussed above, MPPT controller provides an indirect connection between the battery bank and the PV array. This indirect connection includes a DC/DC voltage converter (Fig. 58) that takes extra PV voltage and transforms it into an additional current at a lower voltage without necessarily losing power. MPPT controllers are able to pull this off because of an adaptive algorithm that follows the MPPT of the PV array, which then adjusts the incoming voltage in order to maintain the most efficient level of power for the system in place (Fig. 57).

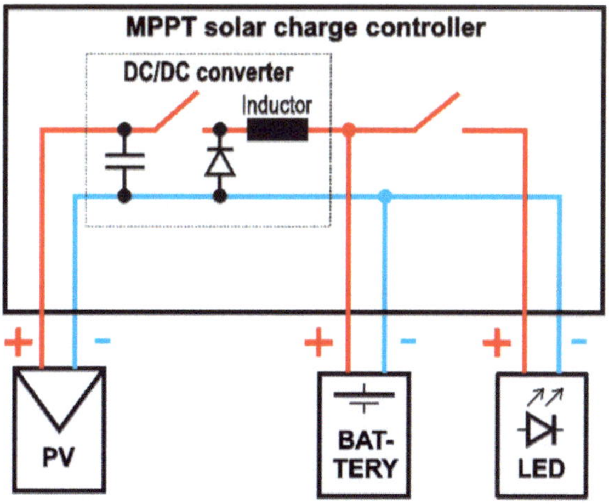

Fig. 58 PV-LED system with MPPT solar controller

The most outstanding feature of MPPT controllers is intelligent tracking input voltage from solar panel, which could let solar panel always working at maximum power point of *V–A* curve. Compared with the PWM solar charge controller, the MPPT controller could increase 10–40% electrical power using full efficiency from solar panel, especially when the solar cell temperature is low (below 45 °C), or very high (above 75 °C), or when irradiance is very low.

6.6 MPPT Charging Algorithm

The MPPT solar controller has a three-stage battery-charging algorithm for rapid, efficient, and safe battery charging—Fig. 59.

In *bulk charge* stage, the battery voltage has not yet reached boost voltage and 100% of available solar power is used to recharge the battery.

When the battery has recharged to the boost voltage set point, constant voltage regulation is used to prevent heating and excessive battery gassing. The *boost stage* remains 120 min and then goes to float charge. Every time when the controller is powered on, if it detects neither over-discharged nor overvoltage, the charging will enter into boost charging stage.

After the boost voltage stage, solar controller will reduce the battery voltage to float voltage set point. When the battery is fully recharged, there will be no more chemical reactions and all the charge current transmits into heat and gas at this time. Then, the solar controller reduces the voltage to the floating stage, charging with a smaller voltage and current. It will reduce the temperature of battery and prevent the gassing, also charging the battery slightly at the same time. The purpose of *float stage* is to offset the power consumption caused by self-consumption and small loads in the whole system, while maintaining full battery storage capacity.

In float stage, loads can continue to draw power from the battery. In the event that the system load(s) exceed the solar charge current, the controller will no longer be able to maintain the battery at the float set point. Should the battery voltage remain below the boost reconnect charging voltage, the controller will exit float stage and return to bulk charging.

Table 8 shows the main advantages and disadvantages of both PWM and MPPT-type solar charge controllers discussed above.

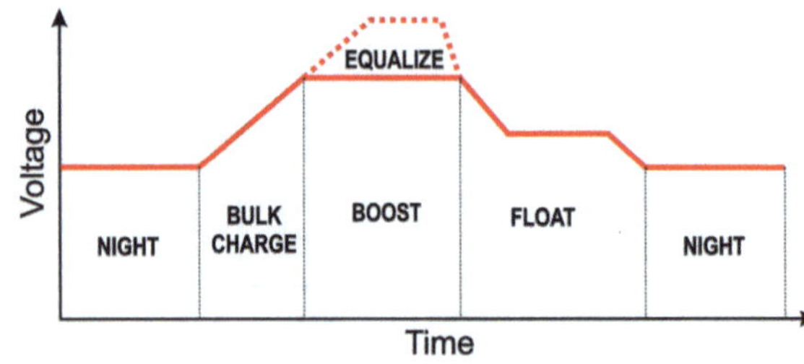

Fig. 59 MPPT charging algorithm

Table 8 Pros and cons of both PWM and MPPT-type solar charge controllers

PWM-type solar controllers	MPPT solar controllers
Advantages	
– Built on a time-tested technology. They have been used for years in solar systems and are well established – Inexpensive – Available in sizes up to 60 A – Durable, most with passive heat sink-style cooling – Available in many sizes for a variety of applications	– Offer a potential increase in charging efficiency up to 40% – Offer the potential ability to have an array with higher input voltage than the battery bank – Available in sizes up to 80 A – Warranties are typically longer than PWM – Offer great flexibility for system growth – The only way to regulate grid-connected modules for battery charging
Disadvantages	
– The solar input nominal voltage must match the battery bank nominal voltage – There is no single controller sized over 60 A DC as of yet – Many smaller PWM controller units are not UL listed – Many smaller PWM controller units come without fittings for conduit – PWM controllers have limited capacity for system growth – Cannot be used on higher-voltage grid-connected modules	– More expensive, sometimes costing twice as much as a PWM controller – MPPT units are generally larger in physical size – Sizing an appropriate solar array can be challenging without MPPT controller manufacturer guides – Using an MPPT controller forces the solar array to be comprised of like photovoltaic modules in like strings

As temperature has a considerable influence on the electrical parameters and life span of the battery, contemporary charge controllers have connected a temperature sensor, which is attached to the battery housing. This sensor allows the charge controller to adjust the electrical charge parameters of the battery to the temperature. The charge controller thus keeps the operating range of the battery within the optimal range of voltages [17].

7 Inverter

The inverter is a device that converts the DC voltage of 12 or 24 V into the AC voltage of 110 V/220 V. Inverters, which are used in photovoltaic system to supply AC power to the consumers, use MOSFET (unipolar transistors), whose output power ranges from 100 W up to 32 kW. Selecting the right inverter to use in photovoltaic solar systems depends on the waveform of the output voltage, the load requirements of the system, the efficiency of the system, etc. Depending on the waveform of the output voltage (output signal), inverters can be divided into: a rectangular output signal inverters, modified rectangular output signal inverters, modified sine output signal inverters, and pulse-width modulated output signal inverters (Fig. 60).

Fig. 60 Square wave voltage

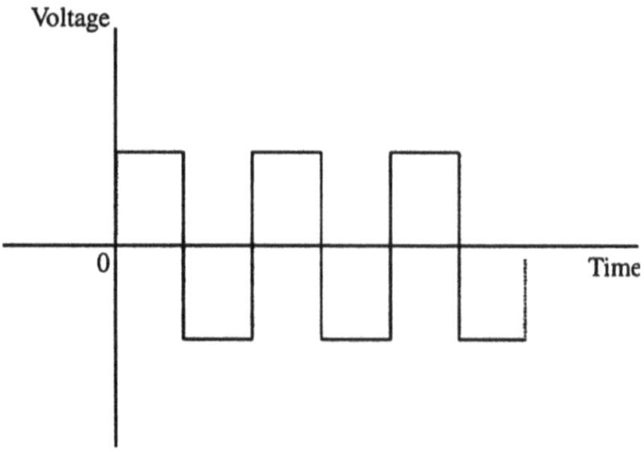

Rectangular output signal inverters fall into the group of the simplest and cheapest inverters, their efficiency ranging from 70 to 98%. This type of inverter converts DC voltage at its input into an alternating voltage of a rectangular waveform at the output and is typically used when consumers are not so demanding in terms of the purity of the input signal, because the higher harmonics can cause problems in engines and lampfluorescents. They are manufactured to the nominal power of 1 MW, with a very high capacity of the impact loads (up to 20 times). The use of the rectangular output signal inverters is limited due to the presence of higher harmonics that cause disfiguration (distortion) of the output up to 40% (Fig. 61).

Modified sine output signal inverters have a distortion of the output signal about 5% and the efficiency of over 90%. The negative features of these inverters are small nominal powers of 300–2500 W (Fig. 62).

Pulse-width modulated output signal (voltage) inverters give an output signal that is most similar to the shape of a sinusoidal signal. The operation of this type of the inverter is based on the pulse-width voltage modulation (PWM signal). The inverter

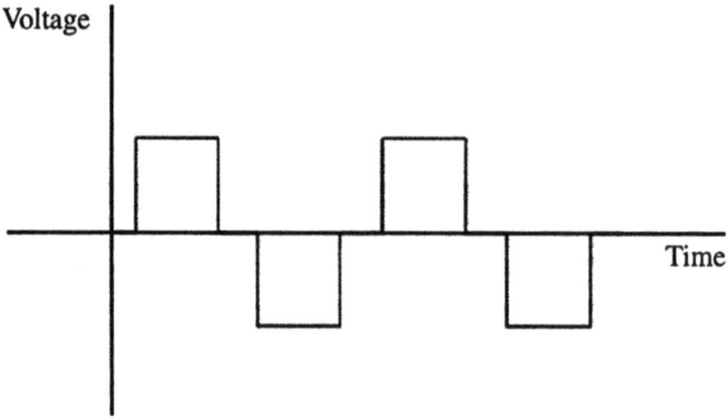

Fig. 61 Modified sine wave voltage

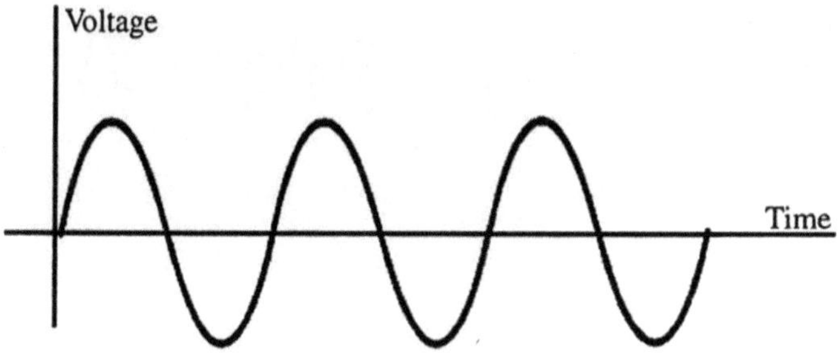

Fig. 62 Sine wave voltage

of this type is of a more complex structure, with the smallest distortion, less than 5% and the efficiency of over 90%. This type of inverter has good possibilities in terms of the nominal powers ranging up to 20 kW per unit and is used in a lot of cases that need good precision sine wave.

The highest quality, but also the most expensive are the inverters with pure sine wave. Usually, these are the inverters that are used for network applications of the photovoltaic systems and belong to the group of *line-commutated inverters*. Cleanliness of the sine wave is obtained by using complex filters that reduce the efficiency of the system below 80%. Nominal powers range up to 2 kW, while the distortion is less than 1% (Fig. 63).

With modified sine wave signal can be run fax machines, laser printers, equipment with variable speed motors, oxygen concentrators, high-voltage cordless tool chargers, electric shovels, garage door openers, lighting etc.

Pure sine wave inverters are the most expensive, and they deliver the most reliable and consistent wave output.

In solar power plants, one central inverter of higher power or several smaller, less power inverters connected in a string can be used.

In private households, an inverter of the corresponding power is usually used [12].

8 Solar Cells' Application

Solar modules are the basics of the photovoltaic system and contain a certain number of serial- or parallel-connected solar cells, whereby serial-connected solar cells increase the output voltage and parallel-connected solar cells increase the output current.

Solar modules are usually fabricated from monocrystalline, polycrystalline, and amorphous silicon. On the market, solar modules of different powers can be found, most frequently those power of 50 W, 100 W, 150 W, and 200 W.

Solar cells can be used for: the lighting, functioning of the audiovisual and refrigeration equipment, signaling devices on roads, tunnels, airports, and lighthouses, operation of telecommunication equipment and systems, operation of solar power

Fig. 63 SMA Sunny Boy 2000HF inverter

plants, with PV systems, electricity supply to facilities, ships, aircraft, cosmic stations and satellites, etc.

The photovoltaic solar system means the system by which solar radiation is converted into electrical energy and consumers are supplied by DC and/or AC power. Photovoltaic solar system can operate independently of the power grid or can be attached to it. Depending on the components of which it is composed, photovoltaic solar system that is not connected to the power grid can supply consumers with DC or AC power.

8.1 Stand-Alone Photovoltaic Systems

Stand-alone (off-grid) photovoltaic systems denote systems which are not connected to the power grid and can supply consumers with direct current (DC) or alternating current (AC). These systems are commonly used as a source of electricity for remote homes, caravans, sailing boats, boats, telecommunication repeaters, etc. Schematic of the stand-alone photovoltaic system which supplies consumers with DC power is given in Fig. 64.

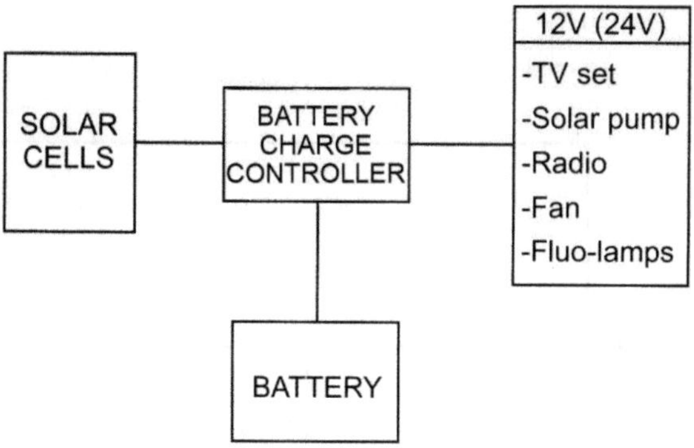

Fig. 64 Schematic of the stand-alone photovoltaic system which supplies consumers with DC power

Stand-alone photovoltaic system for the supply of consumers with AC power consists of solar modules, battery charge controller, battery, and inverter of DC to AC power.

Schematic representation of stand-alone photovoltaic system that supplies consumers with AC power is shown in Fig. 65.

With this system, using the charging controller battery, DC power consumers can be connected to the battery.

The photovoltaic system for supplying the consumers with AC and DC power is given in Fig. 66.

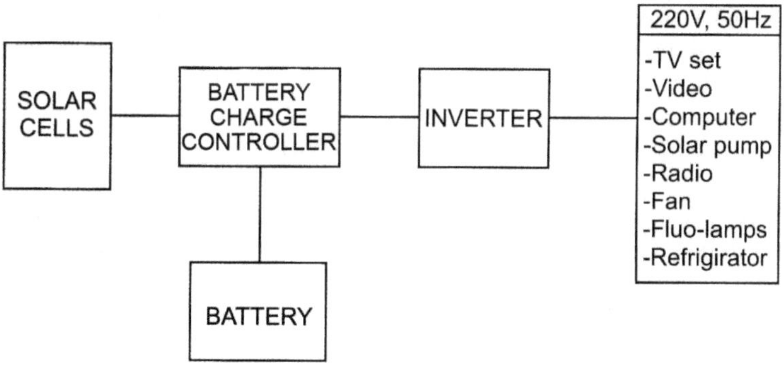

Fig. 65 Schematic of stand-alone photovoltaic system that supplies consumers with AC power

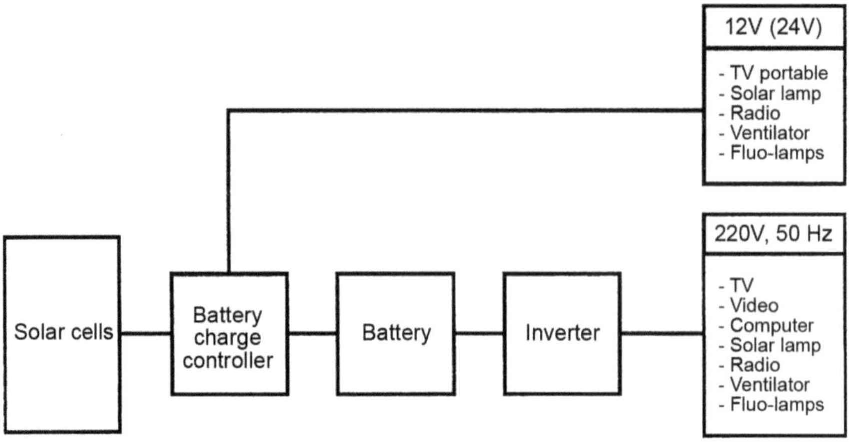

Fig. 66 Photovoltaic system for supplying the consumers with AC and DC power

8.2 Determining Characteristics of the Stand-Alone Photovoltaic Solar System

When designing a photovoltaic solar system for electricity supply to the consumers, it is necessary to take into account the amount of electrical energy which is generated during the day by one solar module, the daily consumption of electrical energy, the geographical position of the object, regulator and converter power, battery capacity, etc.

Electrical energy generated by one solar module during the day E_{m} (Wh) is determined by multiplying the power of the solar module P_{m} (W) and the geographical factor G_{F} (h):

$$E_{\mathrm{m}} = P_{\mathrm{m}} \cdot G_{\mathrm{F}}\,(\mathrm{Wh}). \tag{12}$$

The geographical factor G_{F} represents the mean daily insolation duration (sunshining) in hours in a given geographical area within one year. The values of the geographical factor G_{F} are taken from the solar map given in Fig. 67.

The amount of electrical energy, which is generated by solar cells, is influenced by:

- Local weather conditions, which can be quickly changed in the very narrow geographical area,
- Installation of solar cells—orientation and position of the solar cells can greatly affect the amount of the generated electrical energy,
- Seasons—the calculations are made on the basis of the annual mean values of the solar radiation intensity.

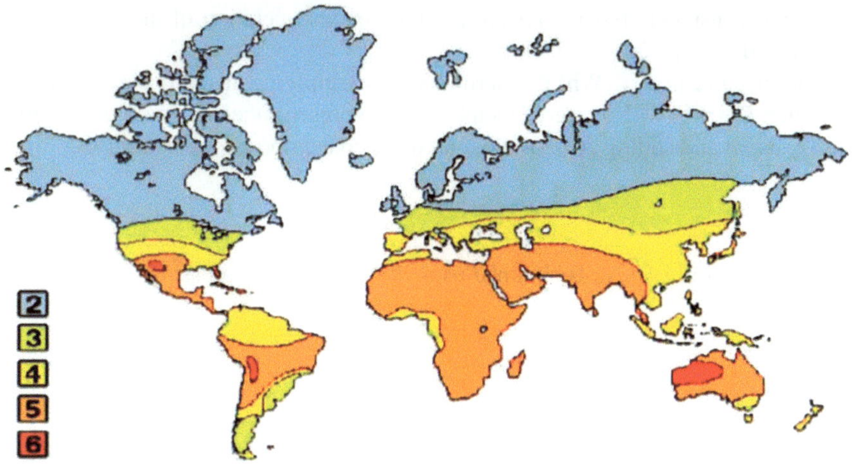

Fig. 67 Solar map for determination of geographical factor G_F

Daily consumption of the direct current E_j (Wh) is calculated by multiplying the individual power of the direct current consumer P_j (W) with the median use time of the consumers t_j (h) and adding to the obtained value 30% to compensate for the losses in the solar system devices:

$$E_j = 1.3\,\Sigma\,P_j \cdot t_j\,(\text{Wh}) \tag{13}$$

Daily consumption of the alternating current E_n (Wh) is calculated by multiplying the individual power of the alternating current consumer P_n (W) with the median use time of the consumers t_n (h) and adding to the obtained value 40% to compensate for the losses in the solar system devices:

$$E_n = 1.4\,\Sigma\,P_n \cdot t_n\,(\text{Wh}) \tag{14}$$

The total daily consumption of electric energy E_p (Wh) is equal to the sum of the daily consumption of the direct current E_j (Wh) and alternating current E_n (Wh):

$$E_p = E_j + E_n \tag{15}$$

The number of solar modules N is the quotient of the total daily electricity consumption E_p (Wh) and the electrical energy generated by one solar module E_m (Wh):

$$N = E_p/E_m \tag{16}$$

Battery capacity depends on the daily needs of consumers for the electrical energy (daily capacity) and the need for electrical energy for the autonomous operation of certain consumers (spare capacity). Autonomous operation of the consumer is determined in days.

Battery capacity K_A (Wh) is determined by multiplying the daily electricity consumption E_p (Wh) with a reserve factor (which represents the number of consecutive cloudy days) and adding 30% of the total daily consumption of electricity:

$$K_A = 1.3\, E_p \cdot \mathrm{n}\ (\mathrm{Wh}) \tag{17}$$

Battery capacity in Ah is obtained using the following expression:

$$C_A = \frac{K_A}{U_A} (\mathrm{Ah}) \tag{18}$$

where K_A is the battery capacity in (Wh) and U_A is the battery voltage in (V).

8.3 *Grid-Connected Photovoltaic Solar System*

Grid-connected photovoltaic solar systems consist of solar modules, inverters, power meter, and connecting lines for the connection of the solar system to power utility grid. In these systems, the entire amount of generated electricity is transmitted to the power utility grid. These systems include large PV solar power plants and small PV solar power plants installed on private homes, residential areas, and other facilities.

8.3.1 Combined Solar System

The combined solar system means the system which one part of the generated electricity transmits to the power utility grid, while the rest is accumulated in the adequate batteries and is used to supply the customers with the DC or AC current in a given facility (Fig. 68). In this system, the generated electricity transmittance to the power utility grid is performed through the inverter, electric meter, and the appropriate ports. The second part of energy is via the battery-charging controller transmitted to the battery. Consumers of DC can directly connect to the battery. If inverter is connected to the battery, the alternating current can be used as well.

8.3.2 Hybrid Power System

The hybrid power system means a system that simultaneously generates electricity by using two or more independent sources of electrical energy such as photovoltaic systems, wind generators, and diesel generators.

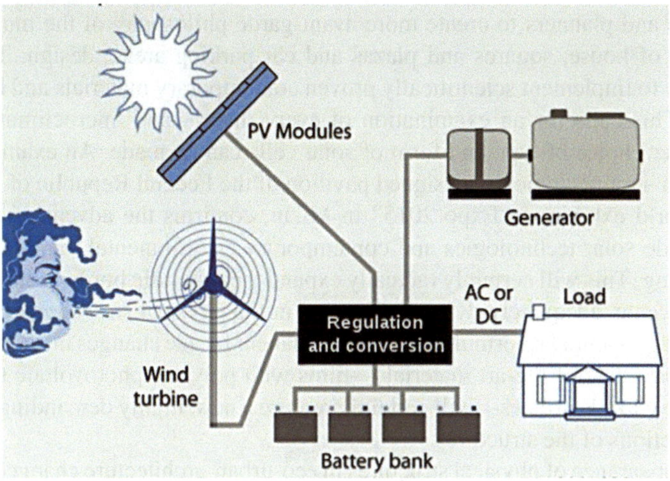

Fig. 68 Hybrid power system [35]

9 Urban Application of PV Systems

9.1 General Remarks

A different, non-globalist pattern in conceptual formation of the buildings' façade planes, coupled with simultaneous observation and support of sensitive esthetic—geometrical differences in urban space, is based on the idea of a rational usage of energy and other natural resources, the new functional technologies providing a reduced, efficient consumption and progressively interesting design of layouts. The contemporary facades with high technical and technological performances provide exquisite potential in conceptualization of the sensitive culture of structure building. Eco-urban-architecture is on a quest for a new identity of "smart" facades and roofs, for an articulation based on multidisciplinary scientific principles and analyses of the material thermal properties in various façade systems. From the theoretical standpoint, this means prioritizing the design of high-performance facades based on the climate conditions in micro-ambient environment as well as on the characteristics of sustainability. Apart from that, thermal behavior and resistance to weather conditions are very important, particularly resistance to humidity. Good contemporary design of solar facades includes thermal comfort in the interior space, quality lighting, and acoustic stability in the environment. The latest city-building experiences of solar cells illustrate the implementation of double facades on the structures with technologically exclusive, artifact materials functioning as energy generators.

Recent studies focus on the environmental and strategic considerations of an impeccable position of the structure layout, adjustment, and orientation of the façade plane forms to observe position of the sun. Energy-efficient facades ask from the designers and planners to create more avant-garde philosophy of the multicultural character of house, squares and plazas and car parking areas' design. Thus, it is necessary to implement scientifically proven contemporary materials and high technology. This calls for an examination of every macro- and microclimate area so that proper choice of type and form of solar cells can be made. An example of the eco-urban–architectonically designed pavilion of the Federal Republic of Germany, at the world exhibition "Expo 2015" in Milan, confirms the advent of new, very avant-garde solar technologies and contemporary environmental materials in civil engineering. This will certainly radically expand the scientific borders and announce radical, scenic, adequately dynamic changes in the physical structures of the cities worldwide. Essential reformulations and permanent facade changes introducing flexible, robust, state-of-the-art materials—films with polymer photovoltaic solar cells to replace glazed surfaces—will certainly require a new, highly demanding architectonic functions of the structures.

The appearance of physical structures in eco-urban-architecture changed substantially at the turn of the century. Integration of solar cells in the architectonic designs of the structure façade planes brought about strategic energy and historical–environmental changes. There were significant innovations concerning energy efficiency in using exterior and interior space and immediate environment. The interpolated solar cells significantly and strategically defined a new non-stereotypic identity of the ambient entities and their esthetics. Their impact generates a different city-building, material-synthetic philosophy, and the urban art culture. Dramatic technical–technological changes in the formation of the houses and installations in them occurred. The innovations can be interpreted from the vertical and horizontal volumes, in spiritual and material-texture changes of the appearance of streets, squares, plazas, and cities. A new age ensues of global creation of the designing–planning values, within the matrixes of environmental–urban–architectonic agglomerations with a new historicity and an artifact, attractive morphology in a community fostering different spatial-conceptual forms of communication. The use of solar cells brings new challenges and creative conceptual solutions and scientific and professional intervention in redefining urban structure of the cities [20].

9.2 BIPV Systems

A building-integrated photovoltaics (BIPV) system consists of PV modules integrated into the building envelope, such as the roof or the façade. This technology provides architects with completely new possibilities to incorporate solar technology into buildings. Photovoltaic (PV) systems and architecture can now be combined into one harmonious mixture of design, ecology, and economy. Wide variety of elegant forms, colors, and optical structures of cells, glass, and profiles enhances creativity and modern architectural design. Solar cells can be incorporated into the façade of a building, complementing or replacing traditional view or spandrel glass. Photovoltaics incorporated into awnings and sawtooth designs on a building façade increase access to direct sunlight while providing additional architectural benefits such as passive shading. PV shade screens provide a large area for generating electricity and also reduce solar heating in the summer, cutting cooling loads, and glare. PV shade screens can be retrofitted onto existing buildings or integrated into a new building's design. Using PV for skylight systems is an exciting design feature. By simultaneously serving as the building envelope material and power generator, BIPV systems provide savings in materials and electricity costs; reduce use of fossil fuels and the emission of ozone-depleting gases; and add architectural interest to the building. The possibility of installing PV generators directly at the point of energy use, and the development of PV modules suited for building integration, makes PV an ideal technology for deployment in the urban environment.

BIPV generate considerable fractions of urban electricity without the need of dedicating exclusive surface areas for PV plant installations. The building envelope provides the surface area for the PV plant at premium urban locations. In many cases, building's electrical installation provides PV plant electrical interface to the public utility grid. Energy is generated at very close proximity to the end user, thus avoiding infrastructure investments and losses in transmitting and distributing electrical power.

When considering BIPV systems, various factors must be taken into account such as shading, installation angle, and orientation but the most important ones are: available solar irradiation and local weather conditions.

Worldwide BIPV systems have been increasingly used for the energy independence of the residential and other objects. Therefore, an issue of the electrical energy generated by BIPV systems in relation to their orientations on the objects and local climate conditions is of a vital importance [21].

9.3 Examples of BIPV Systems

9.3.1 Photovoltaic Honeycomb Glass Screen Façade, New York

This design from 2008 of a building in one heavy-traffic street of New York is based on the honeycomb model according to the concept of the Japanese architects Daisuke Nagatomo and Minnie Jan, inspired by the natural world. The frontal façade wall has a glazed, undulating photovoltaic glass panel with a regular hexagonal geometry, while another panel on the honeycomb structure was installed facing the interior, carrying an efficient LED system for lighting, and displaying of variable colored images and information toward the exterior—the street. The honeycomb structure is very rigid, so that it can statically support the façade screen which can be used for entertainment. This attempt to interpolate a new eco-urban–architectonic structure and bring a metamorphosis in space is a new city-building stimulating strategy of permanent inauguration of contemporary materials and solar technologies in the micro-ambient space (Fig. 69).

Fig. 69 Photovoltaic honeycomb glass screen Façade by Daisuke Nagatomo from Japan [40]

9.3.2 Arizona State University's Photovoltaic Installations

The Arizona State University's photovoltaic installations are one of the largest commitments to solar energy by a university in the USA. In the period 2009–2013, 11 installations, total capacity of 1231 kW, were fitted. Particularly indicative is the solar installation on the roof of the garage with three floors for 420 cars, which produces energy equivalent for 275 households. The advantages of such installation are considerable. The university has a number of financial benefits in the society because of the rational and efficient energy status. Apart from that, the eco-urban-architecture appearance of artifact structures is considerably different. It is a harbinger of new hybrid designing era in engineering, which is a globalizing, non-static view of the world, with harmonic connection of structural elements (Fig. 70).

9.3.3 All Valley Solar, Beijing, China

The increasing number of the environmental hybrid and electric cars challenged planners, urban planners, and architects in shaping of parking areas. Covering of stationary traffic areas using photovoltaic panels opens up new potential for charging batteries of the users of such cars. Industrial Designer and Architect Neville Mars tried to design an interesting car reserve, where the cars apart from being in the cool shade will have a possibility to recharge from the source of generated electric energy.

Fig. 70 Arizona State University's photovoltaic installations are one of the largest commitments to solar energy by a university in the USA [41]

Solar panels are fitted on the posts with power sockets. After parking a car, a driver can easily connect the battery system and recharge it with environmental energy. The "solar forest," with a set of photovoltaic panels, apart from being useful has a very good, eco-urban–architectonic–esthetic, radically new geometry and recognizable non-rigid-angled geometrical composition similar to the natural forest forms, thus representing a good platform for the future changes in the growing urban environments. It is particularly interesting for the region of China, having an accelerated economic growth, under the market condition pressure, where Neville Mars opened in 2003 the foundation (DCF) in Beijing, a center for solar research (Fig. 71).

Fig. 71 Pin by Ted Bavin on all valley solar [42]

9.3.4 Expo 2015 in Milan—German Pavilion

The key motto of the world exhibition "Expo 2015" in Milan is Feeding the Planet, Energy for Life. The Universal Exposition brought together 145 countries representing 94% of world's population. They participate either via self-built pavilions or in the context of clusters. Three international organizations, 13 NGOs, and 5 corporate pavilions are also participating in the event. The largest and most attractive eco-urban–architectonic–technological pavilion in the complex is 2680 m^2, belonging to the Federal Republic of Germany, Architect Schmidhuber. On behalf of the Federal Ministry of Economics and Energy, Messe Frankfurt has entrusted the German Pavilion Expo 2015 Milan Consortium (ARGE) with the realization of the German Pavilion. The ARGE, as general contractor, is responsible for designing, planning, and construction of the German Pavilion and the exhibition. The Schmidhuber architectural office in Munich is responsible for the spatial concept, architecture, and general planning, Milla & Partner in Stuttgart for the content concept as well as the design of the exhibition and media. Nussli in Roth (near Nuremberg) is responsible for the project management and construction. The exterior warped façade surfaces were formed by the flexible, lightweight robust, state-of-the-art material—film with polymer photovoltaic solar cells, which provides exquisite air-conditioning of the building and energy generation. Science fiction came true. The sunlight is transformed into a stream of charged particles generating electric power. The new phenomenal procedure with the printed electronic components was made possible by combining several types of materials, including application of high-quality plastics. All the materials have different electric and optical properties. The implemented innovative technology has an important function in reduction of the external usage of energy in construction industry and over time proves to be a considerable resource-saver. For the architecture of tomorrow, this discovery is of an extraordinary importance. This triggers the fascinating potential of the sustainable physical structure and provides a number of functional advantages in eco-urban–architectonic equipping, remodeling, reconstructions, and modernizations of urban areas (Fig. 72).

9.3.5 AG Facade in Paris

An example of a façade design with solar panels in Paris demonstrates how one interpolated urban–architectonic physical structure in Quai de Valmy street in 10th arrondissement produced a complex esthetic–visual, engineering designing, and historical discrepancy with the surrounding structures. The form of the newly designed structure was seen as physical space and a new technological work of art, whereby the simplified content was taken away from the historical and culturological coordinates. The photovoltaic panels efficiently provide thermal and electric energy for the users. Since December 2011, this has been the first housing building in Paris with photovoltaic panels containing 7200 solar cells and over 170 m^2 panels on the frontal face protecting users in the interior space from the noise and weather effects. There are no solar panels at the ground level (Fig. 73).

Fig. 72 Expo 2015: field of polymer solar cells in the German Pavilion [43]

Fig. 73 A Sunways AG Facade—Quai de Valmy 179 in Paris [44]

Fig. 74 Dubai's vertical village has a skirt of photovoltaics [45]

9.3.6 Dubai's Vertical Village

The vertical district-village in Dubai has prominent sequences of solar panels in façade planes. They are an artifact means for collection of solar radiation and conversion into the energy required for the multifunctional structures. The designers think that such manner of building formation, apart from the esthetics, can result in good, new sustainable structures. Most of collectors are located on the south side of the location, to most efficiently receive solar rays with a potential of the automatic adjustment of solar collector angle, tracking the sun. The building roofs have similar solar surfaces, kinetically adjustable for the optimum operation conditions. The buildings have a nonstandard, futuristic, environmental–architectonic appearance with the spider web of solar panels. The future users will have the potential to enjoy nicely shaped, innovative, and abundant modern interiors of the hotels, cinemas, trade centers, and theaters. They will also enjoy the exterior landscaping of micro-ambient areas (Fig. 74).

9.3.7 Saint-Etienne, France, International Design Center

The city of design in Saint-Etienne, France, has the state-of-the-art photovoltaic panels in triangular pattern on the vertical and roof façade planes. This is a good example of the environmental architecture on the location of Saint-Etienne, where there was once a munitions factory. The historical complex is completely renovated and functionally interpolated into the historical tissue of the city with the new appearance of solar panel façade planes. The replenishment of night lighting, heating, and circulation of air in the great, magnificent International Design Center is thus solved by transformation of the solar energy. The building is a monolithic, attractive, esthetic–visual–physical structure, on a rectangular layout, length of 200 m. The façade skin is composed of 11 different equilateral, modulated triangles. The "Cite du Design" is an international design center and an institution for communication

Fig. 75 Saint-Etienne, France, International Design Center [46]

and research. The project is located on the National Manufacture d'Armes, a site of a former munitions factory, in Saint-Étienne, France. Status: 2004–2009, opening October 1, 2009. Location: Saint-Etienne, Loire, France. Surface: 17,250 m^2 (net), 64,000 m^2 Cite Du Design. Price: Prix Spécial de l'Équerre d'Argent pour la Cité du design de Saint-Étienne, 2010; Prix de la Construction Métallique, Bâtiments à usage tertiaire, and Ouvrages d'art, 2010 (Fig. 75).

9.3.8 Incheon International Airport in Seoul

A triangular pattern with solar panels was implemented in design and realization of the building of Terminal 2, of the "Incheon" International Airport in Seoul on the curved roof surface thus uniquely blending the façade at the ground level, to radically reduce the primary consumption of electric power. The passenger and cargo traffic will be considerably increased in this futuristically conceived terminal with 72 gates, on the location of 7×10^6 m^2. The structure will be completely ready for the Olympic Games in Pyeongchang in 2018. The Terminal 2 hall was designed by the Gensler & HMGY in cooperation with the Heerim–Mooyoung–Gensler–Yungdo (HMGY) Consortium (Fig. 76).

"We designed Terminal 2 with a great central garden in the interior—sizing two football pitches, were motivated by the Korean nature, and we offered an easy and practical usage to the people, wishing to announce new standards in airport design in the world", said Keith Thompson, Head of the Gensler company. The idea that airports are no longer only the aviation infrastructure, but also comfortable energy-efficient generators of the cities, is a new reference and a new concrete orientation in eco-urban-architecture.

Fig. 76 Incheon International Airport in Seoul, South Korea, demonstrates the trend for curved structural forms in buildings [47]

9.3.9 High-Rise Greens Sydney Skyline

The double glass curtain wall for the first high-rise 26-story administrative-business building in Sydney has managed to provide optimum daylight and solar control. The energy efficiency is prominent, showing twice as good results as the standard buildings with a single facade. Its vertical concept essentially differs from the other buildings in the environment. The building has an elliptical layout with 1600 m^2 per floor. It is clad with dark, reflecting glass which looks like "Darth Vader" mask. Darth Vader (born Anakin Skywalker) is a fictional character in the Star Wars universe. He appears in the original trilogy as a pivotal figure, as well as in the prequel trilogy as a central figure. It has a streamline design. In the central part, there is a well-ventilated atrium having ameboid layout, illuminated through the large glass surface, as well as the interior working space ventilated by the air circulation from the floor structures, toward the upper zones, respecting highest environmental standards. The most contemporary materials and technology reducing dependence from the usage of artificial light and mechanical heating and cooling were implemented, thus ranking iconic structure among the "smart" buildings. Particularly important are the solar panels on the roof of the structure, tracking the sun, which yield good energy results and esthetically well define the fifth façade of the building (Fig. 77).

9.3.10 GT Tower East in Seoul

The concept in the design of the attractive, flowing "twist" façade of Seocho Garak Tower East—GT Tower, 130 m high with 24 floors in Seoul, contains solar panels, which apart from the transformation of solar energy bring about a new visual quality

Fig. 77 Sustainable High-Rise Greens Sydney Skyline [48]

and physiognomy of the micro-ambient space with mirrored, deformed visual effects. Technical details: Project Architect: Peter Couwenbergh. Floor area 54,530 m^2. Design and construction: Architect Het Architecten Consort. Developer: GT Construction—Hanbit Structural Engineering (Fig. 78).

Fig. 78 Seocho Garak Tower East. GT Tower East is a 130-m-high contemporary office tower designed by the Dutch architectural firm Architecten Consort and located in Seoul, South Korea [49]

Even though it is located in a cramped space, this archi-sculptural tower of 2010 changes the character of the total geometry of vertical lines in the immediate environment. It is a new philosophy of formation and building of energy-efficient, non-stereotypical physical structures with demanding prefabricated elements. The designers found an inspiration for the appearance of the geometrically fluid façade in the form of the traditional Korean ceramics undulating the frontal glazed façade to create the contrast and dynamics. At night, the tower is illuminated with several thousand LEDs accentuating the unique appearance of the structure through the variation of light coloration [20].

10 Rural Application of PV Systems

Never in the global world was there a need for such a high demand and supply of a cheap, clean, and safe energy as there is today. Moreover, technical and technological changes are required which will not significantly disturb the environment and which will bring noticeable savings. The application of solar panels in rural areas is a potential and priority resource for achieving a long-term energy-saving plan and significantly improving the conditions in the daily life of the population. At first glance, these are small changes, but they produce high content targets in the part of reducing the consumption of energy from oil refineries. The energy efficiency with the use of solar panels in rural, heavily insulated agglomerations of India, Pakistan, Tanzania, Greece, Latin America, Africa, and many European and other countries has shown a reduction in economic costs in the realization of clean energy production and a visible reduction in air pollution and gas emissions, reduction of greenhouse gases, and preservation of existing capacities. In addition, transport costs of energy products are reduced. The greatest effects have been achieved in creating a new sustainable world in which a safe, secure, healthy, and productive life for all the people of the twenty-first century on our planet is ensured. The use of renewable energyenergy sources today is a priority strategy for each country. It is reflected in the savings of non-renewable energyenergy sources, environmental protection, the most rational possible exploitation, and regenerative characteristics. The current use of natural resources reserves involves strict planning and management in the further ecological and economic development of sustainable communities.

Solar panels in rural areas, as well as urban ones, considerably change the ecological urban architecture of diverse and functional–scenic properties in the organization of micro-ambient space. They change the physiognomy of iconological-significant images on the utilitarian contents of façade, vertical and horizontal, especially roof spaces. They bring a different, completely new, positive functional articulation in relation to the artifact–natural physical structure, the change of the historical–esthetic dimensions with which the poor environments become significantly more advanced and organized. Rural areas in today's world of pure energy are in the shortage. This is especially so on those marginal, border and inaccessible mountain-hills and remote areas where the electricity network is still not available. We are dealing with the interpolation of new revolutionary technological solutions in the design of rural environments in which meeting the needs for clean energy is in the focus of interest of hundreds of millions of people.

Changes are visible in the part of the reduction of micro location pollution and the establishment of a better balance with the natural environment in conditions where the global demand for clean and cheap energy, environment friendly, is increasing enormously, especially for renewable, reliable, energy sources and when the use of solar energy in rural areas is the best targeted alternative. Despite the notion that solar energy in rural areas does not always provide enough energy to maintain electricity for a long time and that adverse weather conditions often limit the capacity of electricity produced by modern solar technology, solar panels are a viable and objective potential, a very useful, reliable, and meaningful solution. In the near future, relying on solar energy will increasingly reduce the cost of future generations of rural users in electricity utility companies. The traditional sources of energy production will be replaced by renewable sources. Solar energy is a great advantage for improving the living standards of the rural population, the chance and challenge to overcome conflicting environmental and economic goals in creating their long-term harmony and healthier environment as well as other comfort. In addition, the use of solar panels in rural areas increases the social awareness of creating the necessary quality and importance of the built environment. Although this issue is partially marginalized in some countries, it is an area of national interest in the recognition of the culture of the people. More broadly, the interpolation of this high technology shows an image of the regional and local identity of the rural environment, as well as the quality of the built environment. Such an approach, for future generations of clean energy users, means an organic national strategy for the organization of the rural areas and their development potentials. In such a context, it is necessary to ensure the sustainable installation of new physical solar panel structures as well as their ecological and economic acceptability. Such installations in rural areas must be examples of quality environmental design. It is a building-culture process with a continuous improvement of dialogue, a synthesis in which the balance between artifacts and natural structures has a key, inevitable position (Figs. 79, 80, 81, 82, 83, and 84).

11 PV Solar Power Plants

PV solar power plant denotes a plant using solar cells to convert solar irradiation into the electrical energy. PV solar power plant consists of solar modules, inverter converting DC into AC, and transformer giving the generated electrical energy into the grid net. PV solar power plant is fully automatized and monitored by the applicable software. PV solar power plants mostly use solar modules made of monocrystalline and polycrystalline silicon and rarely modules made of thin-film materials such as amorphous silicon, CdTe, and copper indium diselenide (CIS, $CuInSe_2$). Monocrystalline and polycrystalline silicon solar modules are more suitable for the areas with

Fig. 79 Solar panels in rural areas. Solar panels are installed on the rooftop of a traditional house in the mountain village of Demul in Spiti Valley, India [50]

Fig. 80 Asian solar panels. Solutions and best practices to increase electrification [51]

predominantly direct sun radiation, while solar modules of thin film are more suitable for the areas with predominantly diffuse sun radiation.

Inverter is a device which converts DC generated by PV solar power plants of 12 or 24 V into three-phase AC of 220 V. Depending on the design inverter efficiency is up to 97%. When choosing inverter, it is to bear in mind the output voltage of the solar module array, power of the solar module array, grid net parameters, managing

Fig. 81 Small segment of a German farming village showing how pervasive solar panels are throughout the country [52]

Fig. 82 Beautiful 3.5 kW solar installation in the Poland village [53]

Fig. 83 Common roof types for solar panels in Rochester [54]

Fig. 84 Surge of solar installations in rural America [55]

type of the PV solar power plant, etc. PV solar power plants can use a larger number of the inverters of smaller power or one or two invertors of greater power.

Schematic view of the PV solar power plant is shown in Fig. 85.

PV solar power plant monitoring system comprises central measuring—control unit for the surveillance of the working regime. Monitoring system uses sensors and software to obtain the following data: daily, monthly, and yearly production of the electricity, reduction of CO_2, detailed change of the system parameters, recording of the events after the failure, monitoring of the meteorological parameters, etc.

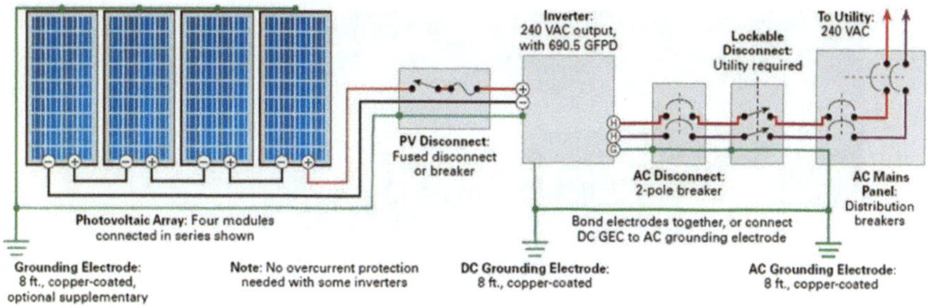

Fig. 85 Schematic view of the PV solar power plant [56]

PV solar power plants in accordance with the power distribution systems' legal regulations use transformers by means of which solar energy generated by PV solar power plant is given to the power grid.

Practice shows that the energy efficiency of PV solar power plant annually decreases from 0.5% to 1%. Lifetime of PV modules depends on the solar cell technology used as well. For monocrystalline and polycrystalline siliconsolar cells, most manufacturers give a warranty of 10/90 and 25/80 which means: a 10-year warranty that the module will operate at above 90% of nominal power and up to 25 years above 80%. The practical lifetime of the silicon-made PV modules is expected to be at least 30 years.

PV solar power plants represent environmentally clean source of energy. PV solar power plant components (solar modules, inverters, monitoring system, conductors, etc.) are manufactured by cutting-edge, environmentally friendly technologies. PV solar power plants operate noiseless, do not emit harmful substances, and do not emit harmful electromagnet radiation into the environment. PV solar power plant recycling is also environmentally friendly. For 1 kWh of PV solar power plant, generated electrical energy emission of 0.568 kg CO_2 into the atmosphere is reduced.

Depending on climate conditions of given location fixed PV solar power plants, one-axis and dual-axis tracking PV solar power plants are being installed worldwide. Fixed PV solar power plants are used in regions with continental climate, and tracking PV solar power plants are used in tropical regions.

11.1 Fixed PV Solar Power Plants

Fixed PV solar power plant denotes plant with solar modules mounted on fixed metal supporters under optimal angle in relation to the horizontal surface, and all are oriented toward south. To install fixed PV solar power plant of 1 MW, it is necessary to provide around 20,000 m^2 (Fig. 86).

Maintenance costs of the fixed PV solar power plants are much lesser than the maintenance costs of the tracking PV solar power plants. Its drawback is that solar

Fig. 86 Fixed PV solar power plant

modules do not follow sun radiation so that on the yearly level one does not gain optimal amount of the electricity.

11.2 One-Axis Tracking PV Solar Power Plant

One-axis tracking PV solar power plant denotes a plant where solar modules installed under the optimal angle are adapted toward the sun by revolving around the vertical axis during the day from the east toward the west, following the sun's azimuth angle from sunrise to sunset. For solar modules, revolving electromotors are used using electrical energy from the batteries of the power grid. For the rotor revolving monitoring a centralized software system is used. In case software system fails, solar modules can be directed toward the sun manually. It is also possible to manually set the tilt of the solar modules in relation to the horizontal surface in steps of 5° from 0 to 45°. One-axis tracking PV solar power plant gives the shadow effect of solar modules situated on neighboring rotors so that for 1 MW installation it is necessary to provide around 60,000 m^2. The available literature reports the efficiency of one-axis tracking PV solar power plant is 20–25% larger than the efficiency of the fixed PV solar power plant.

Installation and maintenance costs of the one-axis tracking PV solar power plants are higher than the costs of the fixed PV solar power plants. Drawback of one-axis tracking PV solar power plant is in that year round there is no automatic adapting of the solar module tilt toward the sun (Fig. 87).

The diurnal and seasonal movement of earth affects the radiation intensity on the solar systems. Sun-trackers move the solar systems to compensate for these motions, keeping the best orientation relative to the sun. Although using sun-tracker is not essential, its use can boost the collected energy 10–100% in different periods of time and geographical conditions. However, it is not recommended to use tracking system for small solar panels because of high energy losses in the driving systems. It is found that the power consumption by tracking device is 2–3% of the increased energy.

Fig. 87 One-axis tracking PV solar power plant

Practice showed that the yearly optimal tilt angle of a vertical-axis-tracked solar panel for maximizing the annual energy collection was almost linearly proportional to the site latitude, and the corresponding maximum annual collectible radiation on such tracked panel was about 96% of solar radiation, annually collected by a dual-axis tracked panel. Compared with a traditional fixed south-facing solar panel inclined at the optimal tilt angle, the annual collectible radiation due to the use of the vertical-axis sun-tracking was increased by 28% in the areas with abundant solar resources and was increased by 16% in the areas with poor solar resources.

11.3 Dual-Axis Tracking PV Solar Power Plant

Dual-axis tracking PV solar power plant denotes a plant where the position of solar modules is adapted toward the sun by revolving around the vertical and horizontal axis. These PV solar power plants follow the sun's azimuth angle from sunrise to sunset, but they also adjust the tilt angle to follow the minute-by-minute and seasonal changes in the sun's altitude angle. Solar modules are oriented toward the sun by means of the appropriate electromotors. Photosensors mounted on the array send signals to a controller that activates the motors, causing the array angles to change as the sun's altitude and azimuth angles change during the day. Efficiency of the dual-axis tracking PV solar power plant is 25–30% bigger than the efficiency of the fixed PV solar power plant (Fig. 88).

For the installation and function of dual-axis tracking PV solar power plant, a substantially bigger surface is necessary than for the fixed PV solar power plant.

Installation and maintenance costs of the dual-axis tracking PV solar power plants are higher than the costs of the one-axis tracking and fixed PV solar power plants.

When designing a large PV solar power plant, it is very important to optimize energy yield and occupation of land. It has been found that the energy gains associated with one north–south axis tracking referenced to static surfaces range from 18 to 25%, and from 37 to 45% for the dual-axis tracker for reasonable ground cover ratios.

Fig. 88 Dual-axis tracking PV solar power plant

Until December 2008, Spain installed 2382 MW_P, Germany 698 MW_P, USA 260 MW_P, Korea 100 MW_P, Italy 70 MW_P, Portugal 60 MW_P, and other countries 102 MW_P PV solar power plants. Worldwide more fixed than tracking PV solar power plants were installed [22].

11.4 Software for the Calculation of the PV Solar Power Plants' Energy Efficiency

In the world market, there is an apparent extent of solar database and software programs available for analyzing solar photovoltaic systems, either commercially available or not. Solar resource information is needed in all stages of the development of a PV project. Reliable solar radiation statistics is required for system siting, design, and financing. In most cases, monthly averages and probability statistics of typical meteorological years (TMYs) are sufficient. This information is sufficient also for the manufacturing industry and for policy makers defining support programs. Some of solar database are NASA Surface meteorology and Solar Energy database, RETScreen solar database, PVGIS solar database, HelioClim-1, Meteonorm, European Solar Radiation Atlas, SoDa Service, Solar and Wind Energy Resource Assessment (SWERA), etc.

Solar PV software simulators on the market are designed with different goals in mind and have various limitations for solving certain problems. The desirable features of software for manufacturing simulation depend on the purpose of their use. Each software works in its specific area of application in solar PV systems. As more PV systems are installed, there will be an increase in demand for software that can be used for design, analysis, and troubleshooting. There are twelve major types of software for simulating solar PV system, they are RETScreen, PV F-Chart, SolarDesignTool, INSEL, TRNSYS, NREL's Solar Advisor Model, ESP-r 11.5, PVSYST 4.33, SolarPro, PV-DesignPro-G, PV*SOL Expert, and HOMER, and many others available are DDS-CAD PV, Polysun, APOS photovoltaic StatLab, PV Designer, SolarNexus, Valentin Software, PV Cost Simulation Tool, PV Potential Estimation Utility, Solmetric iPV, Solmetric SunEye, Blue Oak Energy and Solar Pro Magazine's Solar Select, Seneca Software & Solar, Inc., Sombrero, Horizon, Panorama Master, Meteonorm, GOSOL, , Shadow Analysis, Spyce, Ecotect, Tetti FV, KeryChip, PV Professional, PVCAD, Meteocontrol, etc.

11.5 PVGIS

Quantity of sun radiation intake on the surface of earth is influenced by numerous factors such as: geographical latitude of the given place, season of the year, part of the day, purity of the atmosphere, cloudiness, and orientation and surface inclination. These data are very important because of their use in calculations of the cost-effectiveness of equipment using sun radiation. Very reliable data can be found in data basis PVGIS (*Photovoltaic Geographical Information System*–PVGIS ©, European Communities, 2001–2008, http://re.jrc.ec.europa.eu/pvgis/apps3/pvest.php) (Fig. 89).

PVGIS is a part of the SOLAREC action aimed at contributing to the implementation of renewable energyenergy in the EU. SOLAREC is an internally funded project on PV solar energy for the 7th Framework Programme. PVGIS has been developed at the Joint Research Centre (JRC) of the European Commission within its Renewable Energies Unit since 2001 as a research GIS-oriented tool for the performance assessment of solar PV systems in European geographical regions. From the very start of its functioning, PVGIS was envisaged to be locally used; however, access to the PVGIS database and estimations was drawn as open system access for professionals and the general European public as well, by means of the Web-based interactive applications. PVGIS provides data for the analysis of the technical, environmental, and socioeconomic factors of solar PV electricity generation in Europe and supports systems for EU countries' solar energy decision-making.

PVGIS methodology comprises solar radiation data, PV module surface inclination and orientation, and shadowing effect of the local terrain features (e.g., when the direct irradiation component is shadowed by the mountains); thus, PVGIS represents immensely important PV implementation assessment tool that estimates dynamics of the correlations between solar radiation, climate, atmosphere, the earth's surface, and the PV technology used. Several fast Web applications enable an easy estimation of the PV electricity generation potential for selected specific locations in Europe.

The methods used by PVGIS to estimate PV system output have been described in a number of papers. The basis for the European part of PVGIS is a dataset with 10 years of data from 566 ground stations in Europe measuring global horizontal radiation and in some cases diffuse radiation. The station data were collected and

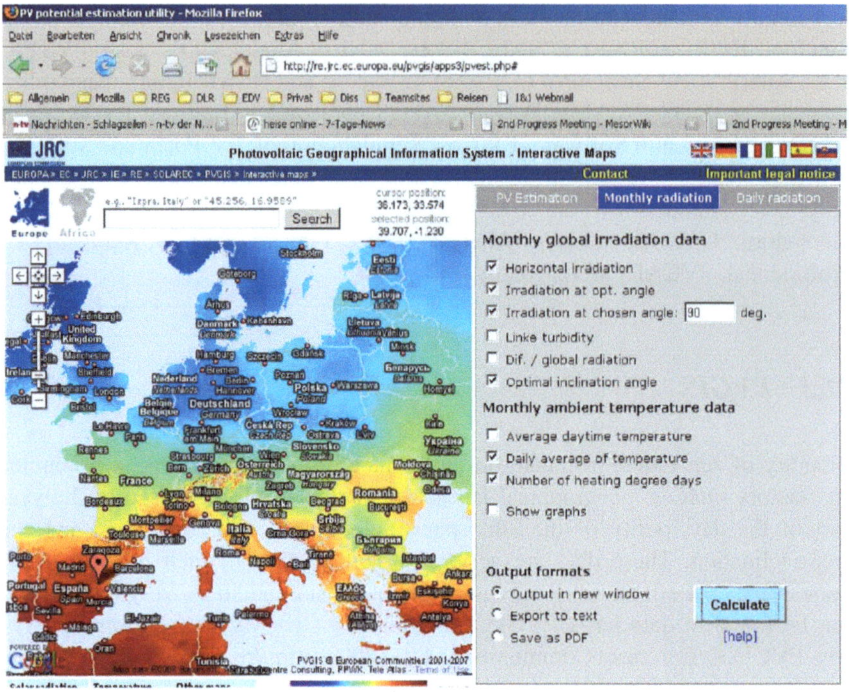

Fig. 89 PVGIS-CMSAF Web site

processed as a part of the European Solar Radiation Atlas and published as monthly averages of daily irradiation sums.

The construction of high spatial resolution datasets for solar radiation has been previously reported. The computational approach is based on a solar radiation model (r.sun) and the spline interpolation techniques (s.surf.rst and s.vol.rst) that are implemented within the open-source GIS software GRASS. The (r.sun) model algorithm uses the equations published in the European Solar Radiation Atlas. This is certainly a powerful tool that can be used for the development of new solar power plants that will obviate climate change and promote sustainable development through poverty alleviation. Other details of the PVGIS methodology and development can be found in some key reference papers.

In order to calculate electricity generated by the fixed PV solar plants and one-axis and dual-axis tracking PV solar plants, today PVGIS software packages easily found on the Internet are used. These programs can produce the following data: average daily, monthly, and yearly values of the solar irradiation taken on square meter of the horizontal surface, or the surface tilted under certain angle in relation to the horizontal surface, change in the optimal tilting angle of the solar modules during the year, relation of global and diffused sun radiation, average daily temperature, and daily, monthly, and yearly electricity generated by the fixed PV solar plants, one-axis and dual-axis tracking PV solar plants, etc. A typical PVGIS value for the performance ratio (PV system losses) of PV solar power plants with modules from monocrystalline and polycrystalline silicon is taken to be 0.75.

This program gives a map which, when appears, activates the program, spots the location of the PV solar power plant to be, sorts out the type of solar cells, and inputs the power and type of PV solar power plant (fixed, one-axis, and dual-axis tracking PV solar plants).

11.6 List of PV Solar Power Plants Larger Than 300 MW

List of PV solar power plants larger than 300 MW is given in Table 9.

Time line of the largest power plants in the world is given in Table 10.

Some of the PV power plants are given in Figs. 90, 91, 92, 93, and 94.

Table 9 List of PV solar power plants larger than 300 MW [65]

Name	Country	Capacity (MWp)	Year
Tengger Desert Solar Park	China	1547	2016
Bhadla Solar Park	India	1365	2018
Kurnool Ultra Mega Solar Park	India	1000	2017
Datong Solar Power Top Runner Base	China	1000	2016
Longyangxia Dam Solar Park	China	850	2015
Villanueva Solar Park	Mexico	828	2018
Rewa Ultra Mega Solar	India	750	2018
Kamuthi Solar Power Project	India	648	2016
Pavagada Solar Park	India	600	2017
Solar Star (I and II)	USA	579	2015
Copper Mountain Solar Facility	USA	552	2016
Desert Sunlight Solar Park	USA	550	2015
Topaz Solar Farm	USA	550	2014
Huanghe Hydropower Golmud Solar Park	China	500	2014
Mount Signal Solar	USA	460	2018
Mesquite Solar Project	USA	400	2016
NP Kunta	India	400	2018
Pirapora Solar Project	Brazil	400	2018
Yanchi Solar Park	China	380	2016
Charanka Solar Park	India	345	2012
Springbok Solar Park	USA	328	2016
Cestas Solar Park	France	300	2015
Stateline Solar	USA	300	2016

Table 10 Timeline of the largest power plants in the world [65]

Name	Country	Capacity (MW_P)	Year
Lugo	USA	1	1982
Carrisa Plains	USA	5.6	1985
Bavaria Solarpark	Germany	6.3	2005
Erlasee Solar Park	Germany	11.4	2006
Olmedilla Photovoltaic Park	Spain	60	2008
Sarnia PV Plant	Canada	97	2010
Huanghe Hydropower Golmud Solar Park	China	200	2011
Agua Caliente Solar Project	USA	290	2012
Topaz Solar Farm	USA	550	2014
Longyangxia Dam Solar Park	China	850	2015
Tengger Desert Solar Park	China	1547	2016

Fig. 90 Solar Star Projects, Rosamond, California, 579 MW [57]

12 Solar Module Efficiency Dependence On Their Orientation and Tilt Angle

In order to determine solar module efficiency dependence of their orientation and tilt angle at the same time in real climatic conditions, the experiment was conducted in Solar Energy Laboratory at the Faculty of Sciences and Mathematics, University of Nis in Serbia (Fig. 95).

Fig. 91 Longyangxia Dam Solar Park, Qinghai. Longyangxia Dam Solar Park is world's largest solar park China, 550 MW [57]

Fig. 92 Charanka Solar Park, Gujarat, India, 345 MW [58]

The experimental system comprises five monocrystalline silicon PV solar modules, each of 60 W_P power and the area of 0.514 m^2. Three solar modules are positioned vertically and oriented toward the east, south, and west, respectively. The fourth module is horizontal, and the fifth is oriented toward the south and tilted at the angle of 32°, which is the yearly optimum angle for a fixed solar module in Niš.

The solar radiation intensity, solar energy, and the ambient temperature were measured by Davis Vantage Pro meteorological weather station also placed on the roof of the faculty building. A Mini-KLA (Ingenieurbüro Mencke & Tegtmeyer GmbH) device was used to measure the current/voltage (I/V) characteristics of each solar module in a rapid succession, thus measuring simultaneous behavior of solar modules.

Fig. 93 Cestas Solar Farm, Bordeaux, France, 300 MW [59]

Fig. 94 California Valley Solar Ranch (CVSR) in San Luis Obispo, 250 MW [60]

The average monthly solar energy measured by Davis meteorological weather station in 2013 and average monthly solar energy calculated by PVGIS-CMSAF software, for the horizontal plane, are shown in Fig. 96.

Based on the values presented in Fig. 96, it can be concluded that the measured values of solar energy received by the horizontal plane are, on average, by 33.4% less than the values calculated by PVGIS-CMSAF software. The difference between the measured and the calculated values arises from the fact that PVGIS-CMSAF software gives 12-year averages for the solar energy received by the horizontal plane and that these values are compared to a single year measurement in this experiment.

Fig. 95 Experimental system composed of five differently oriented PV solar modules [21]

The measured average monthly electrical energy and the energy calculated by PVGIS-CMSAF for five solar modules in 2013 are given in Figs. 97, 98, 99, 100, and 101, respectively.

It was found that the measured values of the electrical energy generated by the solar module oriented toward the south at the angle of 32° are on average by 1% higher than the values of the electrical energy calculated by PVGIS-CMSAF software. The biggest difference in data was observed in January (44%) and November (42.3%), while the smallest difference was observed in October (0.6%).

The measured values of the electrical energy generated by the horizontal solar module are on average by 1.4% smaller than the values of the electrical energy calculated by PVGIS-CMSAF software for the same solar module (Fig. 98). The biggest difference in data was observed in November (47.1%) and the smallest in June (3.1%). From July till October, the measured values were greater than the calculated values of the average generated electrical energy, whereas for all other months these values were smaller.

The measured values of the electrical energy generated by the vertical solar module oriented toward the south were on average by 2.1% higher than the values of the

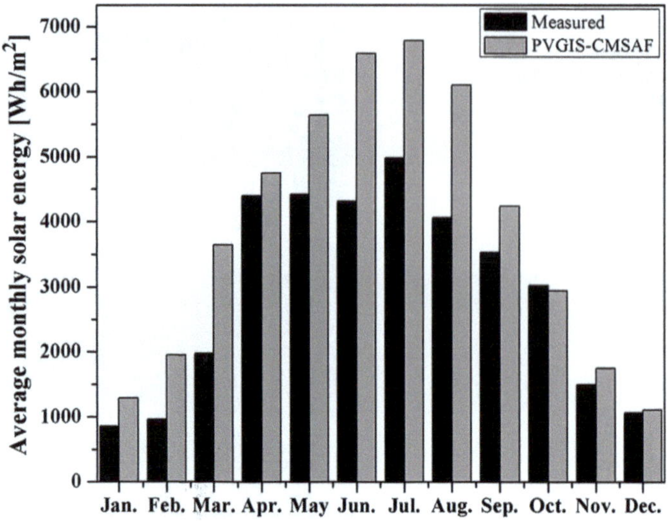

Fig. 96 Average monthly solar energy measured by Davis meteorological weather station in 2013 and average monthly solar energy calculated by PVGIS-CMSAF software, for the horizontal plane [21]

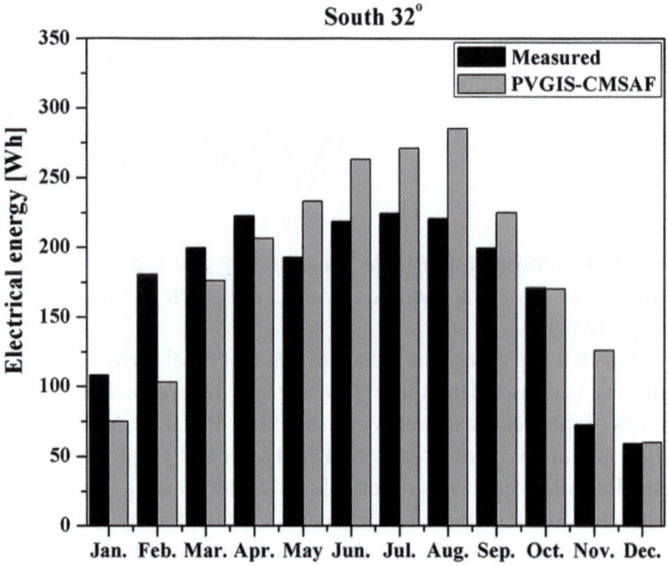

Fig. 97 Measured average monthly electrical energy and the energy calculated by PVGIS-CMSAF software for the solar module oriented toward south at the angle of 32° in 2013 [21]

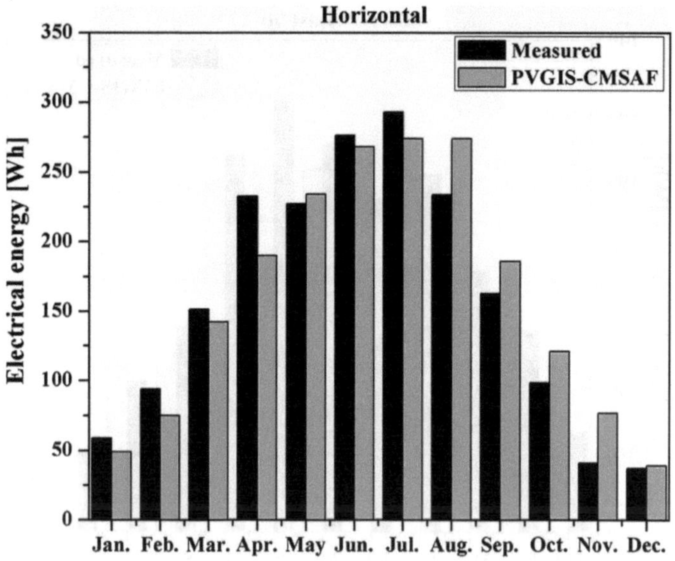

Fig. 98 Measured average monthly electrical energy and the energy calculated by PVGIS-CMSAF software for horizontal solar module in 2013 [21]

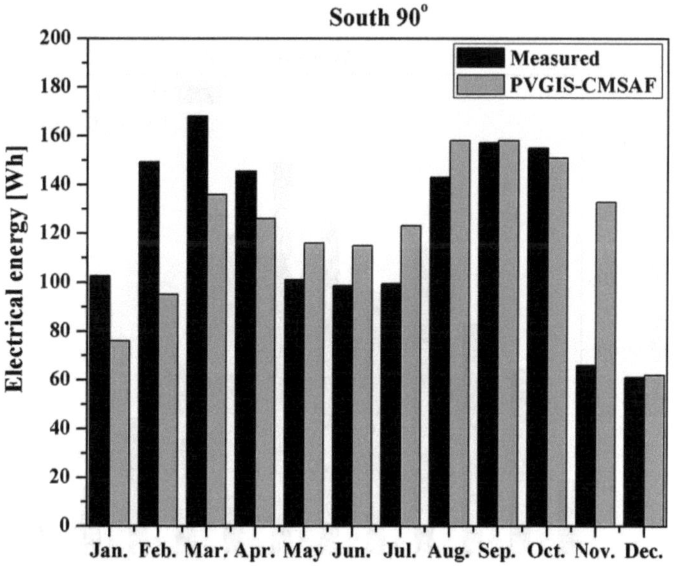

Fig. 99 Measured average monthly electrical energy and the energy calculated by PVGIS-CMSAF for the vertical solar module oriented toward south in 2013 [21]

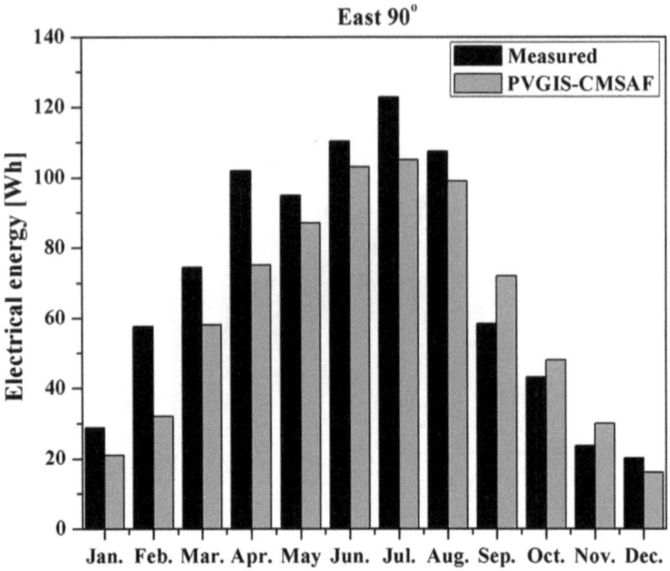

Fig. 100 Measured average monthly electrical energy and the energy calculated by PVGIS-CMSAF for the vertical solar module oriented toward east in 2013 [21]

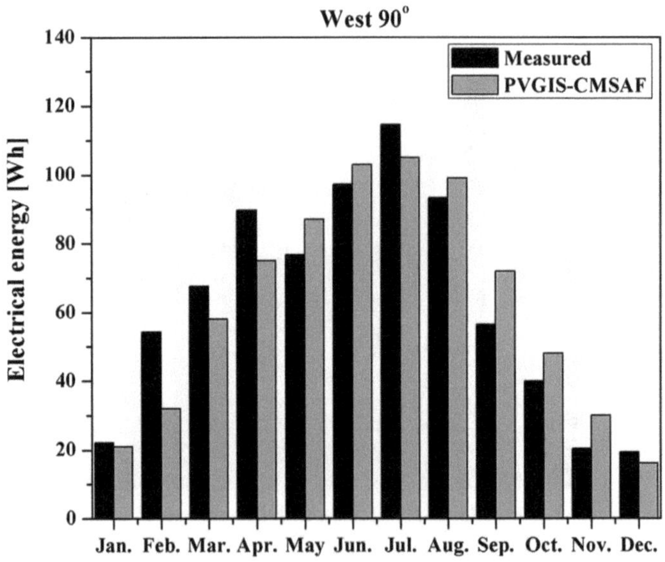

Fig. 101 Measured average monthly electrical energy and the energy calculated by PVGIS-CMSAF for the vertical solar module oriented toward west in 2013 [21]

Table 11 Total electrical energy generated by five differently oriented PV modules in 2013

Orientation of a PV module	E	(kWh)
	PVGIS-CMSAF	Measured
South 32°	66.8	62.9
Horizontal	58.8	58.1
South 90°	44.1	43.9
East 90°	22.7	25.7
West 90°	22.7	22.9

electrical energy calculated by PVGIS-CMSAF software. The biggest difference in data was observed in February (57.1%) and November (50.5%), while the smallest difference was observed in September (0.5%) and October (2.6%), respectively.

The measured values of the electrical energy generated by the vertical solar module oriented toward the east were on average by 16.5% higher than the values of the electrical energy calculated by PVGIS-CMSAF software. The biggest difference in data was observed in February (79.7%) and the smallest in June (7.1%).

The measured values of the electrical energy generated by the vertical solar module oriented toward the west were on average by 3.9% higher than the values of the electrical energy calculated by PVGIS-CMSAF software. The biggest difference in data was observed in February (69.7%) and the smallest in June (5.6%).

Total electrical energy generated by five differently oriented PV modules in 2013 is shown in Table 11.

As shown in Table 11, the most of electrical energy was generated by solar module oriented toward the south at the angle of 32° (62.9 kWh), followed by the horizontal solar module (58.1 kWh). Compared to the optimally oriented solar module, the vertical solar modules oriented toward south, east, and west generated by 30.2, 59.2, and 63.6% less electrical energy, respectively. The difference in generated electrical energy by the vertical solar modules oriented toward east and west can be attributed to the local climate conditions (afternoon clouds and fog).

For cities and regions for which there are no measured data, it is possible to use, as a guide, the information provided by the PVGIS-CMSAF software.

In light of all the above presented, it can be concluded that:

- Experimentally obtained values of the solar energy received by the horizontal plane in 2013 are on average by 33.4% less than the energy values calculated by PVGIS-CMSAF software.
- Optimally oriented monocrystalline solar module of 60 W_P in 2013 generated 62.9 kWh; horizontal module 58.1 kWh; vertical module oriented toward the south 43.9 kWh; vertical module oriented toward the east 25.7 kWh; and vertical module oriented toward the west 22.9 kWh of electrical energy.
- Difference between the theoretical and the experimentally obtained values of the generated electrical energy in 2013 by five differently oriented modules ranges from 1.0 to 16.5% [21, 23, 24].

13 Solar Module Soiling

Soiling is the term used to describe the deposition of dust (dirt) on solar modules. Soiling of solar modules is particularly problematic in dry areas where the amount of dust in the air is high and the amount of precipitation is low. Solar radiation is absorbed and scattered on dust that exists in atmospheric air, but it is also absorbed and scattered on the dust deposited on solar modules, resulting in a decrease in the intensity of solar radiation that reaches solar cells. This may cause the entire PV system to be difficult to operate and that less electrical energy is generated.

Soiling includes not only the deposition of dust, but also the deposition by plant products, salts, bird droppings, the growth of organic species, etc., which negatively affects the performance of solar modules.

Reducing the performance of solar modules due to soiling also depends on the distribution of dust particles deposited on the surface of the module by size. Smaller dust particles have larger specific surface and can be distributed more uniformly and close to each other on the solar module surface, in such a way that the space between them through which light can pass is much smaller, compared to larger dust particles. As a result, smaller particles of dust cause a significant reduction in the performance of solar modules, and therefore greater energy losses, compared to the same mass concentration of deposited larger dust particles.

Reducing the efficiency of solar modules due to their soiling depends on the area in which they are located and on the climate characteristics of the area.

13.1 Dust

Different definitions of dust can be found in the literature, depending on the field of application and field of research. In a geological sense, dust consists of solid inorganic particles that have been caused by the decay of the rocks over time. In the atmospheric sense, dust is particle hovering in the air or aerosols in a solid aggregate state. The organic particles that wind lifts up in the atmosphere can act as dust, which in turn act as an adhesive that connects dust particles or particles of dust with the surface to which they are deposited.

Generally, dust is a mixture of different pollutant characteristic of a particular geographical area. The word dust represents a general term for particles of any matter of diameter less than 500 μm. The important characteristics of dust are the size and distribution of its particles, density, shape, chemical composition, etc. The size and shape of dust particles, as well as the behavior of deposits and the rate of dust accumulation, depend on the geographical location, climate conditions, and urbanization of a particular site. Particles of dust can be divided into: fine, or smaller (with a diameter of less than 2.5 μm), and rough, or larger (with a diameter greater than 2.5 μm). Dust particles have different and irregular shapes. Important ambient conditions that can affect dust characteristics and behavior are air humidity, speed, direction of wind movement, etc. Dust concentration in the air exponentially decreases with an increase in altitude, except in the case of a dust storm.

Dust composition reflects the characteristics of the region from which the dust comes. Depending on the location, the chemical composition of the dust and its quantity on solar modules can vary significantly. Dust most often contains organic minerals (mostly sand (quartz) and eroded limestone (calcite), somewhat less dolomite and clay, etc.) and particles that result from the burning of fossil fuels, but dust can also contain small quantities of pollen and fungi, bacteria, vegetation, microfibers, etc. Air pollution is higher in urban areas due to the high density of population and industrial activities, and there are particularly particles that occur as a product of combustion of fossil fuels and during construction works. In rural areas, dust particles originate from various types of fertilizers, soil particles blown by the wind, or particles originating from plants.

If the relative humidity of the air over a prolonged period of time is high, some biological species may begin to grow on solar modules. Fine dust particles usually represent a mixture of carbon that occurs due to the incomplete combustion of fuels, and secondary particles formed in chemical reactions in the atmosphere (acid condensates, sulfates, nitrates, etc.) and particles of rough dust usually originate from organic materials and soot.

Dust in the air is mainly composed of silicon because it is a chemical element that is most represented in the earth's crust. Dust contains mostly oxygen and calcium, and slightly less aluminum, iron, magnesium, carbon, potassium, sodium, sulfur, chlorine, etc. Calcium in the dust mainly comes from industrial activities (e.g., cement factories) and partly from the permanent construction activities and degradation of building elements. There are many exceptions, bearing in mind that dust composition depends heavily on local conditions.

It can be said that particles of dust pass through the so-called life cycle consisting of four phases:

- Generation (formation),
- Deposition,
- Adhesion,
- Removal.

13.1.1 Generation of Dust

Generation of dust represents the process of getting dust particles into the atmosphere.

Dust is airborne in various ways: It can be lifted and carried by the wind, it can be raised also due to movement of pedestrians and vehicles, and it may emerge from the exhaust gases of a vehicle, due to volcanic eruptions, agricultural activities (such as plowing), animal activities, and in general air pollution.

Dust particles reach the atmosphere mainly due to soil erosion under the influence of wind, after which the wind is further transmitting them. The spread of dust under the influence of wind occurs when the wind has enough power to move the dust from the ground.

The process of dust formation is influenced by a large number of parameters, such as soil properties, dust characteristics, atmospheric conditions (precipitation, for example, reducing the dust concentration in the air).

The influence of the human on dust is only at the local level and is reflected in the emission of exhaust gases from vehicles and agricultural activities.

13.1.2 Dust Deposition

Dust deposition (accumulation) is considered to be the most important process that leads to the soiling of solar modules.

Dust deposition describes the processes by which dust particles reach the surface of solar modules. Dust particles reach the surface of the solar modules mainly under the influence of gravity and due to the diffusion of dust particles in the air.

Factors that influence the deposition of dust on solar modules are: their tilt angle, their location, concentration of dust particles in the air, dust particles' characteristics (size, shape, density, composition), local climate and meteorological conditions (atmospheric stability, speed and direction of the wind, the amount of precipitation, relative humidity of the air), the characteristics of the solar module surface, etc. Most of these factors are characteristics of a particular area. The amount of dust deposited on solar modules depends on the time that the modules spent outside (the period during which the solar modules are exposed to dust) without cleaning.

The tilt angle of solar modules also affects deposition of dust on their surface. With an increase in the tilt angle from 0° (horizontal position) to 90° (vertical position), the amount of deposited dust decreases on solar modules. Most dust is deposited on horizontally installed solar modules, so that the modules thus installed have the biggest losses due to dust accumulation on them.

The basic mechanism for dust deposition is gravity; thus, most dust is deposited when the tilt angle of the solar module is 0°. When the solar module is mounted vertically, the basic mechanism for depositing dust particles is the diffusion of particles in the air.

The deposition of dust under gravity is proportional to the square of the dust particle diameter, so that, as the particle is larger, the speed of the deposition of the particle is higher. Accordingly, most particles deposited on a horizontally installed solar module are larger in size and there are less fine particles. The diffusion is inversely proportional to the diameter of the dust particles, so that the dust deposited on the vertically installed solar module consists predominantly of finer particles.

13.1.3 Adhesion

The adhesion between the particles of the dust on the solar module and its surface is the result of various interactions that occur between them at the nano- and micro-level (Van der Waals forces, capillary forces, and electrostatic forces) (Fig. 102).

Van der Waals forces are poorly attractive forces between two molecules that occur due to the interaction of dipoles of different molecules. Unlike capillary and electrostatic forces, Van der Waals forces are always present between dust particles and the surface of the solar module.

Capillary forces between dust particles and the surface of the solar module occur when the relative humidity of the air is large (greater than 40%) or when the surface of the solar module is covered with a layer of water. When present, capillary force is the dominant force in the adhesion process. Capillary force occurs between two wet, touching bodies and is mainly due to the condensation of moisture from the air surrounding the solar module.

If dust particles are charged, electrostatic forces occur in the process of adhesion between the dust particles and the surface of the solar module.

The adhesion forces between the particles of dust and the surface on which they are located primarily depend on the size of the contact surface between them, so that the adhesion forces grow with increasing dust particles. The adhesion forces between the spherical dust particles and the surface of the solar module are larger because the spherical particles have a larger contact surface with the surface on which they are located. At high relative humidity, the adhesion increases due to water condensation and the occurrence of capillary forces between the dust particles and the surface of the solar module.

Adhesion and removal processes are most often considered in parallel because they represent competing processes whose balance determines which dust particles will remain on the surface of the solar module and which will eventually disappear from its surface.

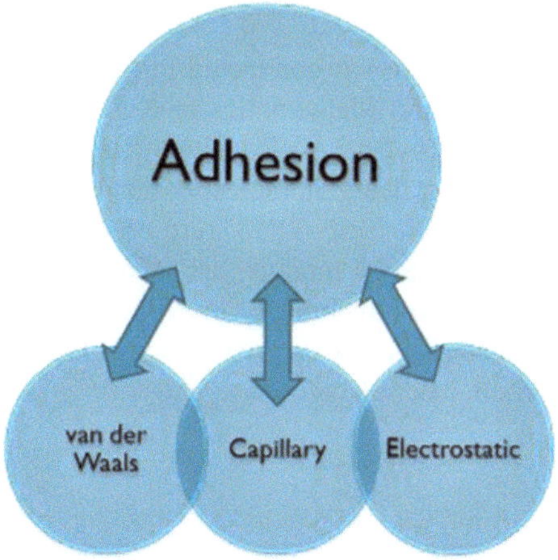

Fig. 102 Adhesion forces acting on a dust particle located on the solar module surface [61]. *Courtesy* Elsevier

13.1.4 Dust Removal

Final stage in the dust life cycle is their removal from solar modules. Except for the cleaning of solar modules by artificial means, dust from solar modules can be removed under precipitation (rain and snow) and wind, i.e., through natural cleaning. Rain and wind are the most effective natural factors for removing dust from solar modules. Various experiments have been carried out to examine the impact of rain and wind on the removal of dust from solar modules. Low rain (less than 1 l/m^2 of rain during the day) is considered to have a negative impact on the process of removing dust from solar modules because it can create a sticky layer that binds dust to the surface of the solar module. It is recommended to clean solar modules immediately after such occurrences in order to restore their efficiency to their original state. The literature indicates a quantity of rain solar module needs to be cleaned, of a value of about 4–5 l/m^2 of rain during the day. It is widely known that intense (and frequent) rain cleans solar modules and in that way improves their performance and restores their efficiency almost to the initial one.

Wind impact studies on dust removal from solar modules have shown that there is no wind velocity threshold to begin removing, but that there is a certain wind speed interval where dust is effectively removed from the solar modules. Basic mechanisms for removing dust from solar modules by wind are: rolling, sliding, and raising dust particles. Rolling dust particles is the dominant process of removing dust from solar modules by wind. Research has shown that, under the influence of wind, larger particles of dust are more easily removed from solar modules compared to smaller particles.

As the tilt angle of solar modules increases, the amount of dust deposited on them decreases because it increases the probability of dust particles (especially larger dust particles) being removed from their surface by wind and rain.

13.2 Solar Modules' Efficiency Dependence on Soiling

The research was conducted in the Laboratory for Solar Energy at the Faculty of Sciences and Mathematics in Niš, Serbia, to compare the efficiency of the clean and soiled solar modules with three pollutants: calcium carbonate ($CaCO_3$), carbon, and soil particles. In the experiment, two identical monocrystalline silicon Isofoton solar modules ISF-60/12 were used, placed side by side on the roof of the faculty (Fig. 103). Solar modules are set at the optimum angle of 32° for the area of Niš and are facing south. Table 12 gives technical characteristics of ISF-60/12 solar modules.

Scanning electron microscope JEOL JSM-5300 was used to observe particles of carbon, calcium carbonate, and soil.

Fig. 103 Two identical monocrystalline silicon solar modules ISF-60/12 on the roof of the Faculty of Sciences and Mathematics in Niš, used in the experiment [26]

Table 12 Technical characteristics of ISF-60/12 solar module

Dimensions (size)	776 × 662 × 39.5 mm
Weight	6.5 kg
Cell type	Si monocrystalline
Power of the module	60 W_P
Module efficiency	11%
Maximum power current	3.47 A
Maximum power voltage	17.3 V
Open-circuit voltage	21.6
NOCT (800 W/m^2, 20 °C, AM 1.5, 1 m/s)	47 °C
Maximum system voltage	760 V

13.2.1 Carbon

Figures 104 and 105 show SEM images of carbon particles used in the experiment at different magnifications.

Figure 104 shows that the carbon particles are distributed homogeneously and densely so that the space between the particles through which the light can pass is very small. Figure 105 shows that the carbon particles have a leaf structure of small thickness, with a diameter of about 30 μm.

The dependence of the efficiency on the soiling of the surface of the optimally installed solar module with different masses of carbon is given in Fig. 106.

Fig. 104 SEM image of carbon particles at 35× magnification [26]

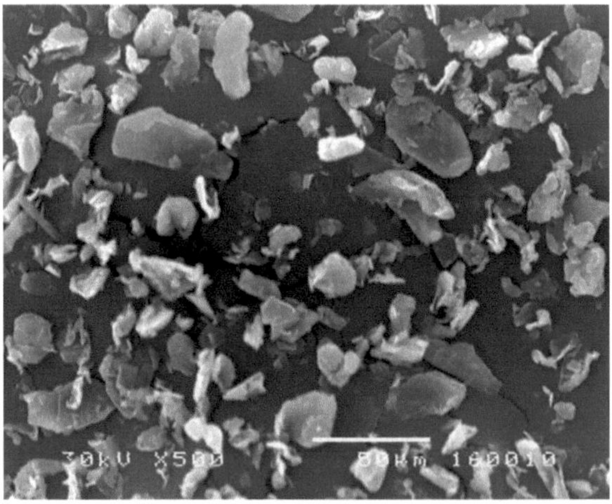

Fig. 105 SEM image of carbon particles at 500× magnification [26]

Figure 106 shows that with an optimally mounted solar module efficiency values are reduced from 10.2% for clean module and up to 5.0% for the module soiled with 3.3 g of carbon.

A comparative display of the efficiency for clean (1) and carbon-soiled (2) optimally installed solar module depending on time during the day is given in Fig. 107.

Figure 107 shows that for optimally installed solar modules the efficiency values range from 10.2% to 10.5% for clean module, and from 9.0% to 5.0% for a module soiled by different carbon masses.

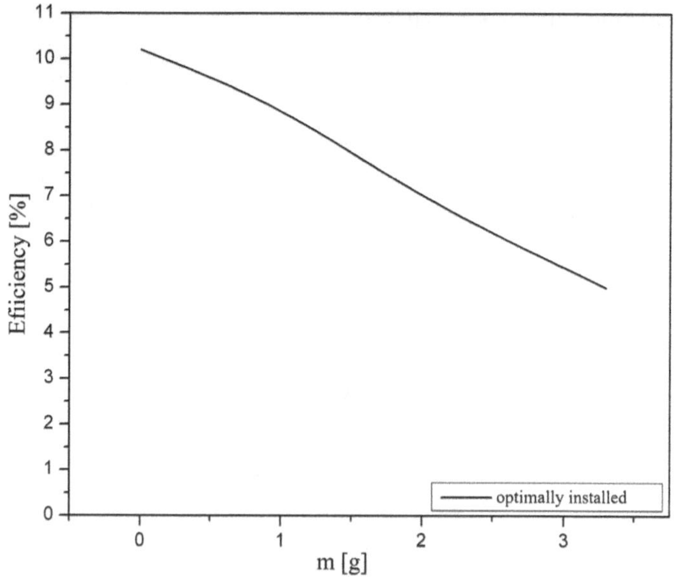

Fig. 106 Dependence of the efficiency on the soiling of the surface of the optimally installed solar module with different masses of carbon [25]

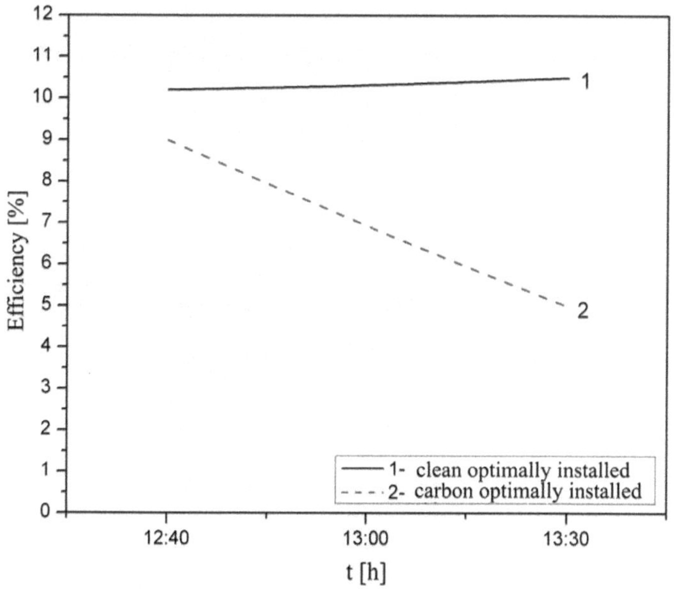

Fig. 107 A comparative display of the efficiency for clean (1) and carbon-soiled (2) optimally installed solar module depending on time during the day [25]

13.2.2 Calcium Carbonate ($CaCO_3$)

Figures 108 and 109 show SEM images of $CaCO_3$ particles used in the experiment at different magnifications.

Figure 108 shows that the particles of calcium carbonate are distributed homogeneously, but the space between the particles through which light can pass is larger than that of carbon particles. Figure 109 shows that calcium carbonate particles have a cubic structure, and their diameter is about 30 μm.

The dependence of the efficiency on the soiling of the surface of the optimally installed solar module with different calcium carbonate masses is shown in Fig. 110.

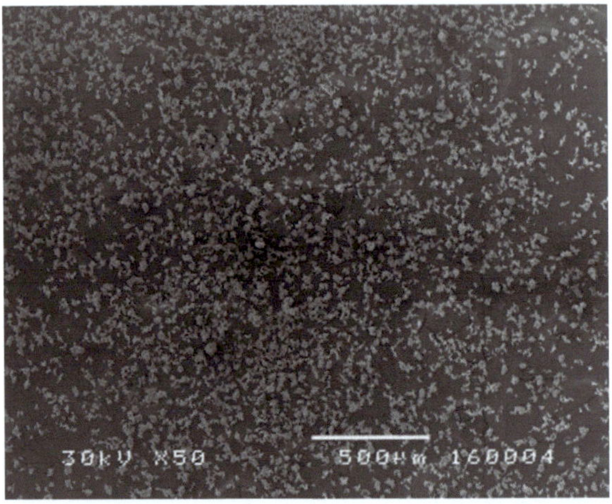

Fig. 108 SEM image of $CaCO_3$ particles at 50× magnification [26]

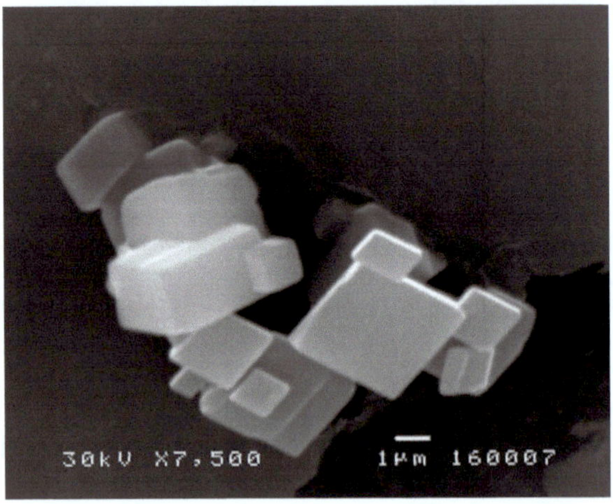

Fig. 109 SEM image of $CaCO_3$ particles at 7500× magnification [26]

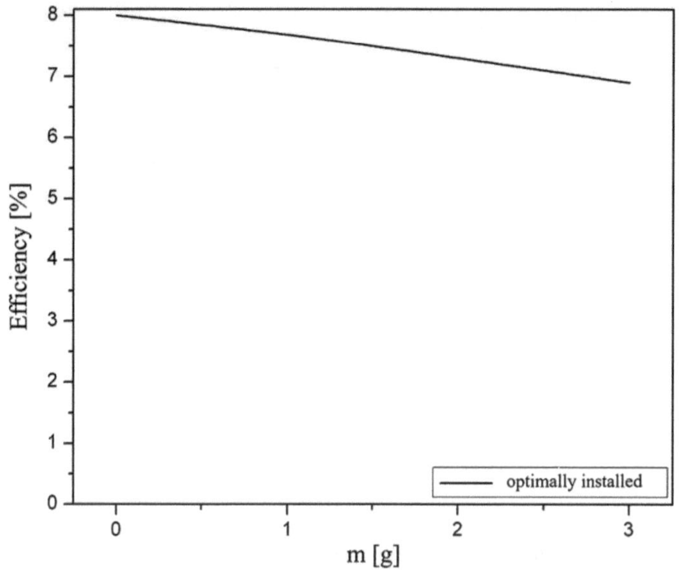

Fig. 110 Dependence of the efficiency on the soiling of the surface of the optimally installed solar module with different calcium carbonate masses [25]

In Fig. 110, it can be seen that with the optimally installed solar module the efficiency values are reduced from 8.0% for clean module and up to 6.9% for the module soiled with 3 g of calcium carbonate.

Comparative efficiency display for clean (1) and calcium carbonate-soiled (2) optimally installed solar module depending on the time of the day is given in Fig. 111.

Figure 111 shows that for optimally installed solar modules, the efficiency has a value of 7.8% and 8.0% for a clean solar module, and that the efficiency values range from 7.7% to 6.9% for a module soiled by different calcium carbonate masses.

13.2.3 Soil

Figures 112 and 113 show SEM images of the soil particles used in the experiment at different magnifications.

Figure 112 shows that the soil particles are considerably different ranging in diameter from 20 μm up to 300 μm, and the small number of particles is of diameter up to 500 μm. The space between the soil particles through which light can pass is larger than that of the carbon particles.

The dependence of the efficiency on the soiling of the surface of the optimally installed solar module with different masses of soil is given in Fig. 114.

Figure 114 shows that for an optimally mounted solar module the efficiency values are reduced from 8.4% for clean module and up to 7.9% for the module soiled with 3 g of soil.

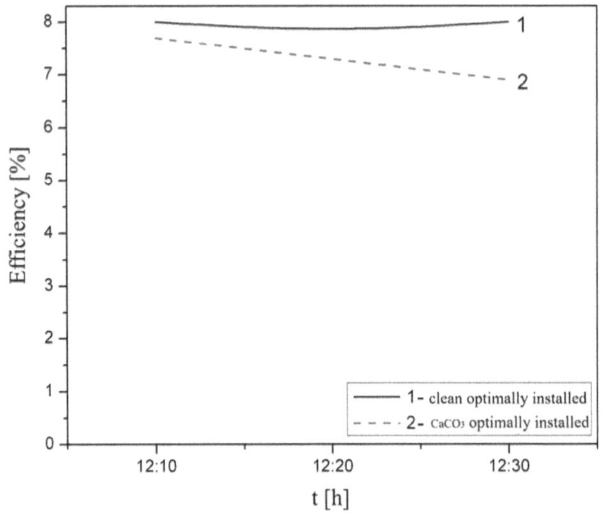

Fig. 111 Comparative efficiency display for clean (1) and calcium carbonate-soiled (2) optimally installed solar module depending on the time of the day [25]

Fig. 112 SEM image of the soil particles at 35× magnification [26]

A comparative display of efficiency for clean (1) and soiled (2) optimally installed solar module depending on the time of the day is given in Fig. 115.

Figure 115 shows that for the optimally installed solar modules the efficiency of the clean solar module during the measurement is constant and amounts to 8.4%, and that the efficiency values range from 8.2% to 7.9% for the module soiled by different masses of soil.

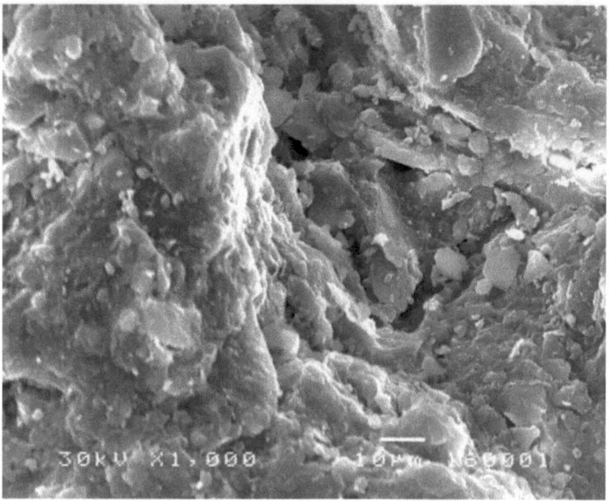

Fig. 113 SEM image of the larger soil particle at 1000× magnification [26]

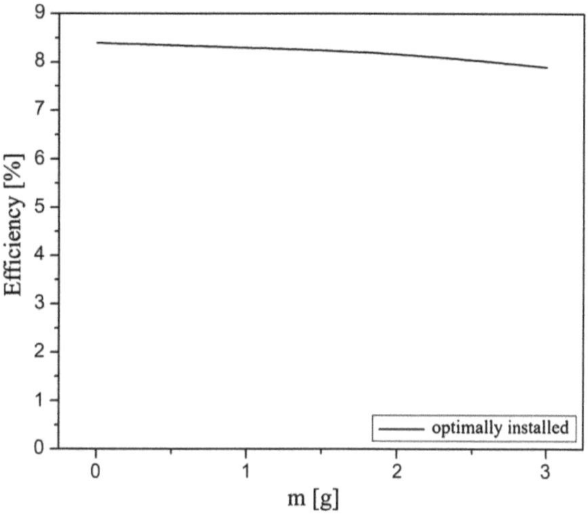

Fig. 114 Dependence of the efficiency on the soiling of the surface of the optimally installed solar module with different masses of soil [25]

13.2.4 Comparison of the Impact of Carbon, Calcium Carbonate, and Soil on Solar Modules

SEM examination of carbon, calcium carbonate, and soil particles showed that the carbon and the calcium carbonate particles are similar in size, while the space between the particles through which light can pass in carbon is less than that of calcium carbonate. The dimensions of the soil particles are different, and the space between the soil particles through which light can pass is similar to calcium carbonate. It can be concluded that solar radiation more easily reaches the surface of the solar modules soiled by calcium carbonate and soil particles than the surface of the solar modules soiled by carbon.

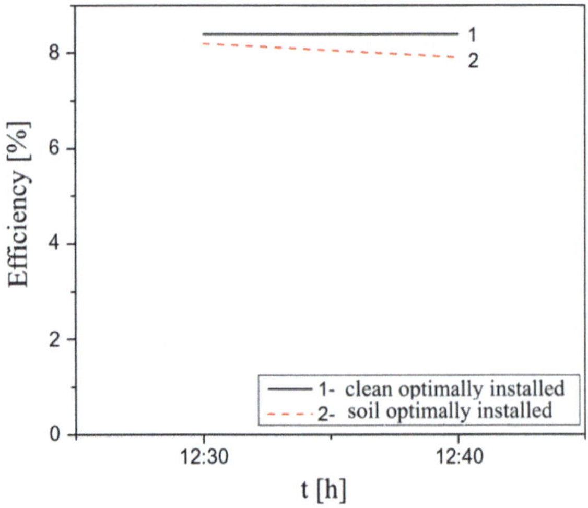

Fig. 115 A comparative display of efficiency for clean (1) and soiled (2) optimally installed solar module depending on the time of the day [25]

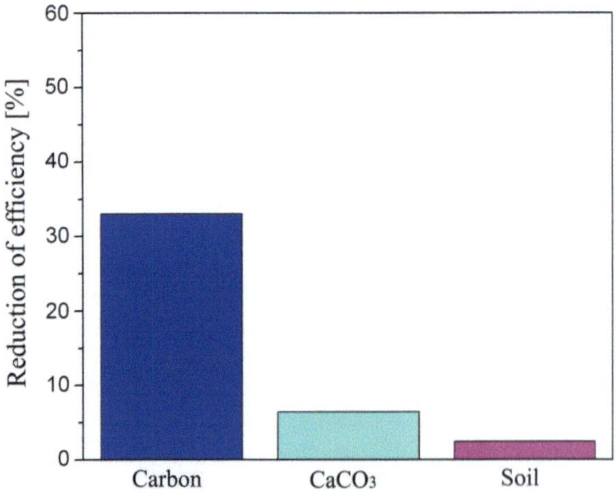

Fig. 116 Reduction of efficiency due to the soiling of an optimally installed solar module with 2 g of carbon, calcium carbonate, and soil [25]

Figure 116 shows that the efficiency of the optimally installed solar module decreases mostly due to the soiling by carbon (by 33.0%), then calcium carbonate (by 6.4%), and the least soil (by 2.4%).

13.3 Cleaning and Maintenance of Solar Modules

Due to the soiling of solar modules, their occasional cleaning must be done.

In commercial PV systems, cleaning of solar modules is an additional cost because a special system for cleaning should be set up or the team of people who will clean them should be engaged. In residential buildings, solar modules are usually cleaned by house owners (Figs. 117, 118, 119, and 120).

Cleaning solar modules is performed in different ways and using different means. The most efficient cleaning agent for solar modules is water. Pressure water or brush water is used to remove adhesive or muddy dirt. In rainy seasons, rain removes all dust from solar modules. However, in periods without rain (e.g., in summer) accumulation of dust on the surfaces of solar modules in some areas of the world can cause daily losses of over 20%.

Various automated systems such as nozzle systems and brush systems are used to clean solar modules with water. Nozzle systems can use fixed or slip nozzles. In systems with fixed nozzles at the top of the solar module, there is a fixed pipeline with several nozzles from which the water comes out at high speeds. Water jets affect the solar module and thus remove dust from its surface.

Fig. 117 Manual cleaning of solar modules [62]. *Courtesy* IEA PVPS

Fig. 118 Cleaning solar modules with water under pressure and brush [63]. *Courtesy* Elsevier

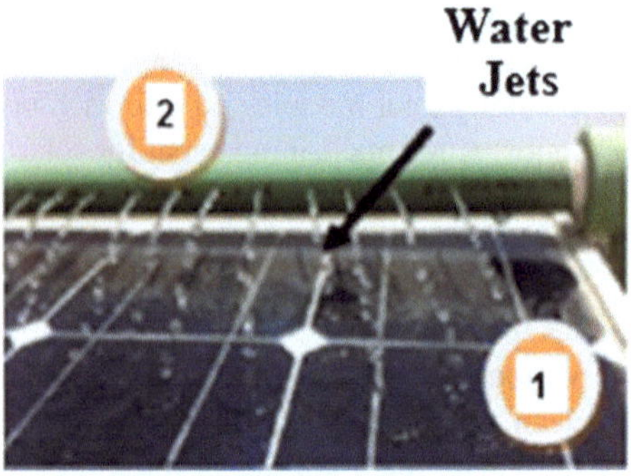

Fig. 119 Nozzle system: (1) solar module and (2) pipeline with nozzles [64]. *Courtesy* Elsevier

Fig. 120 Cleaning solar modules with a mechanical brush system [63]. *Courtesy* Elsevier

In systems with sliding nozzles, the pipeline with nozzles is moved over the entire solar module, and in this way the surface of the solar module is cleaned uniformly. In brush systems, a fixed nozzle system is provided with a movable brush that can slide or rotate around the surface of the solar module.

All these systems are fully automated and contain sensors that control their operation (determine the proper cleaning time and the appropriate amount of water needed for cleaning). These systems simultaneously perform their cooling when cleaning solar modules. Due to the cleaning of solar modules in this way, their efficiency is increased up to 15%. Cleaning solar modules with robotic devices requires great initial investment and incurs high operating costs.

Solar module cleaning can also be done using an air compressor. The air compressor draws dust from solar modules, with some of the dust particles lifted in the air and hovering above the solar modules, so this cleaning method is suitable for small systems.

Solar modules can also be cleaned by a mechanical brush system similar to windshield wipers on vehicles. In this way, the bird droppings cannot be cleaned, nor the adhesive dust that is cemented for the surface glass of the solar module due to high humidity.

Fig. 121 Self-cleaning coating (SCHN107™) [63]. *Courtesy* Elsevier

Some glass manufacturers offer special anti-soiling coatings for cover glasses of solar modules. Coatings can be hydrophobic and hydrophilic.

One of the most effective methods for cleaning solar modules is the use of self-cleaning nano-films from super-hydrophilic substances such as TiO_2 (Fig. 121).

In order to maximize the energy efficiency of solar modules, it is necessary to periodically clean them to remove the dirt deposited on their surfaces. There are general recommendations for proper cleaning and maintenance of solar modules depending on the climate conditions in which they are located.

Various recommendations can be found in the literature on how often solar modules should be cleaned.

Solar modules should not be cleaned too often to prevent damage to their protective glass.

13.4 *Protection of Solar Modules from Bird Droppings*

According to some researches, the presence of bird droppings on solar modules and making nests are much more serious problems than dust. Bird droppings are an organic material that blocks the solar radiation to reach the solar cells of the solar module and thus reduces its performance. Parts of solar modules covered with bird droppings remain in the shade (darkened) until they are cleared. Metal frames of solar modules can corrode under the influence of bird droppings. In order to prevent these problems in practice, various methods for protecting solar modules are used, such as: installation of protective grids, metal spikes, electric fences with electricity, and audio systems for bird frightening in order to keep birds away from solar modules.

Some of these methods may eventually become dysfunctional because birds can get used to them [25–28].

14 Smart Systems and Mini-Grids

14.1 Distributed Generation of Electricity

Distributed energy resource (DER) refers to groups of modular devices located near the consumer. DIS supplies local areas with electricity, which usually reduces overall losses in the electricity system and improves the voltage conditions in it.

Energy resources for distributed generators (DGs) include hydro-energy, solar energy, wind energy, geothermal energy, biomass and biogas, and tidal energy. In the development of small DER, the concept of building large production units and centralized energy systems is given an alternative. In the near future, the DG will play a significant role in preserving the stability and reliability of power systems. DERs are not centrally planned, there is no central transmission, and they are of less power and usually connected with the distribution system.

DER can be divided into several ways:

(a) According to the installed power:

- *Micro*, power less than 5 kW,
- *Small scale*, power from 5 kW to 5 MW,
- *Medium*, power of 5–50 MW,
- *Large*, power higher than 50 MW.

(b) By the type of primary energy:

- Renewable (solar power plants, wind generators, small hydropower plants, biomass power plants and biogas, geothermal power plants, tidal power plants),
- Non-renewable—which include power plants on fossil fuels (coal, oil, and natural gas) and fuel cells.

(c) According to the functional role:

- *Distributed sources for backup power supplies* (diesel power generators, fuel cells, and rechargeable batteries),
- *Autonomous power sources* (diesel power generators, photovoltaic power sources, and low power wind turbines),
- *Resources for supplying remote and rural consumer centers* (small hydropower plants, biomass power plants, wind generators, diesel power generators),

- *Resources for cogenerative production of electric and thermal energy* (thermal power plants—heating plants with gas and steam micro-turbines, diesel electric aggregates, geothermal power plants, and fuel cells),
- *Resources for covering peak load* (fast micro-turbine power plants and accumulation of small hydropower plants),
- *Resources for covering basic production* (small hydropower plants, wind generators, and solar power plants).

The main reasons that contribute to the construction of DER include:

- Reduction of gas emissions (especially CO_2),
- Energy efficiency or rational use of electricity,
- Deregulation of electricity generation and consumption,
- Lack of conventional energy resources,
- National needs for electricity,
- Easier location for small-scale power plants,
- Shorter time and lower construction costs,
- The plant can be closer to the load, which reduces the costs of electricity transmission, etc.

14.2 Connection of a Small-Scale Power Plant to the Distribution Network

Certain criteria must be met for connecting a small-scale power plant to the distribution network. Depending on the power of the small-scale power plant, the mode of its operation, and the distance from the consumer, the indicated generator voltage can be 0.42 kV; 3.15 kW, 6.3 kW and 10.5 kV, respectively.

The maximum permissible voltage deviation ΔU_m at the point of connection to the power distribution network, in relation to the values of the nominal voltage in the stationary state, can be:

- $\Delta U_m = \pm 5\%$, if the connection point is located on the medium-voltage network and
- $\Delta U_m = +5\text{–}10\%$, if the connection point is located on the low power supply network.

In the intermittent mode, when the small-scale power plant is switched on or off to a medium-voltage distribution network, the tolerance of the voltage is $\pm 2\%$, and for the low voltage connection $\pm 5\%$.

A small-scale power plant can be connected to a power distribution network:

- If it meets certain criteria,
- If it is equipped with protection and other devices protecting generators and other equipment of small-scale power plant from damage due to faults in the network

or due to an unauthorized difference in voltage, frequency, and/or phase position in a small-scale power plant and a power distribution network when connected to a power distribution network,
- If it meets the requirements of the regulations on environmental protection.

The criteria for connection and safe parallel operation of a small-scale power plant include:

- Criterion of the allowed power of a small-scale power plant,
- Flicker criterion,
- Criterion of allowed currents of higher harmonics,
- Short-circuit power criterion.

The permissible power rating of the small-scale power plant ensures that in the transient mode (switching on and off the generator) the voltage change (voltage) at the connection point to the distribution network will not exceed the value $\Delta U_{\mathrm{m}} = 2\%$. If a small-scale power plant has more generators, the connection of the next generator to the distribution network must be carried out at least 2 min after the connection of the previous generator.

The flicker criterion is assessed by the flicker factors (A_{fs}) of the small-scale power plant induced by the long-lasting flicker (over 2 h). This criterion is particularly important for connecting solar and wind power plants to the electricity distribution network. The flicker coefficient (c_{f}) indicates the ability of a small-scale power plant to produce flickers. The flicker criterion is satisfied if $c_{\mathrm{f}} \leq 20$. This condition is usually met by the small-scale power plants whose generators are driven by water, steam, and gas turbines. In case of solar and wind power plants, the c_{f} ranges from 20 to 40. Connecting these power plants requires proof that the plant meets the long-term criterion ($A_{\mathrm{fs}} \leq 0.1$); that is, the small-scale power plant will not cause harmful effects to the electricity distribution network.

The criterion of the allowed currents of higher harmonics requires the current of a given harmonic order not to exceed the allowed value, which is determined by the following expression:

$$I_{\mathrm{vhdoz}} = I_{\mathrm{vhs}} \cdot S_{\mathrm{ks}} \tag{19}$$

where I_{vhdoz} is a permitted value of the current of a higher harmonic at the voltage level of the generator in (A), I_{vhs} is a value of the higher harmonic current that is reduced to the short-circuit power at the point of connection to the power distribution network in (A/MVA), and S_{ks} is a power of the three-phase short circuit at the point of connection of the small-scale power plant to the electricity distribution network (MVA).

Currents of the higher harmonics reduced to the short-circuit power at the point of coupling of the small-scale power plant to the power distribution network for certain harmonics are given in Table 13.

If the voltage of a higher harmonic is less than 0.2% of the normal voltage ($U_{\mathrm{vh}} \leq 0.2\% \cdot U_{\mathrm{n}}$) for $\nu = 5$, and for other harmonics less than 0.1% of the normal voltage

Table 13 Currents of the higher harmonics reduced to the short-circuit power at the point of coupling of the small-scale power plant to the power distribution network for certain harmonics

Reduced current of higher harmonics	No. of higher harmonics (ν)							
I_{vhs} (A/MVA)	5	7	11	13	17	19	23	25
	0.7	0.6	0.5	0.3		0.2		

($U_{\mathrm{vh}} \leq 0.1\% \cdot U_{\mathrm{n}}$), the criterion of allowed currents of higher harmonics is satisfied. If this criterion is not satisfied, a filter for the corresponding line of higher harmonics should be installed in a small-scale power plant or a small-scale power plant should be coupled at a point with a higher short-circuit power value, or it should be coupled to a higher voltage level.

The short-circuit power criterion is significant for equipment in power distribution networks. If, when coupling a small-scale power plant to an electricity distribution network, the power (current) of the three-phase short circuit is increased over the values for which the power supply network equipment is dimensioned, the following measures should be applied:

- Limit the short-circuit current in a small-scale power plant.
- Replace the equipment in a power distribution network that does not meet the requirements regarding short-circuit power.
- Make a change of the coupling point to the power distribution network, or change the parameters of the connection line.

Bearing in mind that small-scale power plants up to 1 MVA cannot significantly increase the short-circuit current in the electricity distribution network, the short-circuit test is mandatory for small-scale power plants with a power output greater than 1 MVA.

A connecting switch is used to connect a small-scale power plant to the electricity distribution network. For small-scale power plants up to 63 kVA, generator switch can also be used.

14.3 The Impact of Distributed Generation on the Distribution Network

Distribution networks are usually designed to work radially and are powered from the appropriate high-/medium-voltage transformer station, and the power flow is exclusively from the power transformer station to consumers. When a small-scale power plant is connected to a distribution network, the situation changes significantly.

When connecting a small-scale power plant to the distribution network, there may be a change in voltage and power flows in certain branches of the distribution network. In addition, the connection of the DER to the distribution network can also affect the quality of the electricity, the operation of the relay protection, and the transmission network.

14.3.1 Influence on Voltage Conditions

Each DER is required to maintain the value of the voltage within certain limits. In this case, it is taken into account that at the maximum load the end-consumer voltage is not below the minimum permissible, and that at the minimum load the consumer's voltage nearest to the transformer station does not exceed the maximum permissible value.

When switching to the DG system, the current flows (currents) in the lines will change and thus the voltage values in the nodes. The voltage rise may occur when the DG current flows toward the power transformer station.

In some cases, the increase in voltage can be limited by the return flow of reactive power using an inductive generator or by exciting a synchronous machine.

Switching DG can also affect the size of the switching voltages, which is the result of the transient processes in the event of malfunctioning. This can lead to a breakdown or damage to the insulation of equipment that is installed in the distribution network before connecting a small-scale power plant.

14.3.2 Impact on the Quality of Electricity

Depending on the type of DER in the distribution network, there may be a transient voltage change and a harmonic distortion of the network voltage.

A transient voltage change in the network occurs when a relatively large current change occurs during DG switching on and off. Limiting transient power changes can be largely achieved by the careful design of the DER plant. If synchronous generators are correctly synchronized, they can be connected to the network with minor interference. Significant voltage drops can occur when switching DG off at full load.

If an appropriate generator is not selected, there may be a cyclical change in the output current, which may cause flicker or flicker interference. If a small-scale power plant uses energy (frequency) converters, it may generate higher current harmonics in the network, which can lead to the unacceptable interruptions in the mains voltage. Direct linking of DG to the network leads to a reduction in the harmonic impedance of the distribution network and the conditions for the emergence of harmonic resonances. This is especially important if the power factor correction capacitors are used as compensators for induction generators.

14.3.3 Impact on the Operation of Relay Protection

The connection of small-scale power plants to the distribution network can lead to the occurrence of a large fault current. The presence of DG for parallel connection to the power supply network leads to a significant reduction in the impedance of the fault, which causes large fault currents that could endanger network components not designed to operate under these conditions.

Due to the presence of DGs that supply a fault location, measuring devices do not measure the actual current at the fault site, but only the part of the fault coming from the power supply network.

A particular problem is the occurrence of the distribution network island operation that occurs when due to some malfunction (often transient) the protection reacts and the distribution network is separated from the power supply network, but the DGs remain attached. Since these generators are not supplied with the appropriate voltage and frequency regulators, unexpected voltage and frequency changes occur. If the fault is of a passing character and after a certain time, AR (automatic restart) attempts to turn on the switch on the power supply side, there will be a connection of two non-synchronous systems, which can lead to very severe consequences and damage.

14.3.4 Influence on the Transmission Network

DERs can to a certain extent reduce investments in the transmission high-voltage network. Incorporating DG into the distribution network also leads to a reduction of power losses in the transmission network.

14.3.5 Stability of Distributed Generation

If there is a defect on the network that reduces the mains voltage and the DG switches off, there will be a short interruption in the generation of electricity. The DG will strive to exceed normal speed and will activate its internal protection. The control unit in DG will wait for the network conditions to be restored and will get restarted automatically. Bearing in mind that DG inertia is often small and the time of activation of protection in distribution networks is relatively long, it is often impossible to ensure stability for all disturbances on the distribution network.

If DG is used as a support to the power supply system, its transient stability is gaining in importance. Depending on the current circumstances, transient and static stability can be significant for the proper functioning of the distribution system [29].

14.4 Smart Grids

A smart network is an electricity network that allows monitoring and control of the use of electricity in real time in order to prevent the overload of the power system during peak consumption. Introducing smart grids enables electricity suppliers to actively monitor and manage power consumption and automate these processes.

Increasing use of renewable energy sources (RES) is a challenge for the stability of the electricity system, and its safety and reliability. Currently efficiency of wind power plants is about 20–40% and solar power plants about 10–30%.

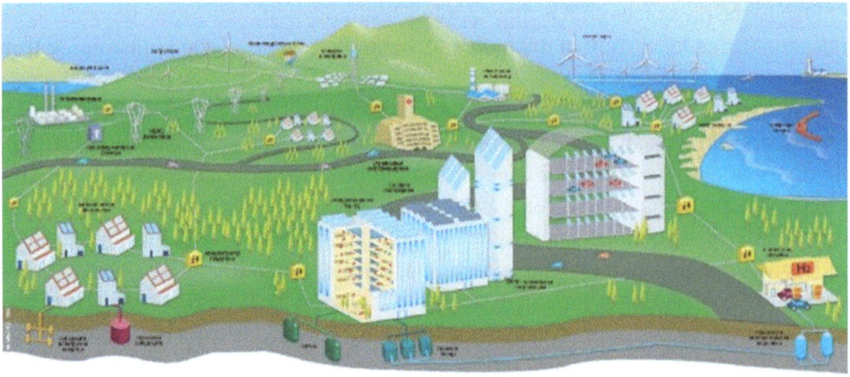

Fig. 122 Power system with smart grid

In order to optimize the production and consumption of electricity, it is necessary to increase the flexibility of the electricity system and introduce new commercial services. This can be achieved if the distribution system operators (DSOs) have at their disposal information about the state of the power system in real time. This can be achieved with smart middle- and low-voltage networks. Smart grids can intelligently integrate the activities of the transmission and power distribution system from generators to electricity consumers to effectively ensure sustainable, economical, and secure electricity supply (Fig. 122).

Smart grids as an upgrade of the existing power distribution system contribute to the realization of desired applications using modern information and telecommunication technologies.

Smart grids consist of:

- IT structures,
- Communication networks,
- An appropriate power distribution system.

The basic applications of smart grids include:

- Smart metering system,
- Automation of SN network,
- Basic distributed management system (DMS) functions,
- Automation of transformer stations.

The basis for the solution for smart grids is the telecommunication system, and the first step in the implementation is the smart measurement of electricity consumption.

Today, worldwide for telecommunication systems in smart grids power lines, wireless or mobile networks, and copper or optical cables are used. As the best solution, optical cables have been used to transmit information at the speed of light and achieve a high degree of reliability and efficiency of the electric power system.

14.5 Power Consumption

Consumption refers to a part of the territory where the electricity distribution takes place. Electricity consumption accounts for about 1/3 of the total consumption of primary forms of energy.

In the document of the Organization for European Cooperation and Development (OECD), the International Energy Agency (IEA), and the European Commission Statistics Office (Eurostat), electricity consumption in EES is divided into the following sectors:

- Households,
- Public and commercial consumption,
- Industry,
- Traffic,
- Agriculture.

In addition, electricity consumption in the EES is divided into the following consumers:

- Households,
- Administrative–commercial consumption,
- Street lighting,
- Traffic,
- Agriculture,
- Industry,
- Other consumption.

Moreover, the following division of electricity consumption is also found in the literature:

- Wide consumption (which includes households, utility, and agriculture),
- Commercial consumption (which includes categories of petty entrepreneurship, crafts, and trade),
- Industry and traffic, etc.

There is also the following distribution of consumption:

- High voltage consumption,
- Consumption at medium voltage,
- Low voltage consumption,
- Wide consumption,
- Public lighting.

The load of a consumer or system is the load on its connectors that represents the mean value over a certain time interval (usually 15 min).

14.5.1 Load Diagrams

The peak power of consumers in a given period of time (day, month, year) denotes the highest registered medium power (most often at intervals of 15 min) during that period. To measure peak power, devices known as *maxigraphs* are used. In households, peak power is estimated by the power of all electrical household appliances.

To find peak loads, the following methods are generally used:

- The method based on the probability of using individual devices,
- Analytical procedures (Rusck's form),
- Methods based on consumed electricity, i.e., power–energy correlations.

Electricity consumption changes during the year, month, and day depending on climate conditions, work activities (working and non-working days), during day and night, etc. The daily load diagram of the EES (load dependency on the time during the day) is given in Fig. 123.

The basic characteristics of electricity consumers are derived from their load diagrams that can be daily, weekly, monthly, and annual. Load diagrams vary in the place where the load is measured and can be viewed for individual categories and consumer sectors and for the entire ESS or its specific parts. In principle, several load diagrams are distinguished:

1. *Generator load diagram*

This diagram represents the gross power consumption of an electrical power system. It includes in addition to net consumption all losses in the system, as well as the own consumption of power plants and other power plants.

2. *Load diagram at the threshold of the generator (power plant)*

This diagram differs from the previous one, insofar as it does not involve the own consumption of power plants.

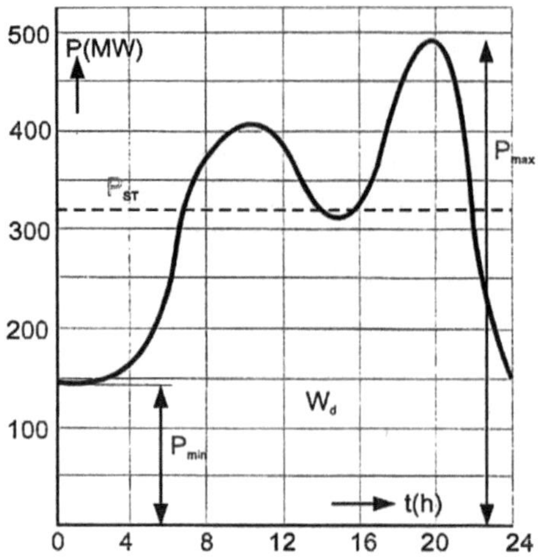

Fig. 123 Daily load diagram of the EES

3. *Load diagram at the threshold of transmission*

In this diagram, from the power consumption on the generator, besides its own power consumption, losses in generator block transformers are subtracted, too.

4. *Load diagram at the threshold of distribution*

This diagram refers to the consumption taken from the transmission network by the distribution organizations. In addition to its own power consumption, it does not include transmission losses (these are losses in transmission lines and all transformers connected to the conveyor network).

5. *Load diagram at the threshold of consumers*

This diagram is a net consumption diagram, which is charged from the electricity consumer. It differs from gross consumption from point 1, as it does not contain either its own power consumption or losses in transmission and distribution (these are all losses in transmission and distribution lines and transformers).

14.5.2 Factor of Simultaneity

The factor of simultaneity is the ratio of the common peak load of all consumers in general and the sums of the peak loads of individual consumers:

$$f_{\mathrm{n}} = \frac{P_{\mathrm{v}}}{\sum_{i=1}^{n} P_{\mathrm{vi}}} \tag{20}$$

where P_{v} is the peak power of a group of n consumers and P_{vi} is the peak power i of that consumer.

14.5.3 Active and Reactive Power

The factor of homogeneity of the household allows determining the peak load at the level of the low voltage line or the corresponding transformer station TS SN/NN. Bearing in mind that TS SN/NN is supplied with medium voltage line, it is necessary to know the factor of simultaneity TS SN/NN for determining the peak load of that line.

The rated power of the electrical energy receiver is the power recorded on each receiver. The rated active motor power P_n is the power the engine develops on the shaft at nominal voltage. The motor's active power is obtained as the ratio of the rated power and the degree of useful effect (P_n/η_n). The indicated active power of other electrical receivers is the active power that these receivers take from the network at a nominal voltage.

The indicated reactive power is the reactive power the receiver receives from the network at the indicated active power.

The mean active and reactive power of the electrical energy receiver at a variable load for a time interval t can be calculated using the following expressions:

$$P_{\mathrm{sr}} = \frac{1}{t} \int_{0}^{t} P \mathrm{d}t \tag{21}$$

$$Q_{\mathrm{sr}} = \frac{1}{t} \int_{0}^{t} Q \mathrm{d}t \tag{22}$$

The average active and reactive power of a group of n receivers can be obtained using the following expressions:

$$P_{\mathrm{rs}} = \sum_{i=1}^{n} P_{\mathrm{sri}}, \quad Q_{\mathrm{rs}} = \sum_{i=1}^{n} Q_{\mathrm{sri}} \tag{23}$$

The average active and reactive power of the receiver group for a specific time interval t can be determined using the electricity meter and the following expressions:

$$P_{\mathrm{rs}} = \frac{W_{\mathrm{a}}}{t}, \quad Q_{\mathrm{rs}} = \frac{W_{\mathrm{r}}}{t} \tag{24}$$

where W_{a} and W_{r} are active and reactive power over time t, respectively.

14.5.4 Household Consumption

In cities that do not have gas pipelines, households account for up to 60% in total electricity consumption and in peak power up to 80%. According to the degree of electrification, there are three categories of households:

1. *Households with partial electrification*

This includes households not using electricity for heating, households with non-electrical food and water preparation (e.g., in settlements with a gas network), and households with non-electric hot water preparation (in settlements with heating and hot water from the heating plant).

2. *Households with "full" electrification*

For these households, the heating is not completely electrical, while the water is heated by means of storage and flow boilers.

3. *Households with total electrification*

For these households, electricity is used for heating, for water heating with storage and flow boilers, for other appliances and devices, etc.

14.5.5 Other Consumers

Other electricity consumption includes consumption in education, health care, sports centers, hotels, administrative–technical and shopping centers, state institutions, and street lighting. Due to the heterogeneity of these areas, the use of electricity, degree of development of the country, density of population, etc., peak power and electricity consumption depend on several factors.

For the peak power forecast in the area of *other consumption*, the following expression is used:

$$P_{\mathrm{m}} = P_{\mathrm{s}} \cdot S_{\mathrm{ob}} \tag{25}$$

where p_{s} is a specific load of a particular activity and S_{ob} is an active surface of the building (s) in which a certain activity is performed.

The values of the specific load of p_{s} for individual activities are given in Table 14.

The electrification of transport greatly contributes to the decarbonization of the economy. It is expected that by 2020, the share of electric vehicles in traffic will be 10%. Electric vehicles will be supplied with electricity from the electricity system. Electric vehicles and the intensification of heating and cooling using electricity will contribute to increasing the consumption of electricity [17, 18].

14.6 Energy Needs

Based on the research conducted in the last few years by the World Energy Alliance, it was concluded that:

- Demand for electricity will be doubled by 2060.
- The need for clean energy sources will require significant investment in infrastructure and the integration of delivery systems for all consumers.
- With the development of new technologies for the production of electricity from renewable sources, it is expected to reduce the share of fossil fuels in primary energy to 50–70% of the current 80%.

Table 14 Values of specific load of p_{s} for individual activities

Activities	p_{s} (W/m^2)
Education	10–25
Health care	10–35
Sports centers	10–50
Hotels with air-conditioning	30–70
Hotels without air-conditioning	20–30
Small business buildings	15–30
Shops	25–60

- Production from the renewable energyenergy sources will create new opportunities and challenges for energy systems.

15 Economy of PV Systems

15.1 Introduction

Global development of society in the future will enormously depend on the situation in the field of energy. Problems that to a greater or lesser extent all countries face in the world are associated with the supply of energy and environmental protection. The increase in human population on earth leads to a steady increase in demand for energy. On the other hand, current structure of conventional energy sources cannot provide an increase in electricity generation. The reason for this is current environmental problems directly related to the combustion of fossil and nuclear fuels, underlying current electricity production in the world. In addition, current dynamics of the exploitation of fossil fuels in the near future will lead to their exhaustion. A direct consequence of the above is a steady increase in electricity prices, which creates an environmentally and economically justified need for the use of renewable energyenergy sources. For this reason, developed countries are making efforts to develop a system for the use of renewable energyenergy sources (solar and wind power, hydropower, geothermal power, biomass and biogas, etc.). As a result of this investment, the technology and industry for technically reliable conversion of renewable energyenergy into electrical and thermal energy were developed. In addition, international protocols and commitments on the reduction of CO_2 emission (Kyoto Protocol) and local environmental problems have forced many governments to use various subsidies so as to encourage the construction of eco-clean power plants using renewable energy sources. These policies have led to a trend of increasing ratio of certain renewable energy sources in the total electric power generation.

Solar energy is the most abundant energy source of renewable energyenergy. Solar energy reaches the earth in various forms like heat and light. Studies revealed that global energy demand can be fulfilled by using solar energy satisfactorily as it is abundant in nature and freely available source of energy. Each day, the sun provides 10,000 times more energy than the energy needed on the planet. Solar energy is obviously environmentally advantageous in comparison with any other energy source, and it is also the milestone of any serious sustainable development program. It does not deplete natural resources, does not cause CO_2 or other gaseous emission into air, nor does it generate liquid or solid waste products. Concerning sustainable development, main directly or indirectly derived advantages of solar energy use are the following:

- No emissions of greenhouse (mainly CO_2, NO_x) or toxic gasses (SO_2, particulates),
- Reclamation of degraded land,

- Reduction of transmission lines from electricity grids,
- Improvement of the quality of water resources,
- Increase of regional/national energy independence,
- Diversification and security of energy supply,
- Acceleration of rural electrification in developing countries, etc.

From the outbreak of the energy crisis in 1973, more attention has been devoted to the use of solar energy for the production of thermal and electrical energy.

One of the most popular techniques of solar energy generation is the installation of photovoltaic (PV) systems using sunlight to generate electrical power. Photovoltaic technology is one of the finest ways to harness solar power.

Solar photovoltaic technology was earlier used mainly in the space programs or in remote locations and was marginalized and exotic. Recently, it has been gaining grounds becoming a basic technology for the production and distribution of the electrical energy in urban areas with the potential to become, in terms of costs, equally competitive to the costs of energy generated and distributed by the conventional technologies.

Starting from 1990 and on, industry of photovoltaic conversion of solar irradiation shows constant annual economical growth of over 20%, and from 1997 over 33% annually. In 2000, total installed capacities worldwide have surpassed 1000 MW and in developing countries have overreached more than million households which are using electrical energy generated by means of the photovoltaic systems. Photovoltaic power is the strongest growing of all technologies examined so far, with recent annual growth rates of around 40%.

Overgrowing number of companies and organizations is taking active part in the promotion, development, and the production of photovoltaic devices and systems. Companies producing and distributing electrical energy in cooperation with the manufacturers of the solar cells, city authorities, and funds are planning and realizing all major projects, thus gaining necessary experiences, mobilizing the public focus, and reducing the cost of electrical energy.

Current market is dominated by the urban (residential) PV systems connected with the electro-distribution grid. PV industry is increasingly represented in the national energy strategies of the large number of countries.

15.2 Solar Cells' Production

PV industry production by region in 1997–2017 is given in Fig. 124.

Annual PV production by technology worldwide in GW_P is given in Fig. 125.

Percentage of global annual production by PV technology is given in Fig. 126.

Annual global PV module production by thin-film technologies is shown in Fig. 127.

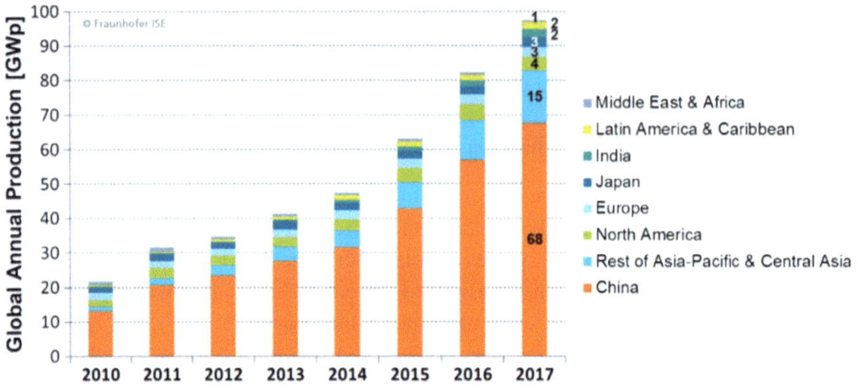

Fig. 124 PV industry production by region in 1997–2017 [38]. *Courtesy* Fraunhofer Institute

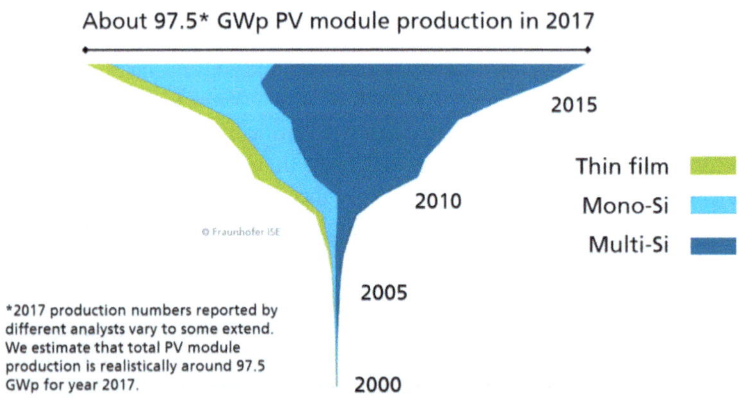

Fig. 125 Annual PV production by technology worldwide in GW_P [38]. *Courtesy* Fraunhofer Institute

15.3 PV Installation

Global cumulative PV installation by region in 2017 is given in Fig. 128.

The total cumulative installations amounted to 415 GW_P at the end of 2017. All percentages in Fig. 128 are related to total global installations, including off-grid systems.

Global cumulative PV installation until 2017 that includes off-grid systems is shown in Fig. 129.

Annually installed capacity of low- and high-concentrator PV systems (LCPV/HCPV) is given in Fig. 130.

LCPV and HCPV have concentration factors below 100 suns and from 300 up to 1000 suns, respectively.

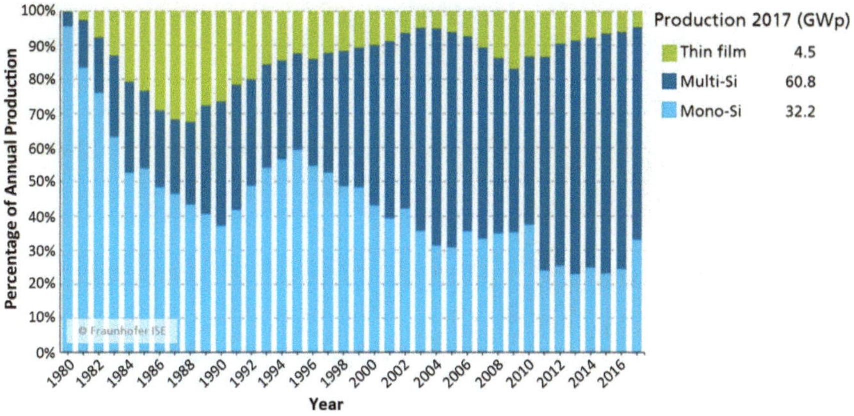

Fig. 126 Percentage of global annual production by PV technology [38]. *Courtesy* Fraunhofer Institute

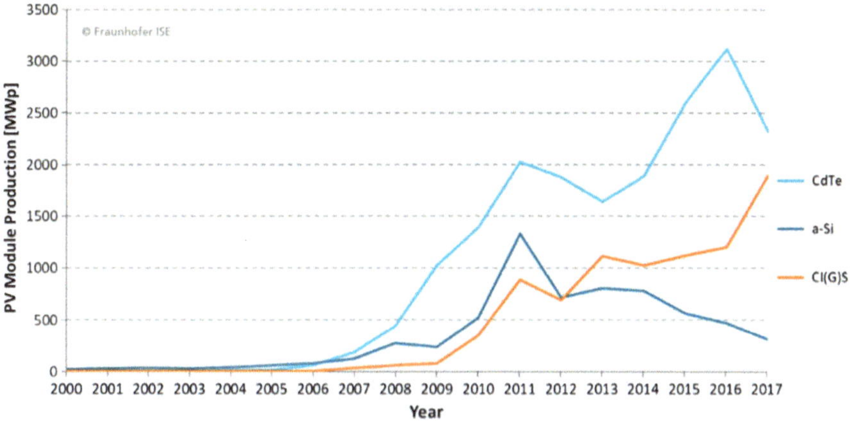

Fig. 127 Annual global PV module production by thin-film technologies [38]. *Courtesy* Fraunhofer Institute

15.4 PV Market

Worldwide growth of photovoltaics has been fitting an exponential curve for more than two decades. During the period between 2000 and 2015, the growth rate of PV installations was 41%. The market for PV systems will likely continue to grow in the future as strongly as so far, due to the thrust of subsidies, tax breaks, and other financial incentives. Support for research and development (R&D) and PV technology change are crucial aspects in accelerating the widespread adoption of PV systems. These two aspects play a key role in climate policy. Some of the largest countries in Europe, such as Germany, Denmark, and Spain, in addition to Asian countries China and

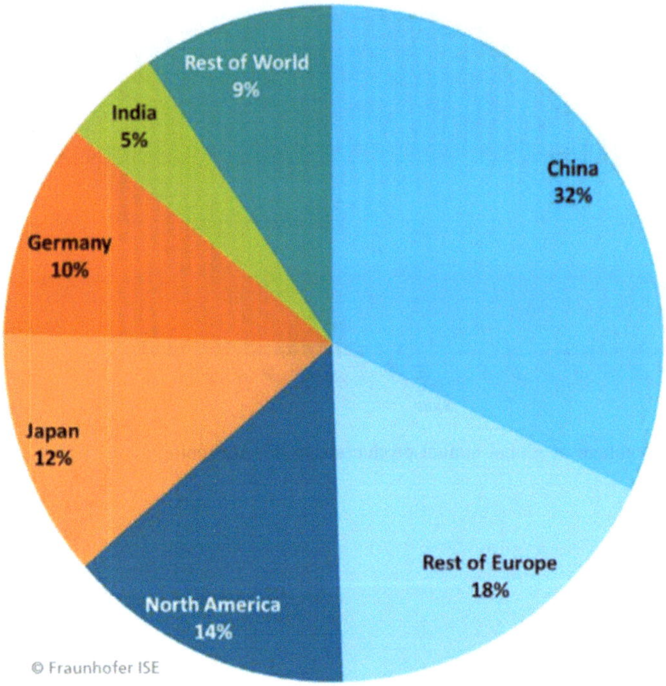

Fig. 128 Global cumulative PV installation by region in 2017 [38]. *Courtesy* Fraunhofer Institute

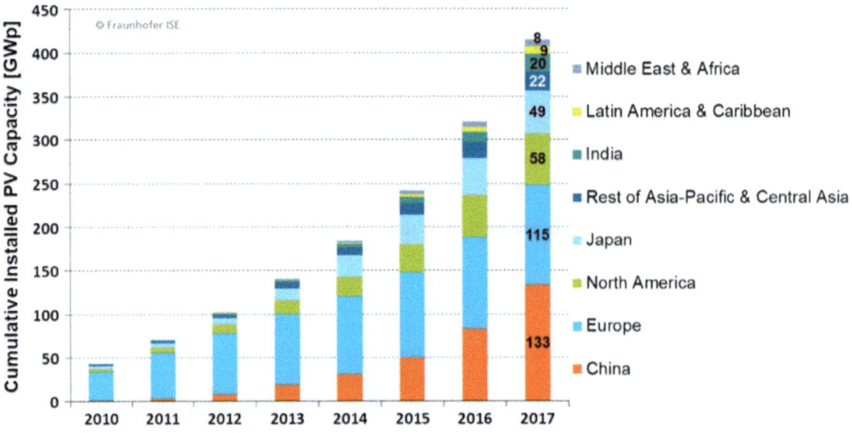

Fig. 129 Global cumulative PV installation until 2017 includes off-grid systems [38]. *Courtesy* Fraunhofer Institute

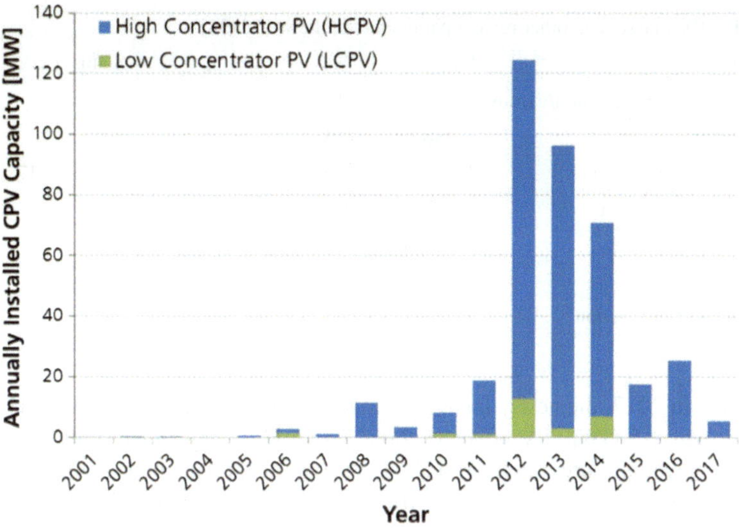

Fig. 130 Annually installed capacity of low- and high-concentrator PV (LCPV/HCPV) systems [38]. *Courtesy* Fraunhofer Institute

Taiwan, have used feed-in tariff (FIT) which is a political mechanism to encourage consumers to invest in renewable microgeneration. On the other hand, the USA, UK, Japan, and Sweden have used the renewable portfolio standard (RPS), which is a regulation that requires that part of the energy consumed comes from renewable sources. Meanwhile, South Korea has changed its plans for renewable energyenergy technologies from a RPS setting to minimize the financial burden on the government.

Photovoltaics is a fast-growing market: The compound annual growth rate (CAGR) of PV installations was 24% between 2010 and 2017. Concerning PV module production in 2017, China and Taiwan hold the lead with a share of 70%, followed by Rest of Asia-Pacific & Central Asia (ROAP/CA) with 14.8%. Europe contributed with a share of 3.1% (compared to 4% in 2016); USA/CAN contributed 3.7%. In 2017, Europe's contribution to the total cumulative PV installations amounted to 28% (compared to 33% in 2016). In contrast, installations in China accounted for 32% (compared to 26% in 2016). Si wafer-based PV technology accounted for about 95% of the total production in 2017. The share of multi-crystalline technology is now about 62% of total production. In 2017, the market share of all thin-film technologies amounted to about 5% of the total annual production.

PV market and other related parameters in the world are given in Table 15.

15.4.1 PV Market in Germany

In 2017, Germany accounted for about 10% (43 GW_P) of the cumulative PV capacity installed worldwide (415 GW_P) with about 1.6 million PV systems installed in

Table 15 PV market and other related parameters in the world [38]

Parameter	Value	Status
Germany/European Union/worldwide		
PV market	1.75/8.6/94.6 GW	2017
Cumulative installation	43/114.6/415 GW	End of 2017
PV power generation	38/120/443 TWh	2017
PV electricity share	7.2% (net)/gross: 3.6%/1.7%	2017
Worldwide		
c-Si share of production	95%	2017

Germany. In 2017, the newly installed capacity in Germany was about 1.7 GW_P; in 2016, it was 1.5 GW_P. PV covered about 7% of Germany's electricity demand in 2017. Renewable sources delivered about 38% of the total net power consumption in 2017 in Germany. In 2017, about 19 Mio. t of CO_2 emissions have been avoided due to 38.4 TWh electrical energy generated by PV in Germany. PV system performance has strongly improved. Before 2000, the typical performance ratio was about 70%, while today it is in the range of 80–90%.

In Germany, prices for a typical 10–100 kW_P PV rooftop system were around 14,000 €/kW_P in 1990. At the end of 2017, such systems cost about 1140 €/kW_P on average. This is a net price regression of about 90% over a period of 27 years and is equivalent to an annual compound average price reduction rate of 8%. The experience curve—also called learning curve—shows that in the last 37 years module price decreased by 24% with each doubling of the cumulated module production. Cost reductions result from economies of scale and technological improvements.

PV energy generated and resulting CO_2 avoided emissions are shown in Fig. 131.

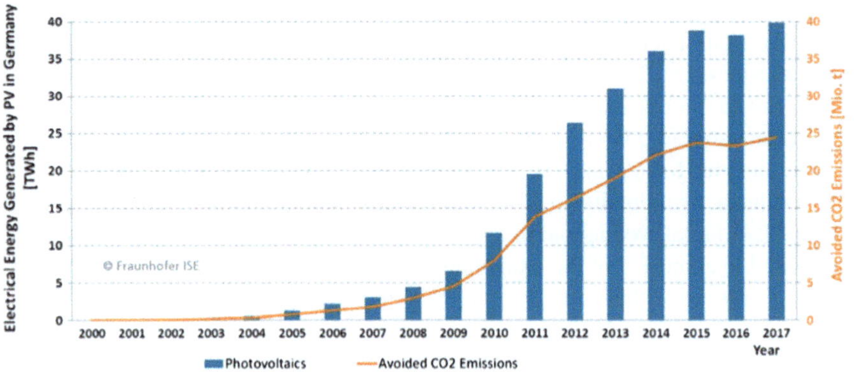

Fig. 131 PV energy generated and resulting CO_2 avoided emissions [38]. *Courtesy* Fraunhofer Institute

Table 16 PV system price and levelized cost of energy in Germany [38]

Price PV rooftop system	~1400 €/kW$_P$	End of 2017
LCOE PV power plant	4–7 ct€/kWh	End of 2017
PV-tender price	4.33 ct€/kWh	February 2018

LCOE levelized cost of energy

In 2017, about 19 Mio. t of CO_2 emissions were avoided due to 38.4 TWh PV electricity consumed in Germany. According to the Federal Environment Agency (UBA), the CO_2 avoidance factor of PV in 2017 is 489 grams of $CO_{2\text{-eq}}/kWh_{el}$.

PV systems' price and levelized cost of energy in Germany are given in Table 16.

15.5 *Installed Cost Trends*

15.5.1 Solar Cells and Modules' Prices

Solar PV module prices in Europe decreased by 83% from the end of 2010 to the end of 2017. Module costs declined 80% between the end of 2010 and the end of 2016, a period over which 87% of the cumulative global PV capacity installed at the end of 2016 occurred. Average monthly solar PV module prices in Europe in 2016 were 13% lower than in 2015, while the decline in average prices across a range of markets was 18% between 2015 and 2016.

In 2016, average selling prices in China were around USD 0.43/W, while California became one of the highest priced major markets with prices of USD 0.61/W, though all analyzed markets experienced a decreasing cost trend between 2015 and 2016. These are average values, and a range of prices around these values occur. In 2017, module prices have dipped as low as USD 0.3/W, but are somewhat higher for modules from Chinese majors and good-quality modules can now be produced sustainably for USD 0.4/W or less.

15.5.2 Total Installed Cost

Utility-scale solar PV total installed cost trends in selected countries are given in Fig. 132.

During the period 2010–2017, utility-scale total installed cost reductions in many markets have exceeded 70%. Between 2010 and 2017, the USA saw utility-scale total installed costs reduce the least, at 52%, with Italy experiencing the largest reduction of 79%.

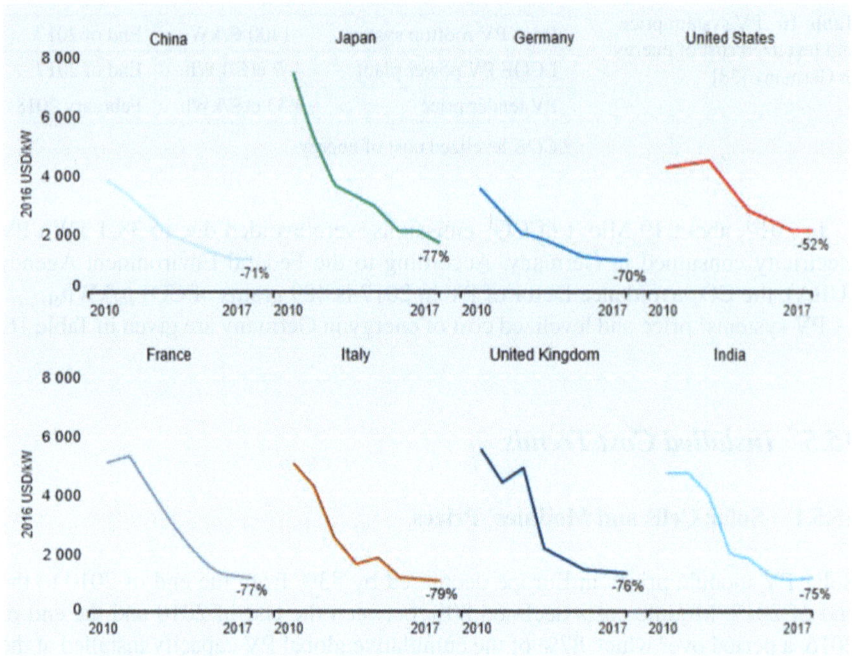

Fig. 132 Utility-scale solar PV total installed cost trends in selected countries, 2010–2017 [9]. *Courtesy* IRENA

15.5.3 Residential PV System Cost

Average total installed costs of residential PV systems by country, 2007–2017, are given in Fig. 133.

Residential PV system total installed costs have declined sharply in a wide range of countries since 2010. The range of residential solar PV total system costs in the markets with the longest historical data decreased from between USD 6700 and USD 11,100/kW in 2007 to between USD 1050 and USD 4550/kW in 2017 (a decline of 47–78%).

15.6 Levelized Cost of Electricity

Utility-scale solar PV: Electricity cost trends in selected countries, 2010–2017, are given in Fig. 134.

Rapid declines in installed costs and increased capacity factors have improved the economic competitiveness of solar PV around the world. The global weighted average LCOE of utility-scale PV plants is estimated to have fallen by 73% between 2010 and 2017, from around USD 0.36 to USD 0.10/kWh.

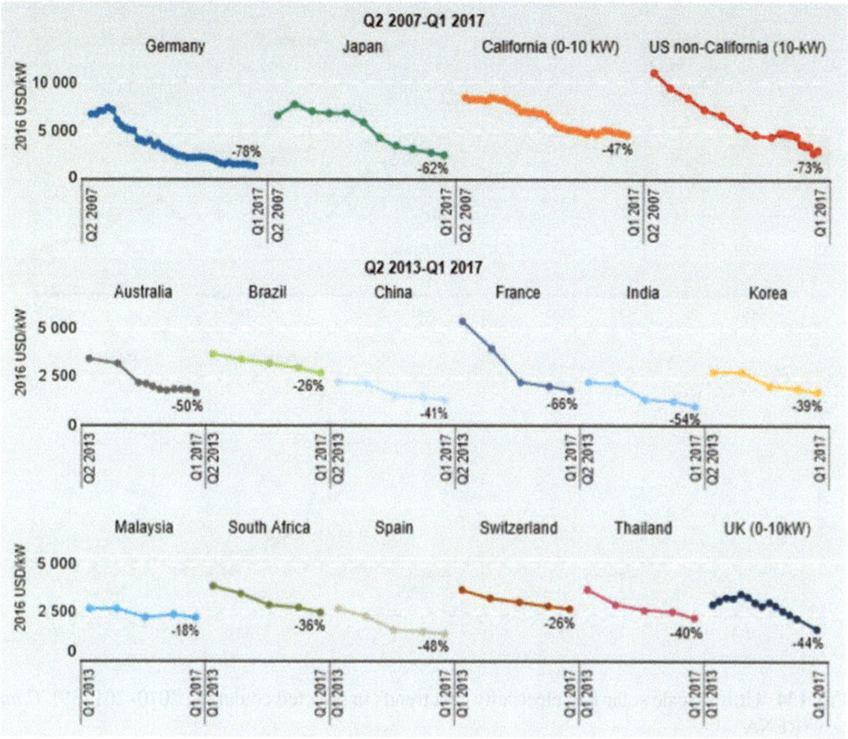

Fig. 133 Average total installed costs of residential PV systems by country, 2007–2017 [9]. *Courtesy* IRENA

15.7 Energy Payback Time

Material usage for silicon cells has been reduced significantly during the last 13 years from around 16 g/W_P to about 4 g/W_P due to increased efficiencies, thinner wafers, and wires as well as larger ingots. The energy payback time of PV systems is dependent on the geographical location: PV systems in Northern Europe need around 2.5 years to balance the input energy, while PV systems in the south equal their energy input after 1.5 years and less, depending on the technology installed. A PV system located in Sicily with multi-Si modules has an energy payback time of around one year. Assuming 20 years life span, this kind of system can produce twenty times the energy needed to produce it. The energy payback time for CPV systems in Southern Europe is less than 1 year.

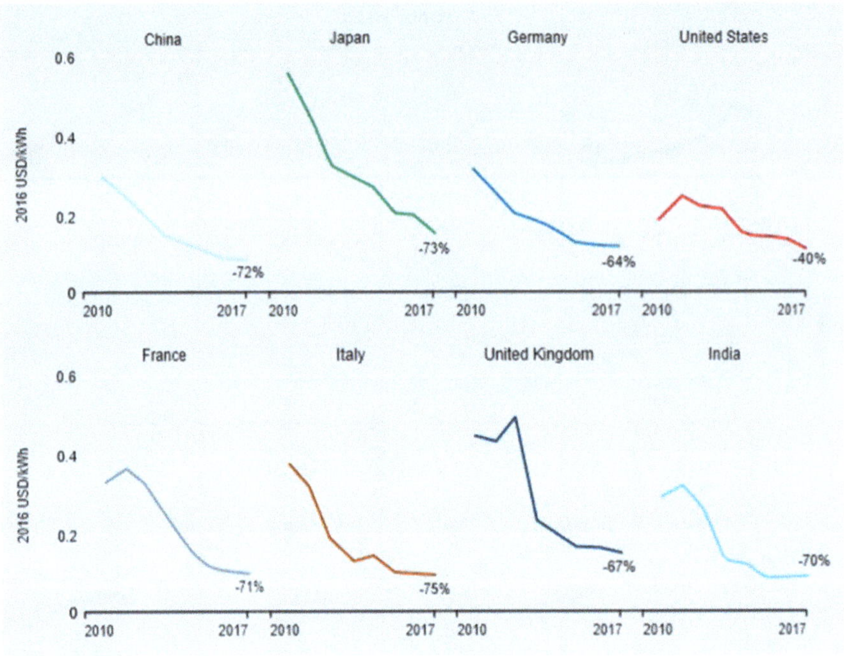

Fig. 134 Utility-scale solar PV: electricity cost trends in selected countries, 2010–2017 [9]. *Courtesy* IRENA

16 Sustainability of the Green Economy

There are numerous approaches to the topic of "sustainable green economy." Key focus is on efficient growth in the use of natural resources, and pollution and adverse environmental impacts are minimized. Indeed, green economy is responsible for all natural hazards. It is a new concept that recognizes the environment, protects and launches national economic development, and redirects society to quality growth and general prosperity through green technology and clean energy on the road to sustainable development.

Solar technologies, wind generators, biomass, biogas, geothermal energy sources, energy products from hydropower plants, and the like are a fundamental driving force for the establishment of a green economy. In this context, on the way to the sustainable development, it is necessary to unconditionally restructure national economic plans for healthy, so-called green solutions based on the strategic assumptions of economic and overall growth. Green energy is the starting point for observing the values of the protection of natural resources and developing strategies for economic growth, for the comprehensive treatment of economic, environmental, social, and technological aspects. Nowadays, more than ever, it is necessary to focus on cost-effective ways of reducing environmental pressures that allow the transition to new models of development and which in turn will avoid the disagreement between local, regional, and global environmental frameworks. According to the assumptions, the implemented strategic green growth options at the national level should encourage ideas for a different, restructured urban–rural identity of agglomerations, for an environmentally sustainable, more acceptable, developmental philosophy of green growth in the behavior of economic organizations and consumers, improving optimal reallocation of labor, to direct technologies toward more green operations, and thus provide motivation for the development and application of the latest, modern eco-innovative solutions. It is necessary to provide greater investment in stimulating efficient use of energy and resources, as well as preventing any damage to biodiversity and the ecosystem.

In this regard, the rapidly growing global population and increasingly noticeable climate changes dominated by increased carbon emissions bring increased risk conditions. That is why urgent, preventive measures and direct planetary agreements are needed on global balancing of national economic growth paths, as well as on the prevention of excessive exhaustion of natural resources and pollution. Only in such circumstances can socio-technological and economic–ecological growth be sustainable as the only rational direction of the development of our civilization. Their presence can be achieved by saving and using energy and resources. Effectively, the impact on climate change and environmental damage must be dramatically reduced. It is necessary to provide new engines for revolutionary growth through the research and development of rational green technologies, creating new jobs and achieving harmony between the economy and environment. It is also necessary to recognize the role of natural capital in the planning processes. It is important that such opinions are refreshed by taking measures and economic initiatives in order to gain support for increasing investments in the green segments of national economies. A global action Green New Deal is needed with concrete proposals of reasonable measures for economic reconstruction and stabilization in which the changes in the environment will be carried out gradually. Changes in space include strategic adaptation and modification to the given conditions, understanding of social, economic, and cultural dimensions in order to achieve the desired goals.

Through green growth, it is possible to solve economic and environmental problems and the realization of potentially new sources of growth by encouraging the increase in the effective use of available resources and natural resources, by reducing the amount of waste and energy consumption. For ecological problems, it is necessary to apply in practice intensive and innovative stimulations for new conditions that precisely define ways of using modern technological solutions, which contribute to the creation of functional development, so that the operational, correlated coexistence strategy between the three green ideas: green growth, green economy, and sustainable development and environmental protection, has a stable character. The goal is to move toward a radical reduction of environmental disadvantages and ecological threats, to reducing crisis situations in the artifact and natural environment as well as to a sustainable green economy. In such an effort, it is important to emphasize the necessity of increasing investments in the economic sectors that create and strengthen the natural capital of the earth and contribute to the reduction of adverse environmental impacts. The sectors in the forefront include renewable energy, low-emission transport, energy-efficient buildings, quality and clean technologies, improved waste management, sustainable agriculture and forests, integrated resource management, and sustainable fishing [9, 38].

References

1. Radmilovic VV (2016) Transparent nanocomposite films for plastic electronics applications. Ph.D. thesis, University of Belgrade, Faculty of Technology and Metallurgy, Belgrade
2. Palz W (1978) Solar electricity-an economic approach to solar energy. UNESCO
3. Angrist WS (1971) Direct energy conversion. Allyn and Bacon Inc., Boston
4. Green AM (1982) Solar cells-operating principles, technology and system applications. Prentice Hall Inc., New York
5. Matsuoka T et al (1990) Solar Cells 29:361
6. Chronar Co. (1997) Technical characteristics of standard photoconversion glass/a-Si module. Princeton, USA
7. Yano M et al (1987) Thin Solid Films 146:75
8. Kalogirou AS (2017) McEvoy's handbook of photovoltaics: fundamentals and applications. Academic Press
9. IRENA (2018) Renewable power generation cost in 2017. International Renewable Energy Agency
10. Boxwell M (2015) Solar electricity handbook. Greenstream Publishing, Coventry
11. Kalogirou AS (2014) Solar energy engineering—processes and systems, 2 nd edn. Elsevier Academic Press, Amsterdam
12. Pode R, Diouf B (2011) Solar lighting. Springer, London
13. Foster R, Chassemi M, Cota A (2010) Solar energy: renewable energy and the environment. CRC Press, Boca Raton
14. Labudovic B et al (2011) Basic applications of photovoltaic's systems. Energetika Marketing, Zagreb (in Serbian)
15. Manimekalai P, Harikumar R, Raghavan S (2013) An overview of batteries for photovoltaic (PV) systems. Int J Comput Appl (0975–8887) 82(12)
16. Tracer 1210RN—maximum power point tracking solar charge controller. Instruction manual, EPSOLAR, utility model patent no. 201120064092.1 (2012)
17. Rajakovic N, Tasic D (2008) Distributive and industrial nets. Akademska Misao, Beograd (in Serbian)
18. Tasic SD, Rajakovic LN, Stojanovic SM (2014) Electro energetic components. University of Nis, Faculty of electronics, Nis (in Serbian)
19. Pavlovic MT, Mirjanic LD, Milosavljevic DD (2018) Electric power industry in Serbia and the Republic of Srpska. Academy of Sciences and Arts of the Republic of Srpska, Banja Luka (in Serbian)
20. Cekić N et al (2015) Application of solar cells in contemporary architecture. Contemp Mater VI–2:104–114
21. Pantić SL et al (2016) Electrical energy generation with differently oriented PV modules as façade elements. Therm Sci 20(4):1377–1386
22. Pavlović TM, Tripanagnostopoulos Y, Mirjanić LD, Milosavljevć DD (2015) Solar energy in Serbia, Greece and the Republic of Srpska. Academy of Sciences and Arts of the Republic of Srpska, Banja Luka
23. Pantić Ranđelović SL (2017) The study of energy efficiency of PV solar modules depending on their geographical orientation, tilt angle and their temperature in real climatic conditions in Nis. Ph.D. thesis, Faculty of Sciences and Mathematics, University of Nis, Nis, Serbia (in Serbian)

24. Pantic SL, Pavlovic TM (2016) Determination of physical characteristics of horizontally positioned solar module in real climate conditions in Nis. Serb Facta Univ 4(1):37–51
25. Radonjic Mitic SI (2018) Investigation of energy efficiency of PV solar modules depending on their soiling in real climate conditions in Nis. Doctoral dissertation, University of Nis, Faculty of sciences and mathematics, Nis (in Serbian)
26. Radonjić SI et al (2017) Investigation of the impact of atmospheric pollutants on solar module energy efficiency. Therm Sci 21(5):21–30
27. Radonjić SI et al (2016) Investigation of solar module energy efficiency depending on their surface soiling degree. In: Proceedings of scientific conference Unitech 2016. Gabrovo, vol 1, pp 147–151
28. Radonjić SI et al (2017) Investigation of the energy efficiency of soiled solar module mounted at the optimal angle. In: Proceedings of international scientific conference Unitech 2017. Gabrovo, vol I, pp 84–88
29. Lasnier F, Ang GT (1990) Photovoltaic engineering handbook. Adam Higler
30. Neville CR (1995) Solar energy conversion—the solar cell, 2nd edn. Elsevier, Amsterdam
31. https://www.slideserve.com/maleah/pv-system-design-and-installlation
32. http://www.daviddarling.info/encyclopedia/
33. Spasojevic AZ, Popovic AZ (1979) Electrotechnical and electronic materials. Naucna Knjiga, Belgrade (in Serbian)
34. Stojanovic M (1986) Primenjena nauka 5, 7
35. https://en.wikipedia.org/
36. www.greenrhinoenergy.com/solar/technologies/pv_concentration.php
37. https://www.osa-opn.org/home/articles/volume_22/issue_1/features/solar_concentrators_using_optics_to_boost_photovo/
38. Photovoltaics report, Fraunhofer Institute for Solar Energy Systems, ISE, Freiburg, 27 Aug 2018
39. www.electronics-tutorials.ws
40. https://www.designboom.com/project/photovoltaic-honeycomb-glass-screen-facade/
41. https://www.asu.edu/fm/images/solarization/tyler.jpg
42. https://www.designboom.com/science/solar-forest-charging-station-for-electric-vehicles/
43. http://www.designboom.com/wp-content/uploads/2015/05/germany-pavilion-expo-milan-2015-fields-of-ideas-designboom-01.jpg
44. http://www.ecoco2.com/images/blog/2011/3F-quai-de-Valmy-big.jpg
45. http://en.stonkcash.com/energy-dubai/
46. http://www.citycool.info/wp-content/uploads/2014/11/Cite%CC%81-du-Design-Platine-Copyright-LIN-Finn-Geipel-et-Giulia-Andi-photographe-Jan-Oliver-Kunze.jpg; https://www.homedit.com/solar-powered-international-design-center-revealed/
47. https://www.pinterest.com/pin/480055641504922067/visual-search/?x=16&y=11&w=530&h=345
48. https://www.urbangardensweb.com/2013/04/02/sustainable-high-rise-greens-sydney-skyline/
49. https://en.wikipedia.org/wiki/Seocho_Garak_Tower_East#/media/File:Seocho_Garak_Tower_East.jpg
50. https://apimagesblog.com/blog/2016/9/25/indias-hidden-valley
51. https://www.irena.org/offgrid
52. https://insideclimatenews.org/news/20121113/germany-energiewende-clean-energy-economy-renewables-solar-wind-biomass-nuclear-renewable-energy-transformation
53. https://twitter.com/SunSol_pl
54. https://greensparksolar.com/tag/residential-solar/page/2/
55. https://www.centuryroofandsolar.com/blog/solar-power-growth-rural-america/
56. https://solarinstallationpanel.blogspot.com/2013/12/solar-schematic-diagram.html
57. http://www.shout92.com/top-5-largest-solar-panel-farms-in-world/
58. https://en.wikipedia.org/wiki/Gujarat_Solar_Park

59. http://www.energia.eiffage.es/en/2014/11/10/eiffage-energia-is-involved-in-the-construction-of-the-biggest-pv-plant-in-europe/
60. https://www.businesswire.com/news/home/20131031006341/en/NRG-Energy-NRG-Yield-SunPower-Commercial-Operations
61. Picotti G et al (2018) Soiling of solar collectors—modelling approaches for airborne dust and its interactions with surfaces. Renew Sustain Energy Rev 81(2):2343–2357
62. Köntges M et al (2017) Assessment of photovoltaic module failures in the field. Report IEA-PVPS T13-09:2017
63. Ghazi S, Sayigh A, Ip K (2014) Dust effect on flat surfaces-a review paper. Renew Sustain Energy Rev 33:742–751
64. Moharram KA et al (2013) Influence of cleaning using water and surfactants on the performance of photovoltaic panels. Energy Convers Manag 68:266–272
65. Knapp V, Kulisic P (1985) New energy sources. Skolska Knjiga, Zagreb (in Serbian)
66. Menniti D (2017) Solar energy and PV systems in smart cities. Hindawi PC
67. Pearsall NM (2017) The performance of photovoltaics (PV) Systems. Elsevier, Amsterdam
68. Ymeri H (2016) Strategic study, the need of an energy strategy for South-East Europe. Balcan Trust for Democracy, Tirana
69. Messenger R, Ventre J (2010) Photovoltaic systems engineering. CEC Press, Taylor & Francis Group, USA
70. Chiras D, Aram R, Nelson K (2009) Power from the sun—achieving energy independence. New Society Publishers, Canada
71. Sen Z (2008) Solar energy fundamentals and modeling techniques—atmosphere, environment, climate change and renewable energy. Springer
72. Masters MG (2004) Renewable and efficient electric power systems. Wiley, Hoboken
73. Green AM, Wenham RS, Watt EM (2006) Applied photovoltaics, 2nd edn. Earthscan Ltd.
74. Tiwari NG, Dubey S (2010) Fundamentals of photovoltaic modules and their applications. RSC Publishing Energy Series No. 2
75. Handbook of photovoltaic science and engineering. Wiley, New York
76. Practical handbook of photovoltaics: fundamentals and applications. Elsevier Science Ltd.
77. Markvart T, Castaner L (2006) Solar cells. Elsevier, Amsterdam
78. Duffie AJ, Beckman AW (1991) Solar engineering and thermal processes, 2nd edn. Wiley, New York
79. Chu Y (2011) Review and comparison of different solar energy technologies. Global Energy Network Institute
80. Altas IH, Sharaf AM (2014) Solar energy and PV systems. Hindawi PC
81. Solanki CS (2013) Solar photovoltaic technology and systems. PHI Leaming Private Limited, Delhi
82. Lorenco E et al (1994) Solar electricity. Progensa, Sevilla
83. Gevorkian P (2008) Solar power in building design, the engineer's complete design resource. Mc Graw Hill, New York
84. Lotsch VKH, Goetzberger A, Hoffmann UV (2005) Photovoltaic solar energy generation. Springer, series in optical sciences, vol 112
85. Mijatovic V (2011) Distributed energy resources. Akademska knjiga, Beograd (in Serbian)
86. Milosavljevic DD (2013) The study of energy efficiency of PV solar power plants in the Republic of Serbia and the Republic of Srpska. Ph.D. thesis, Faculty of Sciences and Mathematics, University of Nis, Nis, Serbia (in Serbian)
87. Parida B, Iniyan S, Goic R (2011) A review of solar photovoltaic technologies. Renew Sustain Energy Rev 15(3):1625–1636
88. Green M (2007) Thin-film solar cells: Review of materials, technologies and commercial status. J Mater Sci Mater Electron 18(1):15–19
89. Ullal SH (2008) Overview and challenges of thin film solar electric technologies. In: Conference paper at the world renewable energy congress X and exhibition. Available online at http://www.scribd.com/doc/58670014/NREL-Thin-Film-Overview-2008. http://www.pvresources.com/en/solarcells.php

90. Klise TG, Stein SJ (2009) Models used to assess the performance of photovoltaic systems. Sandia Report, Sand 2009-8258
91. Angelis-Dimakis A et al (2011) Methods and tools to evaluate the availability of renewable energy sources. Renew Sustain Energy Rev 15:1182–1200
92. Mousazadeh H et al (2009) A review of principle and sun-tracking methods for maximizing solar systems output. Renew Sustain Energy Rev 13(8):1800–1818
93. Súri M, Huld AT, Dunlop DE (2005) PV-GIS: a web-based solar radiation database for the calculation of PV potential in Europe. Int J Sustain Energ 24(2):55–67
94. Súri M (2007) Solar resource data and tools for an assessment of photovoltaic systems, Chapter 7. Status report 2006, EUR 22752 EN, pp 96–102
95. Aelenei L, Goncalves H (2013) From solar building design to net zero energy buildings: performance insights of an office building. Energy Procedia 48:1236–1243
96. Bazilian M et al (2001) Photovoltaic cogeneration in the built environment. Sol Energy 71:57–69
97. Brinkworth JB et al (1997) Thermal regulation of photovoltaic cladding. Sol Energy 61:169–179
98. Brinkworth JB, Sandberg M (2006) Design procedure for cooling ducts to minimize efficiency loss due to temperature rise in PV arrays. Sol Energy 80:89–103
99. Charron R, Athienitis KA (2006) Optimization of the performance of double-facades with integrated photovoltaic panels and motorized blinds. Sol Energy 80:482–491
100. Chemisana D, Ibanez M (2010) Linear Fresnel concentrators for building integrated applications. Energy Convers Manage 51:1476–1480
101. Mallick KT, Eames CP, Norton B (2007) Using air flow to alleviate temperature elevation in solar cells within asymmetric compound parabolic concentrators. Sol Energy 81:173–184
102. Skoplaki E, Palyvos AJ (2009) Operating temperature of photovoltaic modules: a survey of pertinent correlations. Renew Energy 34:23–29
103. Skoplaki E, Palyvos AJ (2009) On the temperature dependence of photovoltaic module electrical performance: a review of efficiency/power correlations. Sol Energy 83:614–624
104. Skoplaki E, Boudouris GA, Palyvos AJ (2008) A simple correlation for the operating temperature of photovoltaic modules of arbitrary mounting. Sol Energy Mater Sol Cells 92:1393–1402
105. Pantić SL et al (2016) The assessment of different models to predict solar module temperature, output power and efficiency for Nis. Serbia Energy 109:38–48
106. Radivojević A et al (2015) Influence of climate and air pollution on solar energy development in Serbia. Therm Sci 19:311–322
107. Milosavljević DD et al (2017) Photovoltaic technology: economical framework. In: Proceedings of 7th scientific conference "Economics and Management—Based on New Technologies—EMoNT-2017". SaTCIP Publisher Ltd., Vrnjačka Banja, Serbia, pp 136–144
108. Solar energy perspectives, International Energy Agency (IEA), OECO/IEA, France, www.iea.org
109. BP energy outlook, 2018 edition, BP energy economics https://www.bp.com/energyoutlook#BPstats
110. Trends 2018 in photovoltaic applications, survey report of selected IEA countries between 1992 and 2017, photovoltaic power systems programme, report IEA PVPS T1-34:2018, http://www.iea-pvps.org/fileadmin/dam/public/report/statistics/2018_iea-pvps_report_2018.pdf
111. Pulfrey LD (1978) Photovoltaic power generation. Van Nostrand Reinhold Co., New York
112. Sanyo Co. (1997) Technical characteristics of photovoltaic products. Sumoto, Japan
113. http://www.pvresources.com/en/solarcells.php
114. http://www.powerfromthesun.net/book.html
115. https://ocw.tudelft.nl/wp-content/uploads/solar_energy_section_19_3_3.pdf
116. https://www.scribd.com/document/34451850/solar-power-based-ups-system
117. https://en.wikipedia.org/wiki/Growth_of_photovoltaics

Solar Lighting

Aris Tsangrassoulis

Abstract In this chapter information about solar lighting, façade systems, light-shelves, blinds, holographic optical elements, and daylight transfer systems are given.

Daylight can be "used" for approximately 4000 h per year. The adoption of daylight savings time offers 210 h/year more daylight during the evening, due to the fact that it lasts an hour longer. Possible energy savings from the adoption of this measure are controversial with a large number of limited and contradictory reports [1]. Daylight use can be competitive against artificial lighting [2] even at cost level as shown in Fig. 1.

Today, all the building's energy and environmental assessment tools (such as LEED and BREEAM) include design prompts not only to increase daylight levels

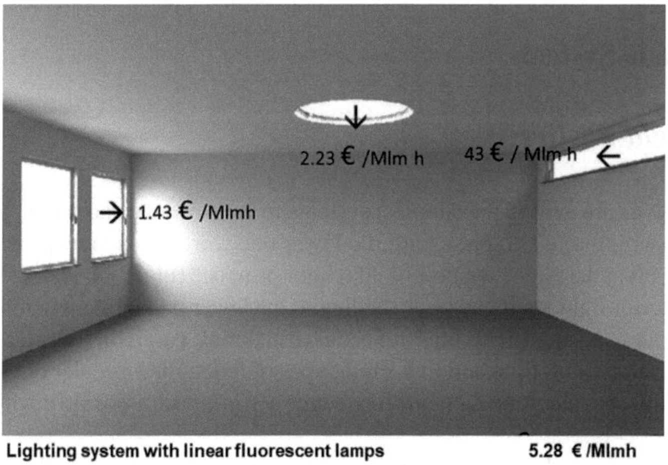

Fig. 1 Lighting cost per million l mh

A. Tsangrassoulis (✉)
Department of Architecture, University of Thessaly, Volos, Greece
e-mail: atsagras@arch.uth.gr

T. Pavlovic (ed.), *The Sun and Photovoltaic Technologies*,
Green Energy and Technology, https://doi.org/10.1007/978-3-030-22403-5_3

(up to a limit), but to ensure that the majority of building users have view contact with the exterior environment. Maximizing building's area receiving daylight can be realized by selecting an appropriate narrow plan together with window sizes and occupied spaces. The targets of such a procedure are:

- The increase of daylight levels without an excessive increase in solar gains, especially during the cooling period.
- The increase of daylight in areas far away from the openings.
- The minimization of glare.
- The ability to maintain view to the external environment.

The use of solar radiation for lighting in the interior of buildings can contribute positively, given the circumstances in their energy balance. Transfer of solar radiation can be accomplished either by a system that is placed in the façade or directly to the core of the building and this is a first classification of daylight systems [3]. An excellent taxonomy of these systems together with their operational principle is presented in the work of Kischkoweit-Lopin [4]. Nevertheless, this chapter focuses only systems using sunlight.

It is true that there is limited use of the aforementioned systems and this is due to the accuracy needed during the manufacturing procedure, which increases initial cost and the difficulty to identify the economic benefits of the application. Increasing daylighting levels in areas away of the window is accompanied by an increase of the cooling load. Until recently, there were very few papers analyzing the overall impact of these systems to a building's energy balance [5].

1 Façade Systems

1.1 Light-Shelves

Light-shelves are among the simplest systems to control solar radiation. They separate the opening into two parts with the lower one working as view provider while the upper one (clerestory window) as daylight provider (Fig. 2).

An exterior light-shelf can offer shading to the lower part of the window while its upper surface, covered with high-reflectance material, can redirect solar luminous flux into the interior of the building. Both types of reflection, specular or diffuse, can be used with specular reflection offering better performance associated with possible glare issues (Fig. 3).

Usually light-shelves are placed in south-oriented windows while the parameters affecting their performance are shown in the following Fig. 4.

The idea of the dynamic movement of the reflective light-shelves according to the position of the sun is quite old and proposed in an effort to increase the lighting levels in areas away from the façade openings. These areas are located next to the perimeter zones (Daylight zones as these are defined in EN 15193 [8]).

Fig. 2 Exterior and interior light-shelf

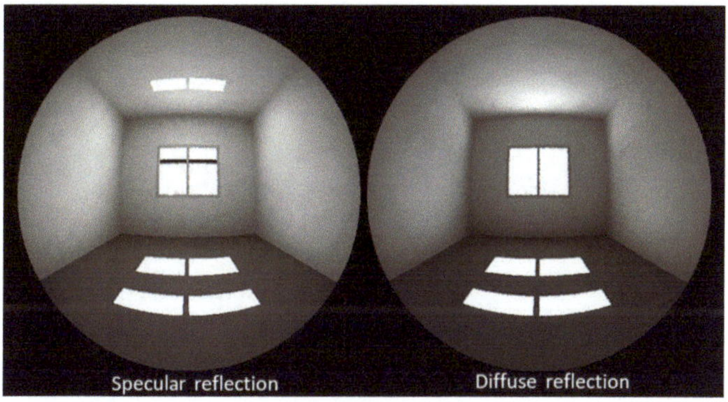

Fig. 3 Two cases with light-shelves using different reflection modes

Clerestory windows are associated with the following problem. When the Sun's position is relatively low, the depth of sunlight penetration is increased; and therefore, shading has to be used (e.g., internal blinds) canceling the luminous flux increase caused by the reflection on the upper surface of the light-shelf. Alternatively, an interior light-shelf can be used. An interesting approach that was proposed during 1986 is the VALPRA system (variable area light reflecting assembly, [9]). Its operational principle is based on the seasonal modification of the inclination of a reflective membrane. The whole system is protected in a transparent closed cavity. This is due to the reduction of surface reflectance when exposed to environmental conditions, which leads to a significant performance reduction (Fig. 5).

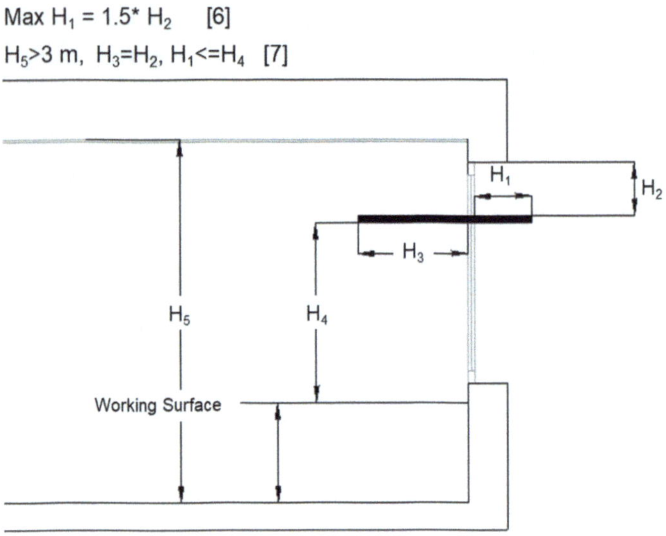

Fig. 4 Parameters affecting light-shelf performance [6, 7]

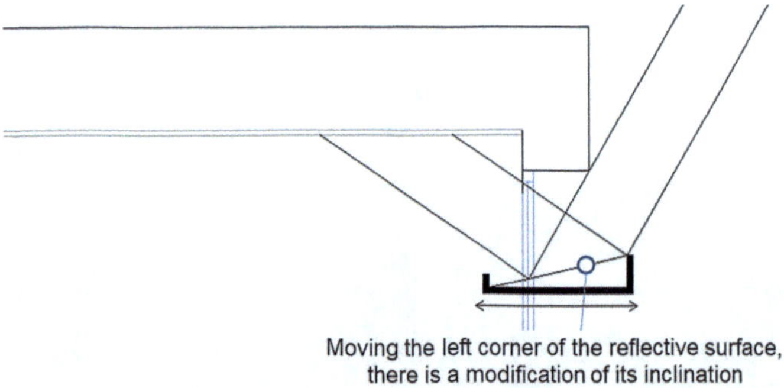

Fig. 5 Schematic representation of the variable area light reflecting assembly system

Anidolic light-shelves, using parabolic or elliptical reflective surfaces, have also been tested [10]. Again, the whole system is protected from exposure to external conditions with glazing. It is capable to increase daylight levels in areas away from the opening, especially in cases of dense urban environment.

1.2 Blinds

The simplest form are the venetian blinds. It is a low-cost system which mainly is used to reduce solar gains and to provide privacy, which in turn leads to a significant reduction of the visual contact with the external environment. Perforation improves this situation, in the sense that even when the blinds are closed, visual contact with exterior environment is maintained. What is interesting is the design of blinds which can satisfy all the aforementioned needs together with sunlight redirection ability and seasonal selectivity (i.e., greater input of solar energy during winter than summer). This can be accomplished using static reflective blinds which have specifically designed profile (Fig. 6).

Modification of blinds' redirection properties according to their height from the floor is generally a good idea to manage reflected radiation. Blinds near the floor redistribute reflected solar flux in an area near the opening, and thus, direct glare is avoided while blinds located further up redirect sunlight at the rear of the room, as presented in the following figure. The concept can be realized either by splitting blinds into two sections having different operation or using blinds having profile geometry depended on their floor distance (Fig. 7).

There are interesting designs of static blinds achieving the seasonal variation either by the appropriate design of their profile or by using micro-structured mirror surface. Typical examples are RetroLux systems and RetroFlex of Koester LichtPlannung [12, 13] (Fig. 8).

The use of mirrored blinds for improved solar control, presupposes their protection to avoid performance reduction. The most typical solution is to place them in the space between the panes of a double glazing unit. This arrangement has the advantage to adjust the overall solar heat gain coefficient and visible transmittance according to the properties of the selected glazing. In many cases, the mid-pane system geometry can be modified before being placed in the gap, satisfying different requirements. The following figures show the operating principle of two such systems (Figs. 9 and 10).

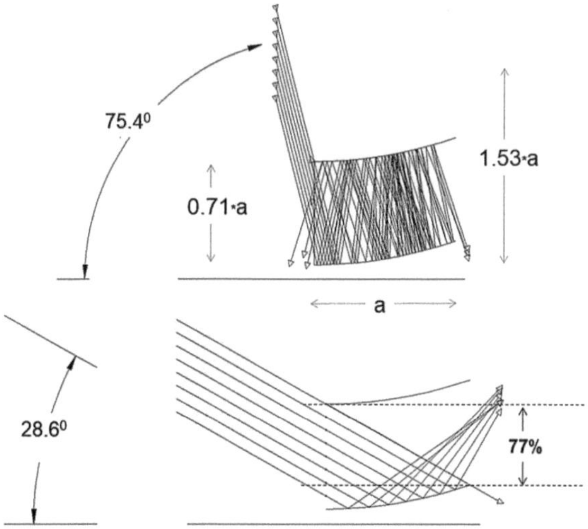

Fig. 6 Static reflective parabolic slats. Due to their design solar transmittance is strongly depended on the elevation angle [11]

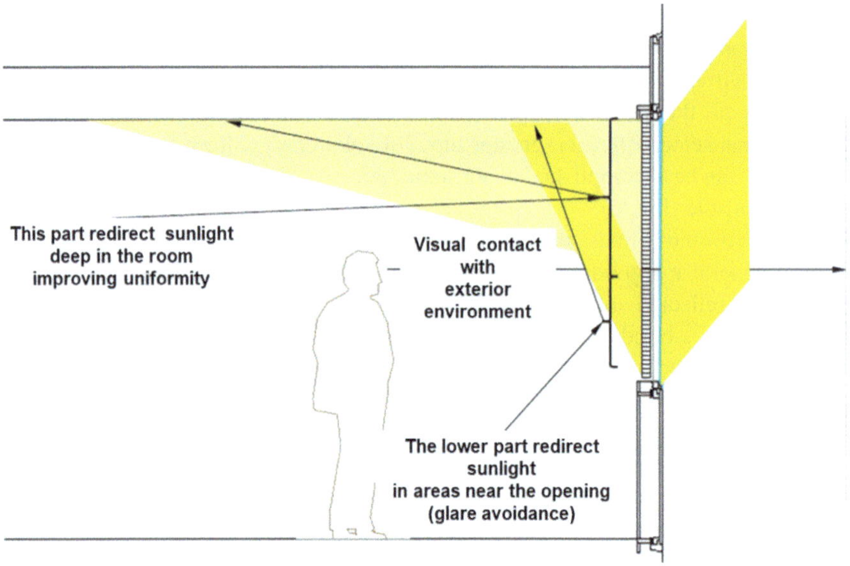

Fig. 7 Blinds and redirection of solar radiation

Fig. 8 Profile shapes for Retroflex (upper image) and Retrolux (lower image) designed by Koester LichtPlannung

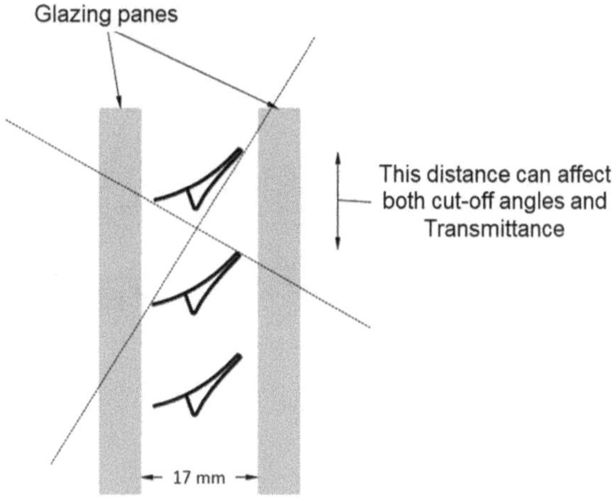

Fig. 9 Operational principle of the daylighting-shading system OKASOLAR W. Visible transmittance varied between 4 and 57% while solar heta gain coefficient—when a low emissivity glazing unit is used—varies between 17 and 45% [14]

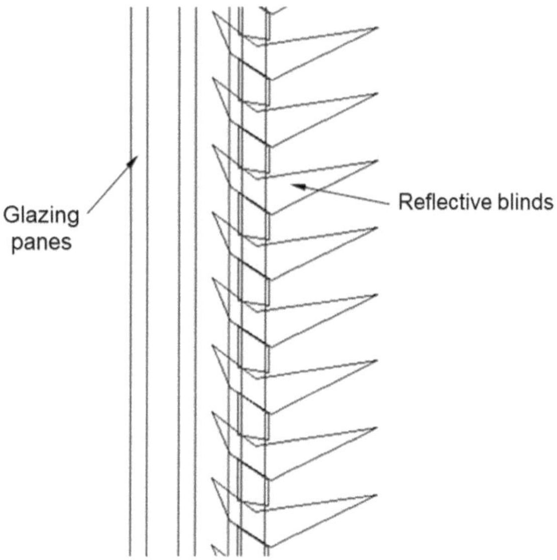

Fig. 10 Static interior daylighting/shading system. Due to the profile shape has the ability to redirect sunlight to the ceiling. LightLouver system [15]

The Micro Sun Shielding Louver from Siteco [16] based its operation on a similar to the above-mentioned sunlight redirection properties with solar heat gain coefficient 12%. It is placed in horizontal openings and aims to significantly reduce the incoming solar radiation while at the same time leave virtually unobstructed views of the northern part of the sky (Fig. 11).

Sunlight redirection can be achieved by simpler means. For example, by making an array of parallel cuts partly through a transparent sheet of acrylic plastic as presented in Fig. 12 [17, 18].

The following Fig. 13 shows the behavior of this panel when placed in a south-facing clerestory window (simulation for Athens and Greece).

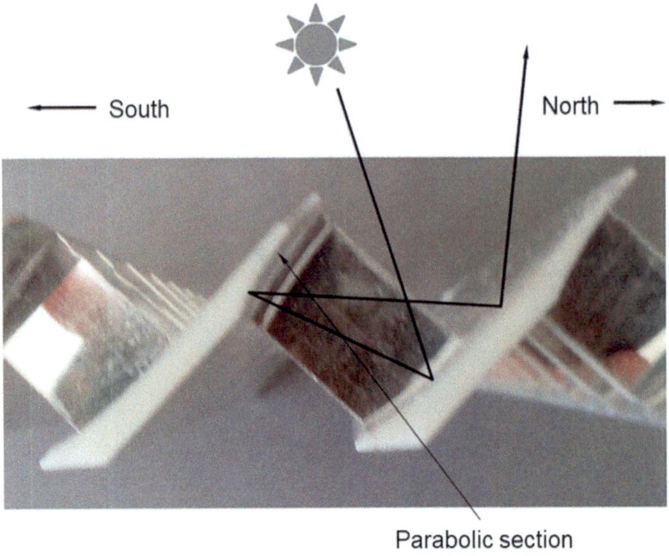

Fig. 11 Micro-sun shielding louver operational principle

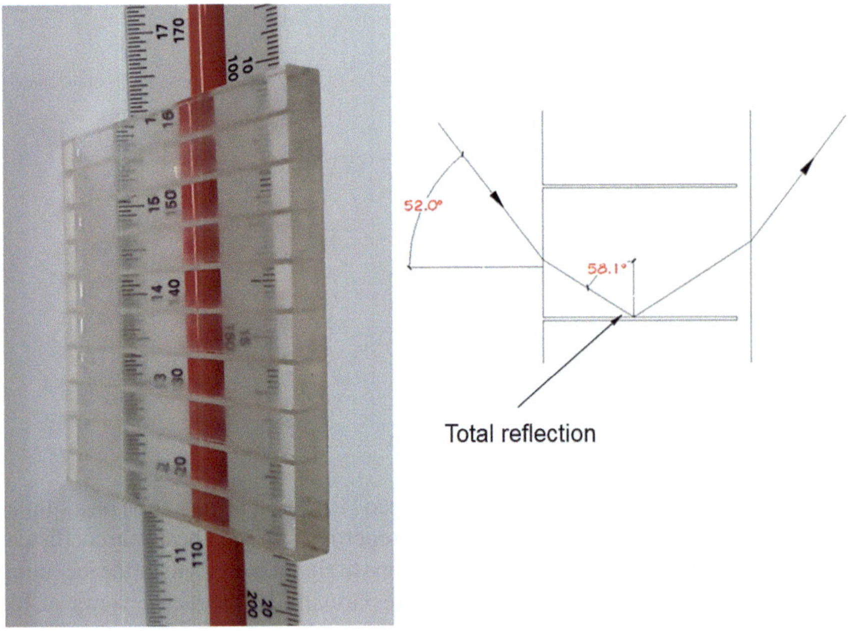

Fig. 12 Laser cut acrylic panels

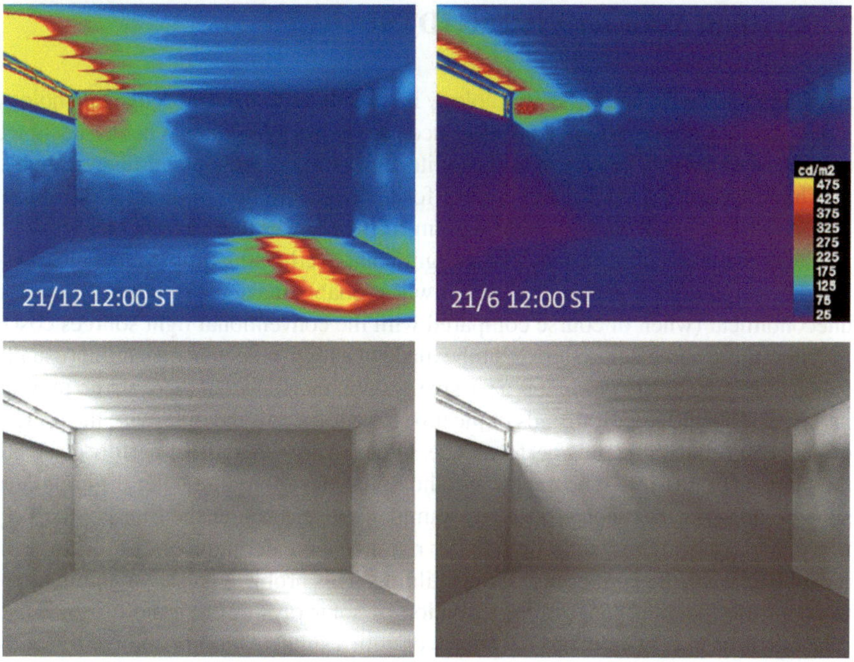

Fig. 13 Due to the sunlight redirection, there is a characteristic increase on ceiling's luminance

The manufacturing process of prismatic structures requires particular attention since slight defects in the edge of the elementary prisms can lead to the appearance of colors on room's surfaces. There is a number of similar systems which base their operation on a microstructure able to modify the direction of the solar rays. Such as system is the Lumitop SGG (g-value = 30%) [19].

1.3 Holographic Optical Elements

These systems when used in buildings are consisted of holographic films laminated between two sheets of glass. These films are produced using the principle of holographic imaging and are capable to redirect sunlight on to ceiling. However, they are adversely affecting visual contact with exterior environment [20].

2 Daylight Transfer Systems (DTS)

Although the daylight transfer technology is considered as a cutting edge technology today, the first prototype system was conceived in 1881. William Wheeler designed a daylight transmission system that is quite similar to current optical fiber systems.

The first part of a DTS is responsible for the concentration of direct solar illumination. This can be achieved either with mirrors or lenses (e.g., Fresnel lenses).

Excessive concentration can lead to significant overheating problems of the transfer material while a low concentration it will probably make daylight transportation uneconomical (when of course compared with the conventional light sources cost).

The simplest transfer system is a tube made of a high-reflectance material (usually called light pipe or sun pipe). It is a passive system, meaning that there are no moving parts to follow the sun. Both ends of the tube have to be covered otherwise the ingress of dust can reduce system's performance. Other factors that affect performance are the number of possible tube angles and the Sun's elevation. The latter differentiates sunpipe behavior between winter and summer. During winter the Sun is relatively low in the sky, and thus, sunlight transfer requires a greater number of reflections in the tube to reach the emitter inside the building. This problem has led manufacturers to adopt the idea of changing the direction of solar rays through the use of either prismatic transparent elements or mirrors or with a suitable modification of the outer transparent cover (e.g., creating a kind of lens) (Fig. 14).

A simple equation for the calculation of a sunpipe direct sunlight transmittance is the following [21]:

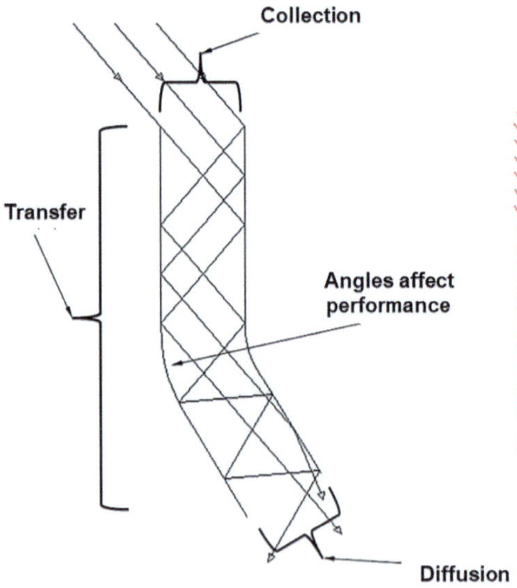

Fig. 14 Characteristic parts of a sunpipe

$$T = \rho^{2L \tan\theta/\pi R} \tag{1}$$

where ρ is the sunpipe's interior surface reflectance, θ the angle of incidence, L is tube's length, and R its radius.

One of the advantages of the sun pipe is related to the height of the emitter inside the building. Decreasing this height increases illuminance decreasing at the same time uniformity. The designer has to optimize diameter, length, and grid in order to achieve proper illuminance levels for a time period during the year.

Alternatively, instead of using reflective material in the interior of the sun pipe, transparent plastic film with prismatic structure may be used (Optical Light Film by 3M). The thickness of the material is small enough and it can be shaped to form a square or circular section.

Its operation is based on achieving total internal reflection when sunlight rays enter the sunpipe for a range of incident angles. For a polycarbonate prismatic film all rays that enter the sunpipe and form with pipe's longitudinal axis (i.e., along the grooves) with the film, angle lower than a critical angle (27.6°) undergo total internal reflection. The reflectance of this film can reach up to 99% (depends on the angle of incidence). The condition of total internal reflection can be interrupted either by using diffuser panels in the interior or by opening holes or modifying its surface. This causes the light to exit from the pipe and thus transform it into a luminaire (Fig. 15).

In general, sunpipes are placed vertically in an attempt to transfer daylight in areas of the building with limited or no access to it. But there are also systems in which this transfer takes place horizontally through the false ceiling. However, obstacles

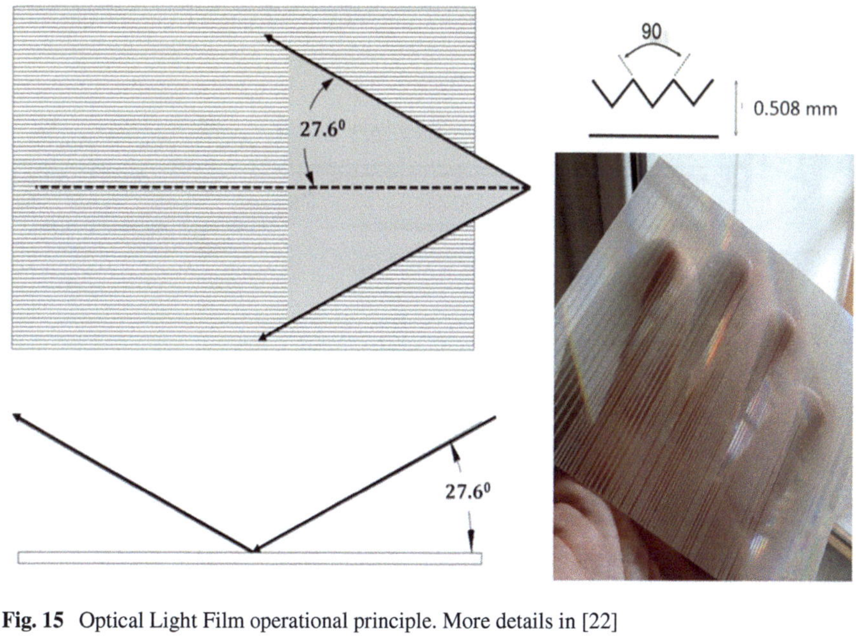

Fig. 15 Optical Light Film operational principle. More details in [22]

due to urban geometry (dense urban environment) can reduce their performance by reducing available sunlight in their entrance. Such systems are presented in the following Figs. 16 and 17:

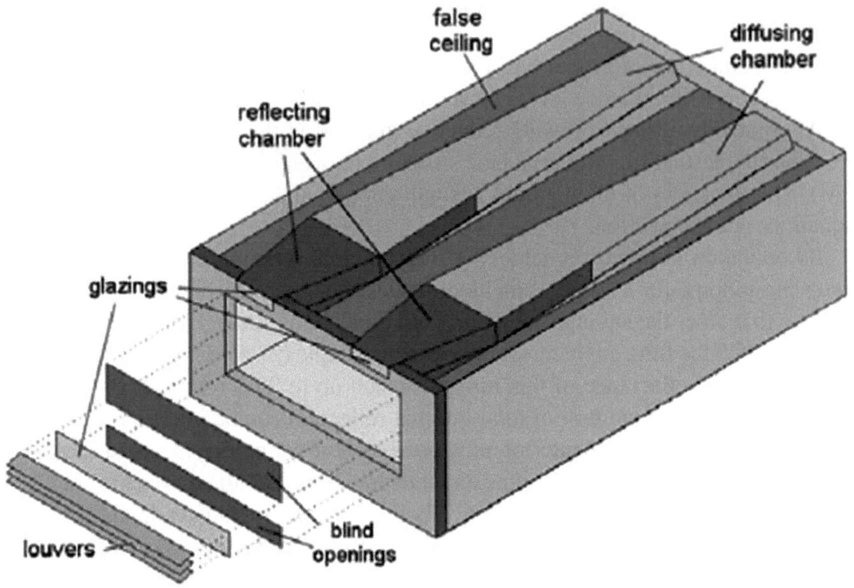

Fig. 16 Daylighting transfer using a horizontal system through the false ceiling [23]. *Courtesy* Elsevier B.V.

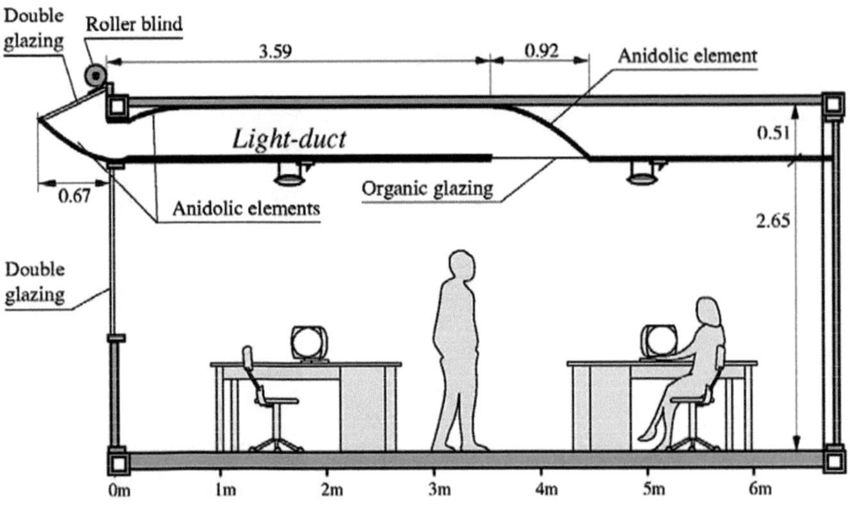

Fig. 17 Daylight transfer using anidolic systems [24]. *Courtesy* Elsevier B.V.

In the aforementioned DTS, optical fibers should be added. An optical fiber is consisted of a transparent core material (plastic or glass) surrounded by a material having a lower refractive index and thus total internal reflection can be achieved. Of course, several optical fibers can be joined together offering greater flexibility in the design of a lighting system.

Optical fibers used for illumination transfer have a diameter between 0.25 and 5 mm. The diameter determines the flexibility of the fiber which practically defines the smallest curvature radius. Various transparent materials can be used, however, acrylic (PMMA) is prevalent due to its high transmittance and heat resistance (−55 to 70 °C in a low humidity environment), with a protective tubing, e.g., PVC.

Glass fibers (with diameter between 0.05 and 2 mm) manufactured using borosilicate glass extrusion. The use of silica glass in the core of the fiber may reduce the attenuation significantly. Acrylic (PMMA) fibers are more transparent at short wavelengths, while the opposite occurs with the glass fibers (due to Rayleigh scattering).

A characteristic magnitude of the fiber is the acceptance angle, which represents the largest incident angle—relative to the axis of the fiber—with which total internal reflection can be achieved. The acceptance angle is defined as:

$$\theta_A = \sin^{-1}\sqrt{(n_1^2 - n_2^2)} \tag{2}$$

where n_1 is the refractive index of the core material and n_2 that of the cladding material (which should be lower that of n_1). Typical values for sin θ_A are 0.2–0.5 meaning that θ_A angles varied between ~11 and 30°. It is evident that light rays with incident angles greater than the acceptance angle cannot be guided through the optical fiber (Fig. 18).

Obviously, the light flux exiting from the optical fiber is contained in a cone which defined by the acceptance angle. Greater acceptance angle means greater transmission capacity. Using Eq. (2) if $n_2 = 1$ then larger values of θ_A can be achieved. This of course in theory because in practice, cladding removal can affect core external surface quality adversely, and thus affecting total reflection conditions.

The transmittance of a fiber optic with length L to monochromatic radiation and rays parallel to its axis is given by the following equation:

$$T = T_{\text{in}} \cdot T_{\text{out}} \cdot e^{-\gamma L} \tag{3}$$

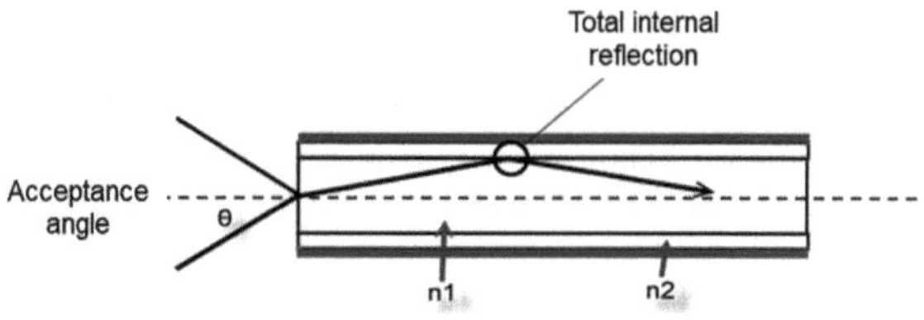

Fig. 18 Optical fiber operation

where T_{in}, T_{out} represent transmittances in relation to luminous flux input and output from the fiber optic. Typical values for T_{in} are 0.94–0.96 while for T_{out} 0.96. These values are due to Fresnel reflection as well as due to the slight irregularities of fiber's exit and entry surface. The attenuation coefficient γ depends on the radiation wavelength and the angle formed between the light rays and the axis of the optical fiber. This attenuation is mainly due to scattering and absorption by the core material of the fiber. The average transmittance of the fiber for solar radiation spectrum is given by the equation [25]:

$$T_{ave} = 10^{-0.1tL} \tag{4}$$

where L is fiber's length and t the attenuation in dB/m (this value is given by the manufacturer. Typical value for 1 mm PMMA fiber is 0.2 dB/m).

Since fiber diameter is quite small, it is difficult to transfer large quantity of luminous flux. Thus, a group of fibers is consolidated to form a bundle. This can be achieved by either gluing together or fusing the fibers. A fiber bundle exhibits the following losses: (a) Losses due to Fresnel reflection in the input and output ~8%, (b) Losses due to the fact that each optical fiber has a cladding ~15–17%, and (c) Losses due to the fact that not all of the bundle cross-sectional area is filled with fibers ~11.9%.

In order to increase the efficiency of the fiber, a concentrator is needed (Fresnel lens or a parabolic mirror). Two factors have to be taken into account; the first one is the resistance of the fiber to increased concentration while the second one is that the radiation should be within fiber's acceptance angle.

In addition, the following factors have to be considered during the design phase of a daylight fiber optic system, which are related to the overall system performance:

1. The Sun's image on the concentration point is not the point but a surface. For example, when a parabolic mirror is used, the radius of the Sun image in the focal point is calculated by multiplying the focal length with the Sun's angular size. If the mirror used has a long focal length it produces large solar images and thus a large diameter fiber is needed.
2. The size of the solar image has to be smaller than the cross-section of the optical fiber.
3. Large fiber diameters can cause problems with the mirror movement while tracking the Sun.
4. The use of high concentration reflectors requires an extremely accurate tracking system.
5. Reflector material and manufacturing process is crucial to overall system's performance. The use of mirror glass increases weight which in turn increases the cost of the Sun tracking system (Figs. 19 and 20).

An interesting improvement albeit quite expensive is the use of fibers with liquid core. With this technology, losses associated with ordinary fiber bundle forming are avoided with diameters be much greater than a typical optical fiber (~10 mm). The system presented in figure below, concentrates solar radiation using a Fresnel lens

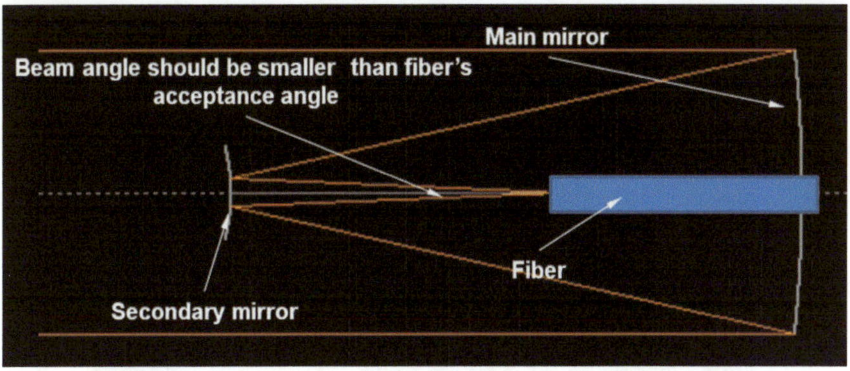

Fig. 19 Concentrating solar radiation into an optical fiber using reflection in parabolic surfaces

Fig. 20 PARANS system [26]. *Courtesy* Parans Lighting AB

(SOLUX system developed by Bomin Solar Research) to a liquid fiber after passing through a water filter. At the other end of the fiber a Prismex sheet is connected to diffuse emitted luminous flux (Fig. 21).

Sunlight transfer can be achieved by using movable mirrors which can reflect sunlight to a predefined direction as the Sun moves across the sky. Their operational principle is presented in the following Fig. 22.

For shading applications, sunlight rays are considered as parallel without this affecting calculation accuracy. But when it comes to sunlight transfer over a long distance, the fact that the rays are emitted by a source (the Sun) of certain dimensions have to be taken into account.

Fig. 21 Sunlight transfer using liquid fiber optic [27]. *Courtesy* Elsevier B.V.

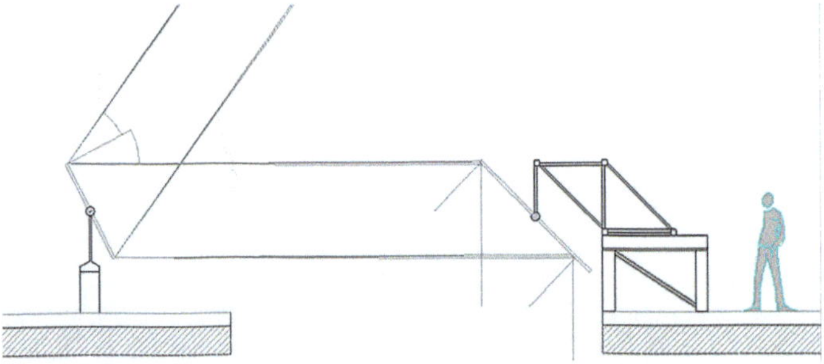

Fig. 22 Heliostat operation with a secondary static mirror

Thus, the error in the calculations due to the solar disc size is proportional to the angle subtended by the Sun ($2 \times \varepsilon$ ($\varepsilon = 0.260$)). During the design phase of a heliostat, the divergence of the reflected beam has to be estimated. This is due to the roughness of the reflector surface, its local curvature and of course the tracking errors. In total, this divergence ($2 \times \omega$) varied ~3–5°. Interesting ideas have been conceived in order sunlight to be used effectively, as these presented in Fig. 23.

Heliostats can be installed at an atrium top generating a fixed light beam. Then, by using secondary reflectors luminous flux can be directed in adjacent rooms.

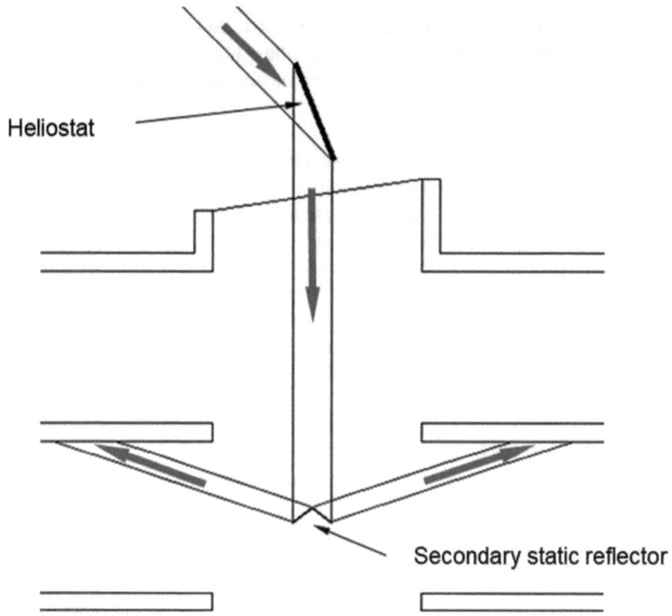

Fig. 23 Sunlight transfer through atrium top

References

1. Myriam CB, Aries G, Newsham R (2008) Effect of daylight saving time on lighting energy use: a literature review. Energy Policy 36(6):1858–1866
2. Fontoynont M (2009) Which daylight in 2050? Daylight Archit Mag 12: 75–81. (VELUX)
3. McCluney R (1998) Advanced fenestration and daylighting systems, Paper presented to daylighting 1998 international conference on daylighting technologies for energy efficiency in buildings, Ottawa, Canada, 10–13 May 1998. Available at http://www.fsec.ucf.edu/en/publications/pdf/FSEC-PF-425-98.pdf
4. Kischkoweit-Lopin M (2002) An overview of daylighting systems. Sol Energy 73(2):77–82
5. Kontadakis A, Tsangrassoulis A, Doulos L, Topalis F (2016) An active sunlight redirection system for daylight enhancement beyond the perimeter zone. Build Environ 113:267–279
6. Selkowitz S, Navvab M, Mathews S (1983) Design and performance of light shelves. In: Proceedings of the international daylighting conference, Phoenix, Arizona, Washington, AIA, pp 267–272
7. Littlefair P (1995) Light shelves: computer assessment of daylight performance. Lighting Res Technl 27(2):79–91
8. EN 15193 (2007) Energy performance of buildings—Energy requirements for lighting
9. Howard TC, Place W, Andersson B, Coutiers P (1986) Variable area light reflecting assemblies, (VALRA), In: Proceedings 2nd international daylighting conference, Long Beach, pp 222–234
10. Courret G, Scartezzini JL, Francioli D, Meyer JJ (1998) Design and assessment of an anidolic light-duct. Energy Buildings 28:79–99
11. Tsangrassoulis A, Maheras V, Axarli K (2013) Simplified design of a specular slat profile curve using 2D ray tracing and genetic algorithms. In: 13th international conference of the international building performance simulation association, BS 2013, 25-28/8/2013, Savoie Technolac, France, pp 3669–3672

12. http://www.koester-lichtplanung.de/pages_gb/news_01.html
13. http://www.koester-lichtplanung.de/downloads/03flex_kl.pdf
14. http://www.okalux.de/fileadmin/Downloads/Downloads_englisch/Infotexte/i_okasolar_f_e.pdf
15. http://lightlouver.com
16. http://www.siteco.com/en/products/daylight-systems/micro-sun-shielding-louvre.html
17. http://www.solartran.com.au/lasercutpanel.htm
18. Edmonds IR, Greenup PJ (2002) Daylighting in the tropics. Sol Energy 73:111–121
19. http://merint.com/site/ourbusiness/SGG%20LUMITOP.pdf
20. http://www.visionoptics.de/index.php?id=6&L=1
21. Zastrow Wittwer V (1986) Daylighting with mirror light pipes and with fluorescent planar concentrators: first results from the demonstration project Stuttgart-Hohenheim, materials and optics for solar energy conversion and advanced lightning technology. In: Holly S, Lampert, CM (eds) SPIE Proceedings, vol 692. Society for Photo-Optical Instrumentation Engineers, Bellingham, WA, p 227
22. Whitehead LA (1998) New simplified design procedures for prism light guide luminaires. J Illum Eng Soc 27(2): 21–27
23. Canziani R, Peron F, Rossi G (2004) Daylight and energy performances of a new type of light pipe. Energy Build 36(11):1163–1176
24. Scartezzini JL, Courret G (2002) Anidolic daylighting systems. Sol Energy 73(2):123–135
25. Grisé W, Patrick C (2002) Passive solar lighting using fiber optics. J Ind Technol 19(1):2–7
26. http://www.parans.com/eng/
27. Final Report (2003) Universal fiber optic project, JOULE programme (ERK6-CT-99900011)

Lighting Technologies

Plamen Ts. Tsankov

Abstract In this chapter, information about lighting fundamentals, nature of the light, light quantities, color characteristics of light sources, light sources, incandescence, luminescence, a brief history of light sources, incandescent lamps, tungsten-halogen lamps, low- and high-pressure mercury discharge lamps, metal-halide lamps, low- and high-pressure sodium discharge lamps, induction lamps, light-emitting diodes (LED), light sources' main parameters comparison, photobiological safety of light sources and PV–LED systems are given.

1 Lighting Fundamentals

1.1 Nature of the Light

The first attempts to study and explain light phenomena and the essence of light date back to ancient times. Democritus (460–370 BC) explains the visual process with the presence of the "atoms" that are emanating from the objects and reaching the eye, creating the images in it. Euclid (300 BC) suggests that "beams of light" are emitted from the eyes. Their naive conception of the nature of light remains unchanged until the Middle Ages.

Interesting are the ideas of Kepler (1571–1630) about light, from each light source emitting endlessly many rays that go into infinity. The speed of light distribution is infinitely large, as the light has no mass. The spread of the beam is not related to its longitudinal extension; it is only a light movement. When moving away from the light source, the density of the light rays decreases back in proportion to the surface of a sphere centered on the light source. The light rays do not interfere with each other and do not illuminate each other.

Galileo (1564–1642) was the first to express doubts about the infinitely fast propagation of light by making unsuccessful attempts to prove this.

P. Ts. Tsankov (✉)
Faculty of Electrical Engineering and Electronics,
Technical University of Gabrovo, Gabrovo, Bulgaria
e-mail: plamen@tugab.bg

T. Pavlovic (ed.), *The Sun and Photovoltaic Technologies*,
Green Energy and Technology, https://doi.org/10.1007/978-3-030-22403-5_4

Descartes (1596–1650) in his Light Treatise assumes that space is filled with particular particles. They transmit mechanically to one another, like a hard rod, the movement of the particles of the illuminating body. The light consists of such accelerated particles that affect the eyes and create light sensations.

Hook (1635–1703) is the first to conjecture that light consists of rapid fluctuations, assuming that they are transmitted instantly at any distance.

Wave theory of light

The actual creator of the overall wave theory of light is Huygens (1629–1695). According to this theory, the propagation of light represents a wave process in a hypothetical environment of ether. The vibrational movement in this elastic medium spreads in all directions as waves. Any particle of the substance in which the wave propagates must convey its motion not only to the nearest particle lying on the straight edge of the luminous point, but necessarily also transmits it to all the other particles that touch it and interfere with its movement. Thus, a wave is formed around each particle. Fresnel (1788–1827) and Young (1773–1829) (Fig. 1) perfected the Huygens wave theory, assuming that the wave propagation of light is periodic in time and space. This gives the opportunity to create a strict theory of interference and diffraction of transverse light waves.

Maxwell (1831–1879) views light not as a change in the position of a physical object, but as a periodic change in the space and time of the intensity of the magnetic and electric fields. An important point in the development of physics, and more precisely in the theory of light in the nineteenth century, is the comparison of the laws of electricity and magnetism with those of light made in 1860 by him. Maxwell shows that the velocity of propagation of electromagnetic waves and the light in the vacuum is the same $c = 3 \times 10^8$ m/s. Maxwell's electromagnetic theory was experimentally confirmed by the experiments of Hertz (1857–1891), Kohlrausch (1809–1858) and Weber (1804–1900), and then she was fully acknowledged.

Particle theory of light

An atomist Pierre Gassendi (1592–1655) proposed a particle theory of light which was published posthumously in the 1660s. Isaac Newton (1642–1727) studied

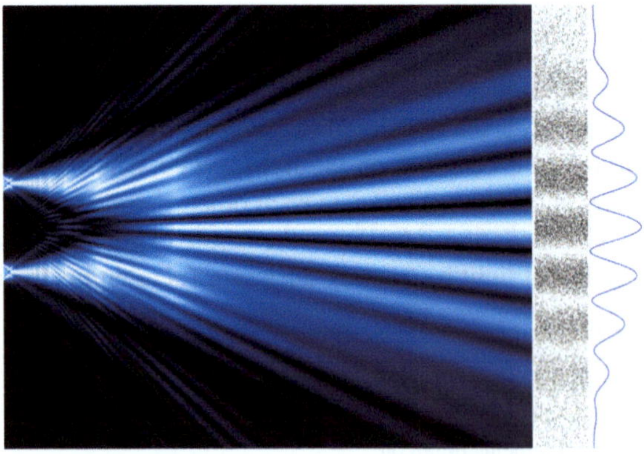

Fig. 1 Young's interference experiment with two slits [1]

Gassendi's work at an early age and preferred his view to Descartes' theory of the plenum. One of Newton's arguments against the wave nature of light was that waves were known to bend around obstacles, while light travelled only in straight lines. According to him, light is a stream of small material particles—corpuscles that emerge from light sources and propagate in space. Refraction of the light rays in the passage through different environments, Newton explains with the different pulling force with which they act on the corpuscles.

At the end of the nineteenth century, new light technic phenomena and laws were found which cannot be explained by the classical electromagnetic theory—the law for the spectral distribution of the radiant energy of an absolute black body, the photoelectric effect, the light pressure, etc. This prompted Planck (1858–1947) to express his hypothesis in 1900 about the discrete nature of light emanation. Light is emitted and absorbed continuously in "portions," with the minimum amount of energy of a "portion" Planck called a quantum of energy ε, quantified by the following way:

$$\varepsilon = h\nu \tag{1}$$

where:

h—Planck's constant, $h = 6.626 \times 10^{-34}$ Js;

ν—frequency of the radiant emission, Hz.

The achievements of physics at the end of the nineteenth and early twentieth centuries, and especially Planck's hypothesis about the discrete nature of the transmissions, enabled Einstein (1879–1955) to create a new photon theory of light. According to this theory, light is a stream of particles—photons. Every photon has a certain energy ε, impulse $h\nu/c$ and a mass $h\nu/c^2$. Radiation, propagation and absorption of light take place in the form of photons, which indicates that it is discrete. Photon theory, by its nature, is in some ways a return to Newton's corpuscular theory, but on a new qualitative basis. Photons are regarded as carriers of both corpuscular and wave properties in accordance with the dualistic nature of light.

Using photon theory, a number of complex phenomena of physical optics are successfully explained: photoluminescence, photoelectric effect (Fig. 2), interference, diffraction, polarization, absorption of light, etc.

Radiometry is concerned with the study of the generation, distribution and redistribution of electromagnetic radiation in the optical area with wavelength $\lambda = 10\text{–}1000 \times 10^3$ nm, which in turn is subdivided into (Fig. 3):

- Ultraviolet waves (UV)—$\lambda = (10\text{–}380)$ nm;
- Visible waves (light)—$\lambda = (380\text{–}770)$ nm;
- Infrared waves (IR)—$\lambda = (770\text{–}1000 \times 10^3)$ nm.

Radiation in the optical field is a flux from photons, which is conditioned by electron excitation of atoms and molecules. From the optical radiation, the ultraviolet radiation photons have the greatest energy—from 5.3×10^{-19} to 2×10^{-17} J, and therefore strongly affect a number of radiation receptors. In lighting technologies, substances called luminophores are used, which when irradiated with ultraviolet

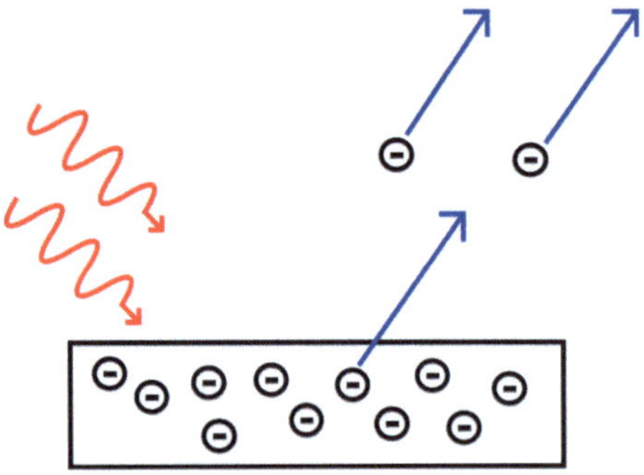

Fig. 2 Photoelectric effect [2]

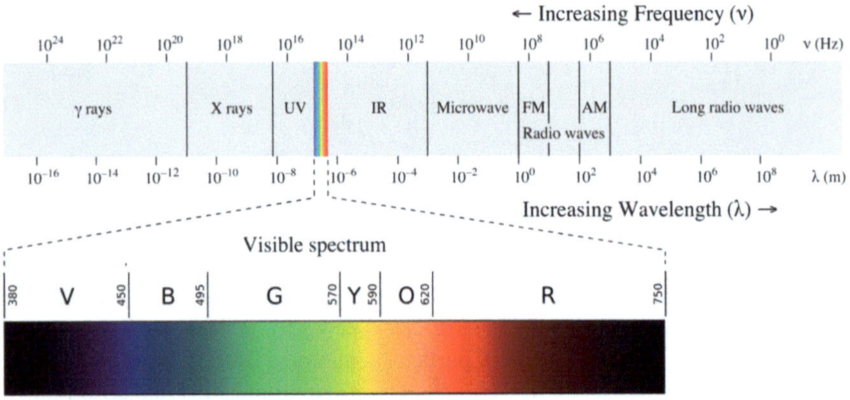

Fig. 3 Electromagnetic, optical and visible spectrum [3]

rays emit energy in the visible spectrum. Based on this effect of luminescence, the fluorescent lamps and some of the light-emitting diodes are built. Ultraviolet radiation also has photochemical and biological activity, which is extremely important for the vital functions of living organisms. The photon energy of infrared rays is considerably smaller—from 5.3×10^{-22} to 2.5×10^{-19} J. They are characterized by their strong thermal and less photochemical and photoelectric effects. Radiations in the visible area of the optical spectrum also have significant photoelectric and photochemical effects. They also have the ability to produce a visual sensation in the human eye and enable us to see our surroundings.

The radiant flux of the light sources is characterized by a different intensity distribution in the visible spectrum. Spectral power distributions (SPD) of some types of lamps, measured in the laboratory of Lighting at Technical university of Gabrovo, are shown in Fig. 4.

The Illuminating Engineering Society of North America (IESNA) defines light as "radiant energy that is capable of exciting the retina and producing a visual sensation." Light, therefore, cannot be separately described in terms of radiant energy or of

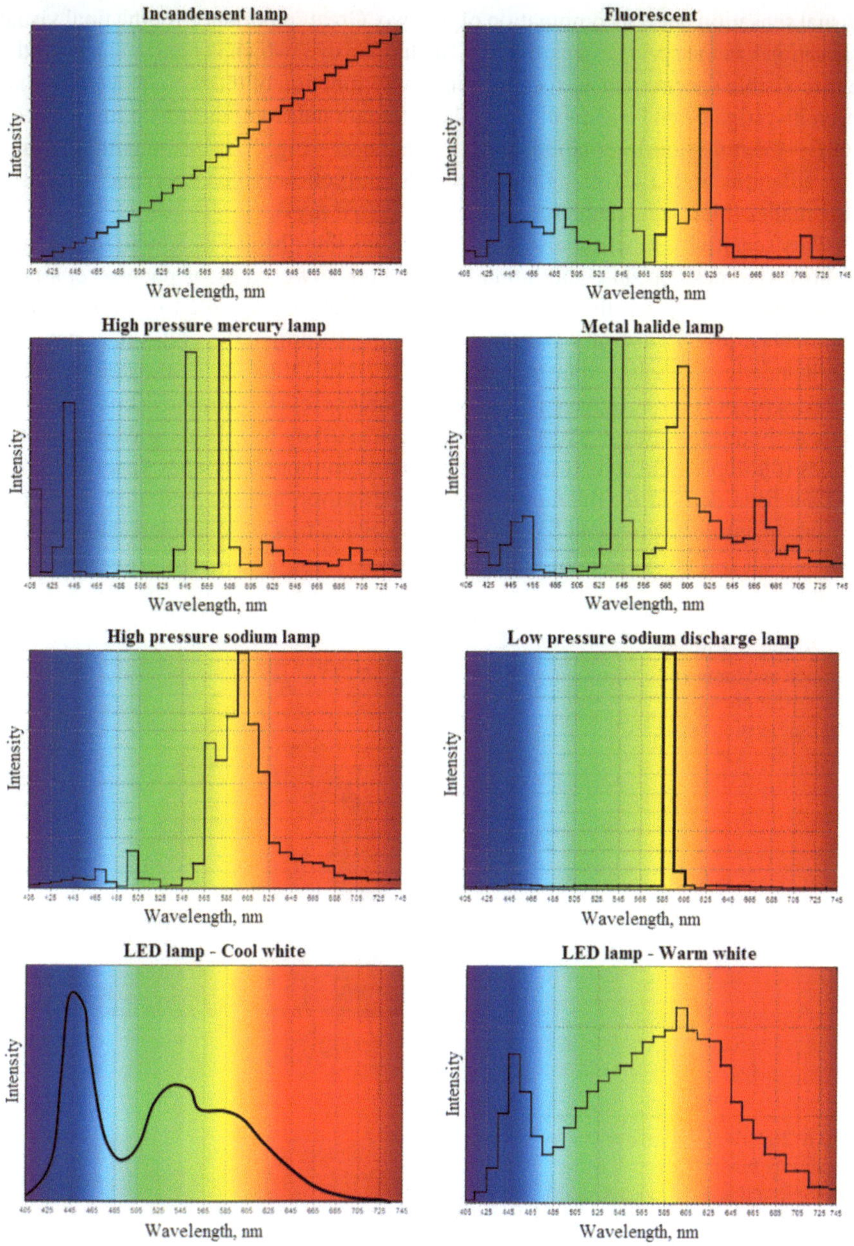

Fig. 4 Spectral power distributions of different types of light sources

visual sensation but is a combination of the two. Great importance for the final visual sensation has the spectral sensitivity of the human eye, which is not the same for the entire visible spectrum. The spectral sensitivity function of the human eye $V(\lambda)$ has been investigated and standardized by the International Commission on Illumination (CIE—Commission Internationale de l'Éclairage)—Fig. 5. It has maximum value for radiations with a wavelength of $\lambda = 555$ nm (yellow–green rays) and decreases for smaller (blue rays) and larger wavelengths (red rays).

This sensitivity function also depends on whether the eye is adapted for bright light or darkness because the human eye contains two types of photoreceptors—cones and rods. When the eye is adapted for bright light, called *photopic vision,* (luminance levels generally greater than about 3.0 cd/m^2), the cones dominate. At luminance levels, below approximately 0.001 cd/m^2, the rods dominate in what is called *scotopic vision*. Between these two luminance levels, *mesopic vision* uses both rods and cones. Figure 6 shows the relative sensitivity to various wavelengths for cones (photopic) and rods (scotopic). Standard luminous efficiency functions have not yet been defined for the mesopic region. However, there is a gradual shift from a peak spectral sensitivity at 555 nm for cone vision to a peak spectral sensitivity at 507 nm for rod vision as light levels are reduced.

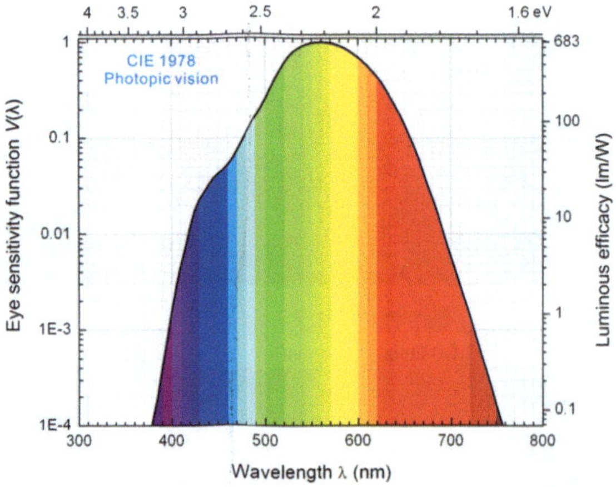

Fig. 5 Eye sensitivity function and luminous efficacy for photopic (daytime) vision [4]

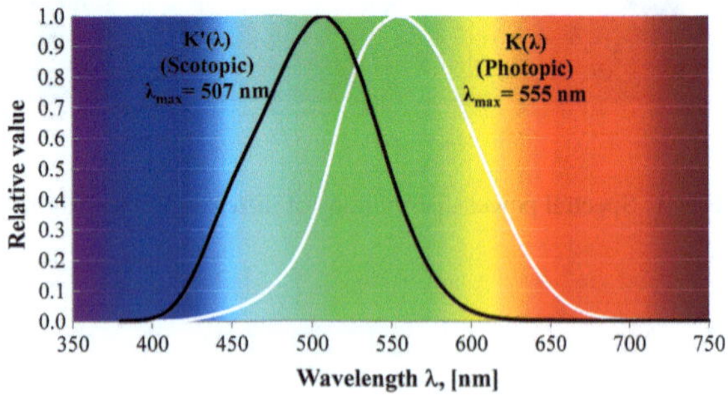

Fig. 6 Relative spectral eye sensitivity functions for photopic and scotopic vision

The different spectral functions of light source radiation and human eye perception in the visible part of the spectrum show that it is important to take them into account when assessing the energy efficiency and color quality of lighting.

1.2 Light Quantities

Basic parameters and dependencies used in lighting engineering are presented in Fig. 7.

Two elements intervene in lighting engineering: both the source of light and the object to be illuminated. The luminous flux and luminous intensity are the parameters of the light sources, the illuminance of the illuminated objects and the luminance could be parameter of both of them.

Luminous flux (luminous output)

Luminous flux Φ is the basic photometric quantity and describes the total amount of electromagnetic radiation (radiant flux) emitted by a source (Fig. 7), spectrally

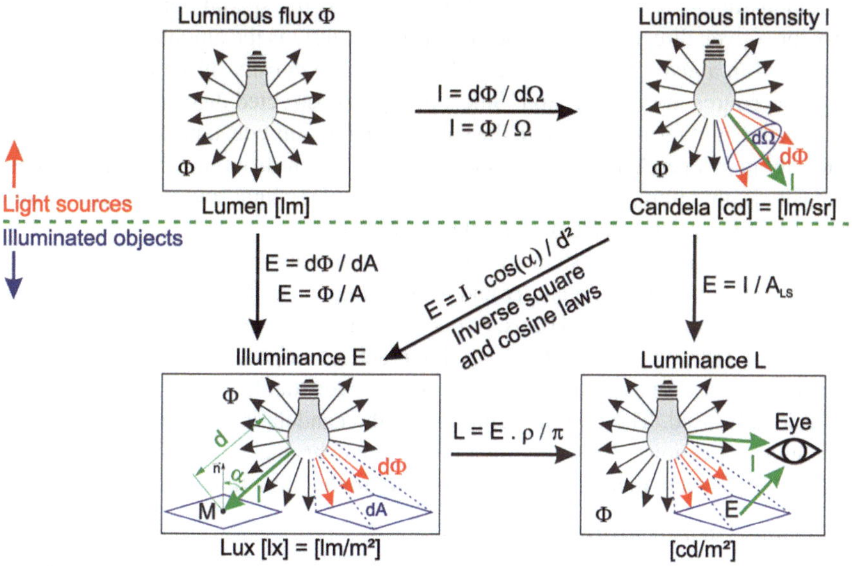

where:
Ω - solid angle into which Φ is emitted
A - area
A_{LS} - visible area of the light source
d - distance from the light source to the point M
ρ - diffuse reflection coefficient of the surface
π = 3.14

Fig. 7 Basic photometric quantities and dependencies

weighted with the human eye spectral luminous efficiency function $V(\lambda)$ (Figs. 5 and 6). Luminous flux is the photometric counterpart to radiant power. The unit of luminous flux is lumen (lm), and at 555 nm, where the human eye has its maximum sensitivity, a radiant power of 1 W corresponds to a luminous flux of 683 lm. Due to the shape of the $V(\lambda)$ curve, the same radiant flux will produce correspondingly less luminous flux at different frequency points.

Luminous efficacy

Luminous efficacy $\chi = \Phi/P$ describes the luminous flux Φ of a lamp in relation to its power consumption P and is therefore expressed in lumen per watt (lm/W). The maximum value theoretically attainable when the total radiant power is transformed into visible light is 683 lm/W. Luminous efficacy varies from light source to light source, but always remains well below this optimum value.

Luminous intensity

An ideal point-source lamp radiates luminous flux uniformly into the space in all directions, and its luminous intensity is the same in all directions. Real light sources do not usually emit their luminous flux equally intensely in the different directions of space. In order to quantify the spatial distribution of the light flux, the magnitude Luminous intensity I is introduced. It is defined as the spatial density of the light flux for a given direction. Luminous intensity $I = \Phi/\Omega$ is the amount of light flux Φ contained in a given solid angle Ω and indicates the density of the light emanating from the source in that direction. Since, at any unobstructed distance from the light source, the light flux contained in the solid angle remains constant, and the luminous intensity is independent of the distance of measurement from the source. Luminous intensity is, therefore, considered to be a property of the light source and can be used to indicate the performance of the light source.

In the International System of Units (SI), luminous intensity is accepted as a fundamental photometric quantity and is expressed in candelas (cd). The magnitude of the candela has a historical basis. At one time called the candlepower, it was defined in terms of flame or filament standards. For practical purposes, the terms candela and candlepower are equivalent and, although no longer standard, the latter term is still occasionally used. The current definition of the candela is the luminous intensity, in a given direction, of a source that emits monochromatic radiation of frequency 540×10^{12} Hz and that has a radiant intensity in that direction of 1/683 W/sr.

Luminous intensity distribution curve

The distribution of the luminous intensity of a light source in all directions throughout a space produces a three-dimensional graph—photometric solid (Fig. 8). A section through this graph results in a luminous intensity distribution curve (LDC—Fig. 8), which describes the luminous intensity on one plane. The luminous intensity is usually indicated in a polar coordinate system as the function of the beam angle. To allow comparison between different light sources to be made, the light distribution curves are based on an output of 1000 lm. In the case of symmetrical luminaire, one

Fig. 8 Photometric solid and light distribution curve

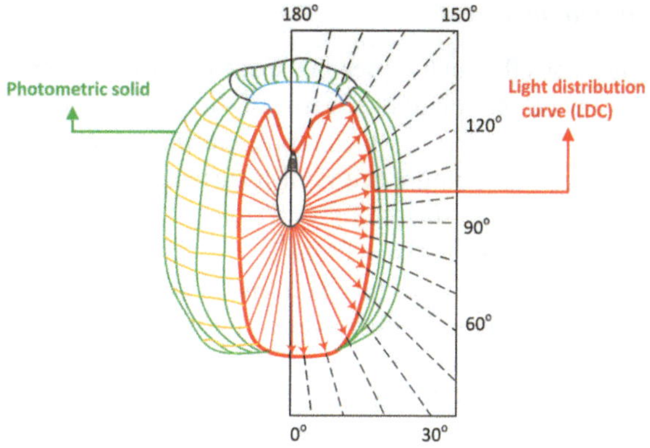

light distribution curve is sufficient to describe one luminaire, and axially symmetrical luminaires require two curves, which are usually depicted in one diagram.

By reviewing the photometric curve of a source of light, luminous intensity in any direction may be determined very accurately. These data are necessary for some lighting calculations. Therefore, spatial directions through which luminous radiation is irradiated may be established by two coordinates. One of the most frequently used coordinate systems to obtain photometric curves is the "$C - \gamma$" represented in Fig. 9.

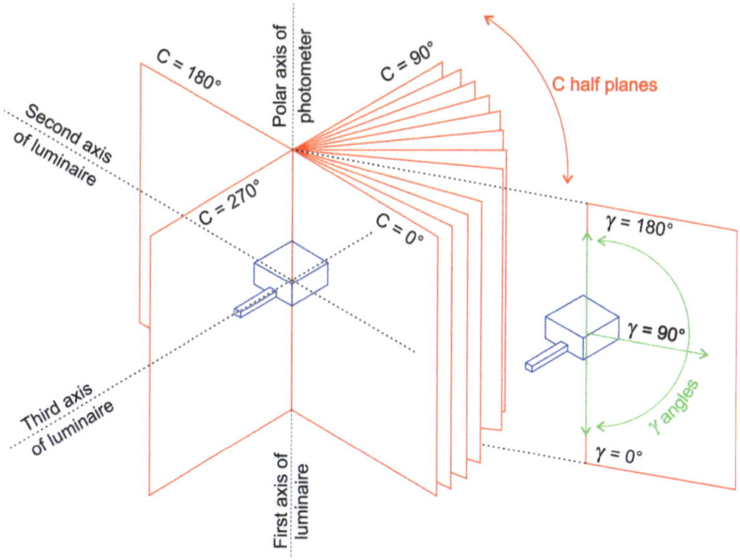

Fig. 9 "$C - \gamma$" for representing of luminous intensity distribution curves [5]

Illuminance

Luminous flux in its spread in space falls on different surfaces and illuminates them. For quantification of the degree of illumination of surfaces, the magnitude illumination E is introduced, which represents the density of the light flux on the illuminated surface. The illuminance E of a surface is the ratio between the luminous flux Φ received by the surface to its area A: $E = \Phi/A$ (Fig. 7). The unit for measuring of illuminance is lux ($\text{lx} = \text{lm/m}^2$).

Illuminance is the most convenient for direct measurement light technical parameter, which assesses the level of illumination of illuminated objects. Therefore, it is the basic quantitative parameter for setting the required level of lighting in light technical standards. The current standards that set the recommendations for the level of lighting of

- work places are
  - EN 12464-1:2011 Light and lighting—Lighting of work places—Part 1: Indoor work places;
  - EN 12464-1:2014 Light and lighting—Lighting of work places—Part 2: Outdoor work places;
- road lighting—EN 13201-2:2015, parts 1–5;
- sports lighting—EN 12193:2007 Light and lighting—Sports lighting;
- energy performance—EN 15193-1:2017 Energy performance of buildings—Energy requirements for lighting, parts 1–2;
- Emergency lighting—EN 1838:2013 Lighting Applications—Emergency lighting.

Luminance

Luminance is the effect which produces a surface on the retina of the eye, both coming from a primary source which produces light, or from a secondary source or surface which reflects light. Luminance measures brightness for primary light sources as well as for sources constituting illuminated objects. This term has substituted the concepts of brightness and lighting density. Luminance is the only basic lighting parameter that is perceived by the eye. The human eye does not perceive colors but brightness, as a color attribute. Light perception is the perception of differences in luminance. Therefore, it may be stated that the eye perceives luminance differences but not illuminance ones (provided that we have the same lighting, different objects have different luminance since they have different reflection characteristics).

Luminance L of an illuminated surface is the ratio between luminance of a source of light in a given direction, to the surface of the projected source depending on such direction and can be defined on the base both light intensity I or illuminance E—Fig. 7. The unit for measuring of luminance is cd/m^2.

1.3 Color Characteristics of Light Sources

Color is a subjective psychophysiologic interpretation of the visible electromagnetic spectrum. Luminous sensations or images, produced in our retina, are sent to the brain and interpreted as a set of monochromatic sensations which constitute the color of the light. The sense of sight does not analyze each radiation or chromatic sensation individually. For each radiation, there is a color designation, according to the frequency spectrum classification.

Objects are distinguished by the color assigned depending on their optical properties. Objects neither have nor produce color. They do have optical properties to reflect, refract and absorb colors of the light they receive. The set of additive monochromatic sensations that our brain interprets as color of an object depends on the spectral composition of the light that illuminates such an object and on the optical properties possessed by the object to reflect, refract or absorb.

Subjective evaluations of object surfaces, in the same way they are perceived by the human eye, are interpreted bearing in mind color attributes or qualities:

- Lightness or brightness: luminous radiation received according to the illuminance possessed by the object. The further from black in the gray scale, the lighter is the color of an object. It refers to intensity;
- Hue or tone: common name for color (red, yellow, green, etc.). It refers to wavelength;
- Purity or saturation: proportion in which a color is mixed with white. It refers to spectral purity.

In order to avoid a subjective evaluation of color, there exists a chromaticity diagram in the shape of a triangle, approved by the CIE—Fig. 10. It is used to treat sources of light, colored surfaces, paints, luminous filters, etc., from a quantitative point of view. All colors are ordered following three chromatic coordinates, x, y, z, whose sum is always equivalent to the unit ($x + y + z = 1$). When each of them equals 0.333, they correspond to the white color. These three coordinates are obtained from the specific potencies for each wavelength. It is based on the fact that when three radiations from three sources of different spectral composition are mixed, a radiation equivalent to another with a different value may be obtained. The result is the triangle in Fig. 10, in which any two coordinates are enough to determine the radiation color resulting formed by the additive mixture of three components.

There are two systems of measurement commonly used to describe the color properties of a light source in lighting engineering. The "color rendering index" (CRI or R_a) suggests how an object illuminated by that light will appear in relation to its appearance under other standardized common light sources, and "color temperature" (CCT or T_c) expresses the color appearance of the light itself.

Color rendering index

CRI is the main quantitative indicator of the ability of a light source to correctly reproduce the test colors R_i of different objects as compared to a reference light

Fig. 10 CIE chromaticity diagram [6]

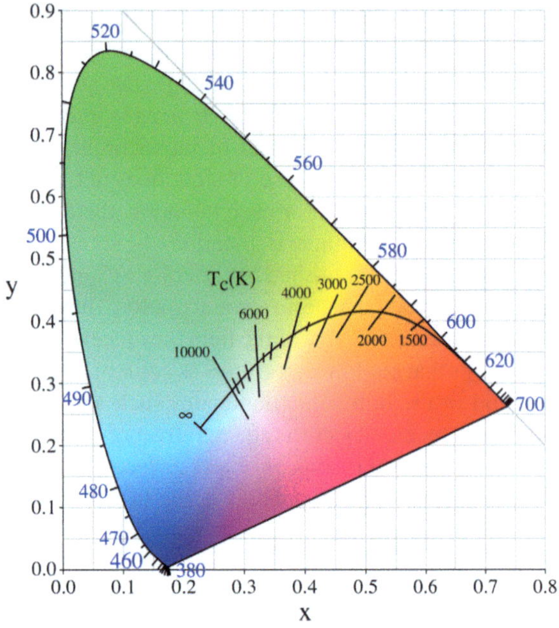

source (D65-sun, A-halogen lamp). It changes from 0 (no color rendering) to 100 (perfect color rendering). A visual comparison of the color rendering at different values of CRI is shown in Fig. 11.

The value of CRI is calculated as CRI = $(\Sigma R_i)/8$, where R_1 to R_8 are standardized test colors—Fig. 12. The saturated test colors R_9 to R_{14} are also used occasionally to

Fig. 11 Visual comparison of the color rendering at different CRI values lamps [5]

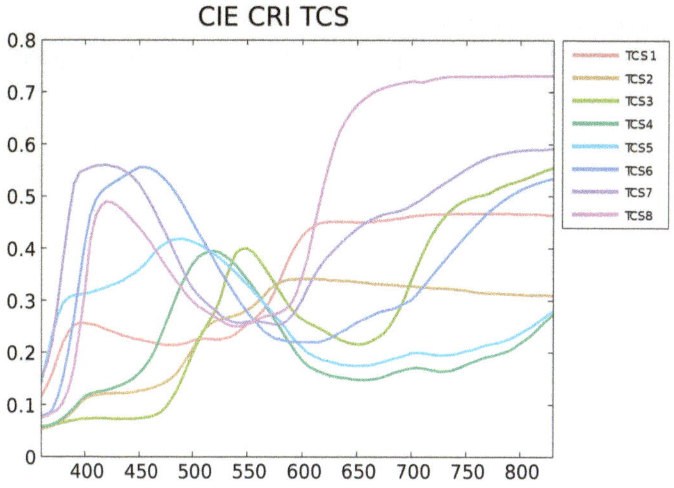

Fig. 12 Standardized test colors for accessing of CRI [7]

describe special functions of a light source. The reproduction of these colors is then quoted separately.

Typical standard test color values for white light-emitting diode (LED) with color rendering index CRI = 82 are shown in Fig. 13.

Light sources are divided up into color rendering groups—Table 1.

According to the latest recommendations, color rendering of less than 80 should not be selected at workplaces. If light sources with a color rendering index below 80 are used in exceptional cases, it has to be ensured that safety colors can be recognized without any problems.

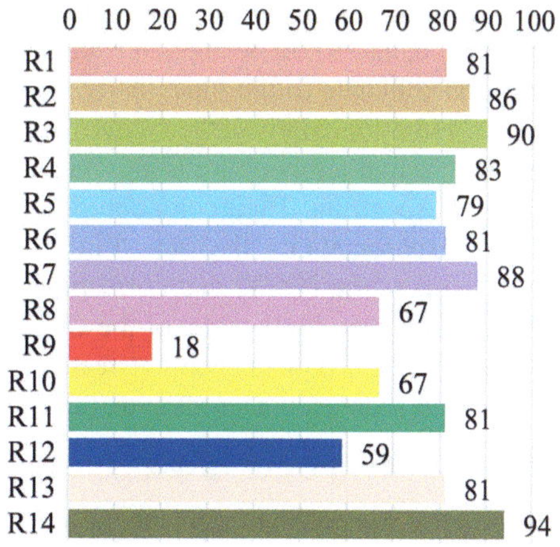

Fig. 13 Typical standard test color values for LED with CRI = 82 (R1–R8)

Table 1 Color rendering groups of light sources

Color rendering group	CRI	Importance	Typical application
1A	CRI $\geq$ 90	Accurate color matching	Galleries, medical examinations, color mixing, TV broadcasting
1B	90 > CRI $\geq$ 80	Accurate color judgement	Houses, hotels, offices, schools, hospitals
2	80 > CRI $\geq$ 60	Moderate color rendering	Industry
3	60 > CRI $\geq$ 40	Accurate color rendering is of little importance	Rough industry
4	40 > CRI $\geq$ 20	Accurate color rendering is of no importance	Rough work, acceptable in traffic lighting

Color temperature

In the CIE chromaticity diagram in Fig. 10, a curve has been drawn representing the color emitted by a black body according to its temperature. It is known as black-body color temperature curve. Color temperature T_C in kelvins (K) is an expression used to indicate the color of a source of light by comparing it with a black-body color, which is a "theoretical perfect radiant" (object whose light emission is only due to its temperature). As any other incandescent body, the black body changes its color as its temperature increases, acquiring at the beginning, a red matte tone, to change to light red later on, orange, yellow and finally white, bluish white and blue. For example, color of a candle flame is similar to the one of a black body heated at about 1800 K. Then, the flame is said to have a "color temperature" of 1800 K. Incandescent lamps have a color temperature which ranges from 2700 to 3200 K, depending on their type. Their fleck is determined by the corresponding coordinates and is located virtually on the black-body curve. Such temperature bears no relation at all with that of an incandescent filament. Color temperature only defines color appearance, and it can be applied exclusively to sources of light which have a great color resemblance to the black body. This is only fulfilled for light sources using thermal radiation. Many of the modern light sources use luminescence to obtain light, and their actual temperature is significantly different from the color appearance of their light spectrum. Their color coordinates do not fall on the line of the black body of the CIE chromaticity diagram—Fig. 10. For these light sources, the term correlated color temperature (CCT) that corresponds to the closest to their color coordinates point of the black-body curve of the CIE chromatic diagram.

The practical equivalence between color appearance and color temperature is established arbitrarily according to Table 2 and Fig. 14.

Light Color and psychic effects

It has been proved that color in the environment produces psychic or emotional reactions in the observer. Hence, using colors in the adequate way is a very relevant topic for psychologists, architects, lighting engineers and decorators. There are no

Table 2 Color appearance groups of light sources

Color appearance group	Color appearance	Color temperature T_C
1	Warm	<3300 K
2	Neutral	3300–5300 K
3	Cool	>5300 K

Fig. 14 Color appearance on color temperature scale in kelvins [8]

fixed rules for choosing the appropriate color in order to achieve a certain effect, since each case requires to be given a particular approach. However, there are some experiences in which different sensations are produced in the individual by certain colors.

One of the first sensations is that of heat or coldness. This is the reason why the expression "hot colors" and "cold colors" are mentioned. Hot colors are those which go from red to greenish yellow in the visible spectrum, cold colors the ones from green to blue. A color will be hotter or colder depending on its tendency toward red or blue, respectively.

On the one hand, hot colors are dynamic, exciting and produce a sensation of proximity. On the other hand, cold colors calm and rest, producing a sensation of distance.

Likewise, color clarity also produces psychological effects. Light colors cheer up and give a sensation of lightness, while dark colors depress and produce a sensation of heaviness.

When two or more colors are combined and produce a comfortable effect, it is said that they harmonize. Thus, color harmony is produced by means of selecting a color combination which is comfortable and even pleasant for the observer in a given situation.

From all the above mentioned, it may be deduced that a knowledge of the spectral distribution curve of sources of light is necessary to obtain the desired chromatic effect.

Color discomfort

Color temperature has an important influence on the environment created as long as coldness or heat sensations go. At the same time, it promotes or reduces object chromaticity in the same way. Moreover, the term T_C cannot be manipulated in an independent way, but it must be combined in an adequate way with illuminance so that the disturbing effects of visual perception are not produced. Kruithof's curves delimit possible combinations between T_C and illuminance level E (Fig. 15).

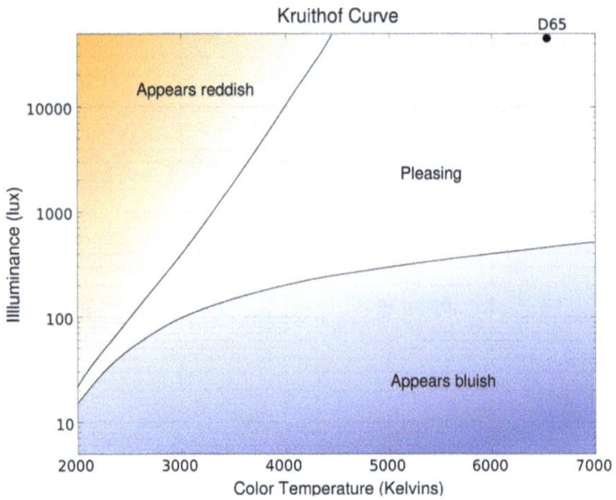

Fig. 15 Kruithof curves for the recommended ratio between color temperature and illuminance level [9]

The experience of using cold-light lamps indicates that when relatively small illumination occurs, unpleasant visual sensations associated with twilight noises often occur. This is partly explained by the fact that the human eye is accustomed to highlighting in daylight, so illumination with cold-white lamps that have a spectrum close to daylight is considered as discomfort if illumination is not high enough—Fig. 15.

Color temperature datum is only referred to the color of light, but not to its spectral composition which is decisive for color reproduction. Thus, two sources of light may have a very similar color and possesses, at the same time, very different chromatic reproduction properties. Therefore, it is important to take into account both the color temperature and the color rendering index when considering the color characteristics of the light sources.

2 Light Sources

As explained in the lighting fundamentals section, light is a form of energy represented by electromagnetic radiation, which may affect the human eye. It is produced by light sources in many ways, depending on the causes that provoke it. If it is due to the radiant body temperature, the phenomenon is called thermal radiation or incandescence. All other examples are considered as luminescence. So, there are two basic types of light production: *incandescence and luminescence*, to which all types of natural and artificial light sources can be classified—Table 3.

Table 3 Classification of the types of light sources

	Incandescence (Thermal radiation)	Luminescence	
		Gas discharge	Solid state (body)
Natural	Sun, lava, stars, combustion	Lightning, electrostatic electric spark	Glowworm, firefly, anglerfish, crystal jellyfish
Artificial	Flame, gaslight, incandescent lamp, halogen lamp	High intensity discharge (HID), low intensity discharge, fluorescent lamps	Solid-state lighting (SSL): light-emitting diode (LED), organic LED (OLED)

2.1 Incandescence

Incandescence involves the vibration of entire atoms, for example when atoms are heated to high optimum temperatures, the thermal vibration is released as electromagnetic radiation. Incandescent light or "black-body radiation" is produced when light comes from a heated solid. Depending on the temperature of the material, the photons released vary in their energies and colors; at low temperatures, the materials emit radiation in infrared wavelengths. In black-body radiation, the trend follows as the temperature increases the peak shifts to shorter wavelengths, firstly produces a red then white and lastly a bluish-white color as the peak moves into the ultraviolet part of the spectrum from the visible part, an example of this is when metal is heated. Incandescent light is the most common type of light; it includes the Sun, fires and light bulbs. Fires involve chemical reactions which release heat and gases, causing materials to reach high temperatures and eventually causes the gases and materials to incandescence. In contrast, light bulbs produce heat, as an electrical current passes through a cable and heats the cable to high temperatures eventually causing the cable to incandescence. Incandescent light bulbs emit approximately 90% of their energy as infrared; whereas, the remainder is visible light [10].

The main characteristic of the thermal light source is its temperature. The higher the temperature, the more intense the radiation is. Also, with the change in temperature, the radiation spectrum changes. At relatively low temperatures, the rotational movement of molecules predominates, creating long-wave infrared radiation. Bouncing is characteristic of higher temperatures and is accompanied by short-wave infrared and long-wave visible radiation. At very high temperatures, when the kinetic energy of the moving particles is large, conditions are created for electronic excitation of the atoms, which emit visible and ultraviolet radiation. At the same temperature, the radiation density of the different bodies may be different if their absorption coefficients are not the same. Depending on the absorbing and radiant characteristics, the bodies are subdivided into three groups: absolute black, gray and selective [11].

The absolute black body absorbs all the energy that falls on it, so its absorption coefficient is $\alpha = 1$. In Fig. 16, a schematic representation of an absolute black-body model is given. The internal surface of the body is painted black. The radiant flux, which enters through a small hole made, is absorbed due to multiple reflections. If

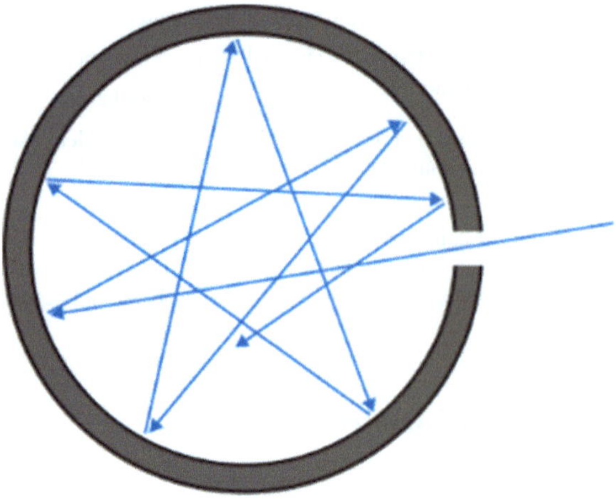

Fig. 16 Approximate realization of a black body as a tiny hole in an insulated enclosure

this body is heated, it will radiate through the opening as an absolute black body. The actual gray and selective radiating bodies have absorption coefficients α, smaller than one.

Research on the characteristics of incandescent light sources is performed on the basis of experimental tests with an absolute black body.

Kirchhoff has established (1859), based on his law "Emissivity = Absorptivity," a quantitative relationship between the radiant exitance M and the coefficients of absorption α of different bodies having the same temperature T_s:

$$\frac{M_1(T_s)}{\alpha_1} = \frac{M_2(T_s)}{\alpha_2} = \cdots = \frac{M_n(T_s)}{\alpha_n} = M_b(T_s) \tag{2}$$

where

$M_1(T_s), \ldots, M_n(T_s)$—radiant exitance of different real bodies at temperature T_s;
$\alpha_1, \ldots, \alpha_n$—coefficients of absorption of the real bodies at temperature T_s;
$M_b(T_s)$—radiant exitance of absolute black body at same temperature T_s.

Based on Eq. (2), the following conclusions can be drawn:

- The radiant exitance of all real bodies is less than that of the absolute black body at the same temperature.
- The spectral intensity of the radiant exitance of the real bodies in each area of the spectrum is always less than the spectral intensity of the radiant exitance of the absolute black body at the same temperature. Therefore, the $M(\lambda, T)$ curve for a gray or selective emitter is always located under the curve $M_b(\lambda, T)$ for an absolute black body at equal temperatures—Fig. 17.

Planck has derived (1900) the following quantitative relationship $M_b = f(\lambda, T)$ between the spectral radiant exitance, the wavelength λ and the temperature T of the black body:

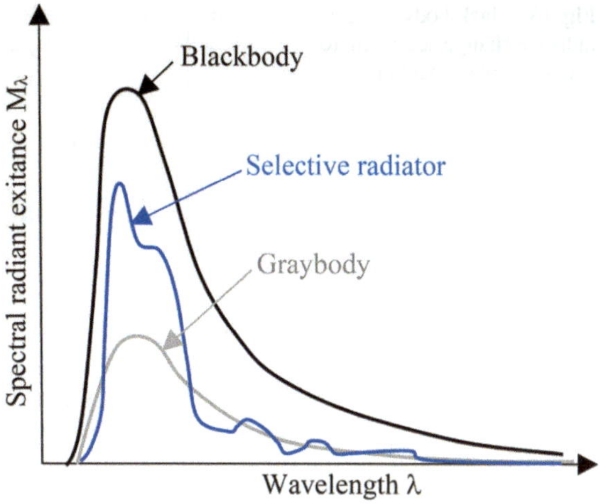

Fig. 17 Comparison of black body, real surface and selective radiator spectral radiant exitance curves at equal temperature *T*

$$M_b(\lambda, T) = \frac{c_1}{\lambda^5} \frac{1}{e^{\frac{c_2}{\lambda T}} - 1} \tag{3}$$

where

$c_1 = 2\pi hc^2 = 3.7410^8$ Wm^{-2} μm^4—first radiation constant;
$c_2 = hc/k = 1.438 \times 10^4$ μm K —second radiation constant;
$H = 6.626 \times 10^{-34}$ Ws^2—the Planck's constant;
$K = 1.3806 \times 10^{-23}$ Ws K—the Boltzmann constant;
$C = 299{,}792{,}458$ m/s—speed of light.

By integrating Formula (3) with the wavelength, the integrated radiant exitance $M_b = f(T)$ can be determined. This dependence was first established experimentally by Stefan, and theoretically by Boltzmann, and is known as Stefan–Boltzmann's law—Formula (4):

$$M_b(T) = \sigma T^4 \tag{4}$$

where $\sigma = \frac{2\pi^5 k^4}{15h^3c^2} = 5.6704 \times 10^{-8}$ Wm^{-2} K^{-4}—Stefan–Boltzmann constant.

In Fig. 18, the curves for different values of the absolute temperature *T* are constructed according to the Plank Eq. (3). The solid lines show the radiation from a black body with the temperature shown in kelvin. The red line shows the radiation for the Sun at 6000 K, the yellow line for the filament of an incandescent halogen lamp at 3000 K and the blue line at 300 K—near to the earth at 288 K. The dotted gray line connects the peak spectral radiances (λ_{max}) for each temperature, a consequence of Wien's displacement law. The dashed gray line shows the visible part of the spectrum.

The solid curves in Fig. 18 show that at low temperatures, the radiance of the black body is entirely in the infrared spectrum. As the temperature rises, the maximum

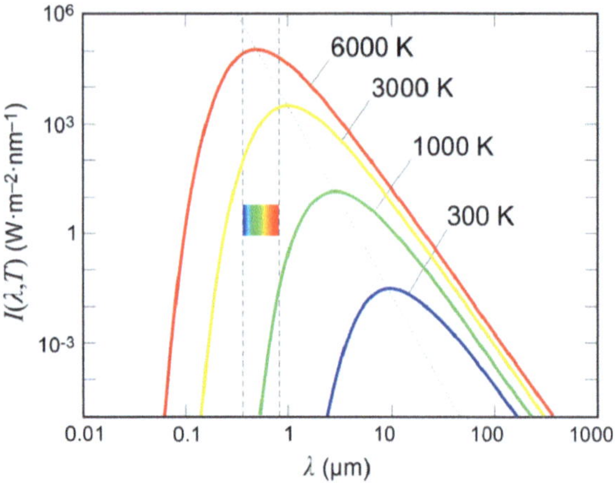

Fig. 18 Black-body spectral radiance change according to Wien's displacement law [12]

values of the curves are rapidly increased while simultaneously shifting to a shorter wavelength.

Interest in lighting technology is the position of the maximum of $\lambda_{\max}$ curves at different temperatures, since depending on this point, a different part of the radiation falls into the visible part of the spectrum. This dependence is known as Wien's displacement law since it was first established by him (1896) based on experimentally captured curves of the spectral radiance of the black-body radiation:

$$\lambda_{\max} T = \text{const} = 2.897773 \times 10^{-3}\,\text{m K} \approx 2900\,\mu\text{m K} \tag{5}$$

From the laws of Wien, Planck and Stefan–Boltzmann, the following essential conclusions can be made:

- upon increasing the temperature of the radiating body, its radiant flux increases sharply, at the same time changing its spectrum;
- when the temperature increases, the maximum spectral intensity of the radiation is shifted to the shorter wavelengths. If the radiator temperature is from 4670 to 10,000 K, this maximum is located in the area of the visible rays.

The light performance of the thermal radiation is quantified by means of luminous efficiency η in percentages. It represents the proportion of radiation falling within the visible spectrum and corrected by the spectral sensitivity of the human eye $V(\lambda)$ to the overall energy emission of the emitter. The luminous efficiency η dependence on the temperature for a black-body radiator is shown graphically in Fig. 19.

The maximum value for η is obtained at a temperature $T = 6500$ K when the maximum of the spectral radiance curve appears at $\lambda = 555$ nm, i.e., coincides with the maximum of $V(\lambda)$—Fig. 5. In this case, the value is $\eta = 14.5\%$, and this represents the theoretically possible maximum value of luminous efficiency of a thermal radiator. This value is relatively low for today's high-energy efficiency requirements for lighting, and therefore incandescent light sources are less and less used.

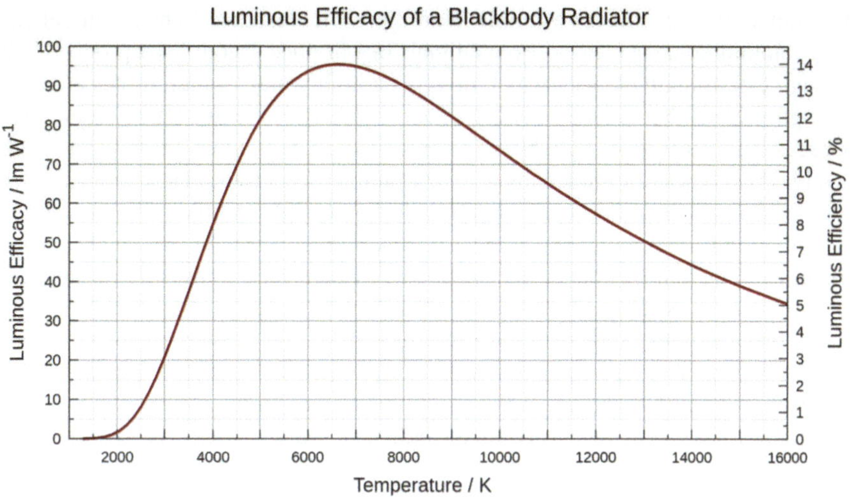

Fig. 19 Variation of luminous efficacy and efficiency of a black-body radiator when changing the temperature [13]

In lighting technologies, the important term luminous efficacy χ is used. Luminous efficacy is a measure of how well a light source produces visible light. It is the ratio of luminous flux to power $\chi = \Phi/P$, measured in lumens per watt (lm/W) in the International System of Units (SI). Depending on the context, the power can be either the radiant flux of the source's output, or it can be the total power (electric power, chemical energy or others) consumed by the source.

The distinction between luminous efficacy and efficiency is not always carefully maintained in published sources, so it is not uncommon to see "efficiencies" expressed in lumens per watt, or "efficacies" expressed as a percentage.

2.2 Luminescence

Unlike incandescence, which is due to the vibration of entire atoms, the luminescence involves only electrons. It generally occurs at lower temperatures, compared to incandescent light and is often referred to as light from different sources of energy that can take place at normal or lower temperatures.

Luminescence light is produced when an electron emits some of its energy as electromagnetic radiation. During specific energy levels, electrons need to have energy. When electrons jump down to lower energy levels, a certain amount of energy which becomes light of a specific color is released. To maintain continuous amounts of luminescence, the electrons require a continuous push to be pushed up to higher energy levels so that the cycle continues. This push or "kick" can be provided for by a range of sources—for example neon lights, fluorescence light, electrical currents, bioluminescence, e.g.,—even animals like fireflies. Neon lights produce light through electroluminescence which involves a high voltage which forces a current through the gas that excites it and eventually causes it to emit light. Bioluminescence is the formation of light by living organisms, e.g., fireflies. Fluorescence light involves two types of luminescence—electroluminescence and photoluminescence, common uses include fluorescent lamps, televisions and computer screens [10].

Radiation from luminescent sources results from the excitation of single valence electrons of an atom, either in a gaseous state, where each atom is free from interference from its neighbors, or in a crystalline solid or organic molecule, where the action of its neighbors exerts a marked effect. In the first case, line spectra result, such as those of mercury or sodium arcs. In the second case, narrow emission bands result, which cover a portion of the spectrum (usually in the visible region). Both cases contrast with the radiation from incandescent sources, where the irregular excitation at high temperature of the free electrons of innumerable atoms gives rise to all wavelengths of radiation to form a continuous spectrum of radiation, as discussed in "Incandescence" above [11].

To clarify the phenomenon of luminescence, the Bohr atomic model (1913) can be used—Fig. 20. According to this model, each atom is formed by a positive atomic nucleus and by a cover of negative electrons. These are distributed in different layers that rotate around the nucleus following certain orbits. Usually, there is an electric balance in the atom, that is to say, the number of positive charges is equal to the number of negative charges (electrons). This balance is known as the fundamental state of the electron, and for electrons in the most internal orbit (Fig. 20). If a certain amount of energy is administered to the electron from the outside, electron is excited and moved from its regular orbit to the next one or to another more external one.

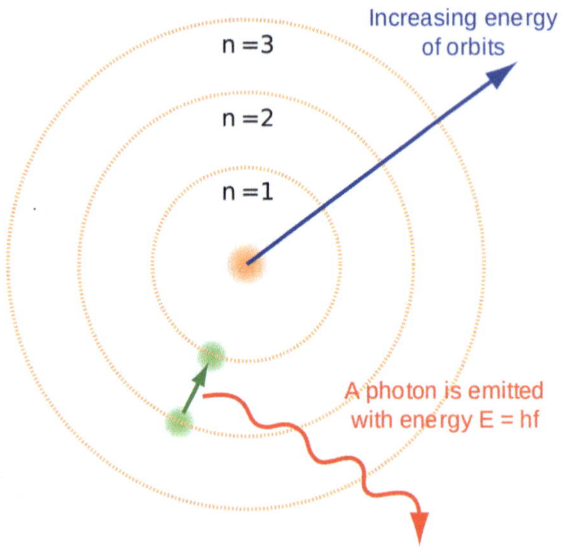

Fig. 20 Bohr atomic model [14]

Thus, the energy supplied is absorbed. The electron is located in a superior energy level (level lines $n = 1$, $n = 2$, $n = 3$, etc., of Fig. 20). After a short time in this level, the electron returns again to its regular initial position (green line in Fig. 20) and emits the amount of energy absorbed at the beginning, usually in the form of a photon (red curve in Fig. 20).

If the amount of energy is greater, the electron may instantaneously reach a more external orbit. As a consequence of the greater range of energy achieved, radiation emitted when the electron returns to base f will be richer in energy.

Therefore, the different layers of energy correspond to a perfectly determined level of energy, and, thus, there are not intermediate levels. Thus, it is deduced that in order to excite an atom, an exactly determined amount of energy is necessary. This is emitted in the form of radiation and/or heat loss when the atom recovers its fundamental shape.

The emission of energy transformed in this process from an atomic point of view takes place in portions or discontinuous parts known as energy quants (Bohr postulated that the atom may not rotate at any distance from the nucleus, but in certain orbits only). However, in the field of practical lighting engineering, light emitted in this transformation is considered to be emitted in a continuous way, in the form of electromagnetic waves, which is acceptable for normal cases of its application.

By means of the theory of energy quants formulated by Plank, it is proved that different chemical elements, when excited, do not emit a continuous spectrum due to the different structure of their electronic layers, but only very particular wavelengths (lines) within all the electromagnetic spectrum. These spectra are known as linear spectra. Each substance or gas has a characteristic linear spectrum. The emission lines of the chemical elements observed through a diffraction grating are given in Fig. 21.

According to the physical technique used to excite atoms, the type of radiation and the form in which it is emitted, several types of luminescence may be distinguished [16]:

A. Photoluminescence:

- Gaseous discharges;
- Fluorescence;
- Phosphorescence;
- Lasers;

B. Electroluminescence:

- Electroluminescent lamps (ac capacitive);
- Light-emitting diodes;
- Cathodoluminescence (electron excitation);

C. Miscellaneous luminescence phenomena:

- Galvanoluminescence (chemical);
- Crystalloluminescence (crystallization);
- Chemiluminescence (oxidation);

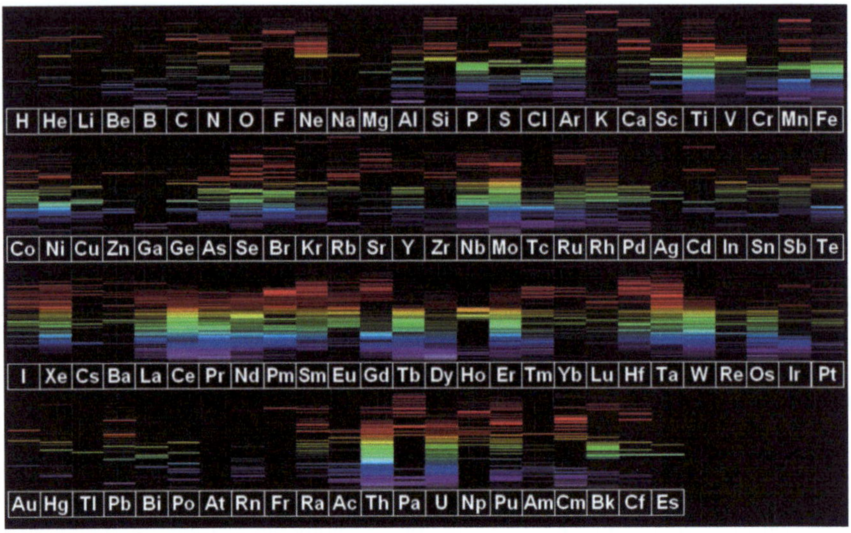

Fig. 21 Visible spectral emission lines of the chemical elements [15]

- Thermoluminescence (heat);
- Triboluminescence (friction or fracture);
- Sonoluminescence (ultrasonics);
- Radioluminescence (α, β, γ, and X-rays).

2.2.1 *Photoluminescence*

Gaseous Discharge

A typical mechanism for generating light (photons) from a gaseous discharge (Figs. 22 and 23) is described below [16]:

1. A free electron emitted from the cathode collides with one of the two valence electrons of a mercury atom and excites it by imparting to it part of the kinetic

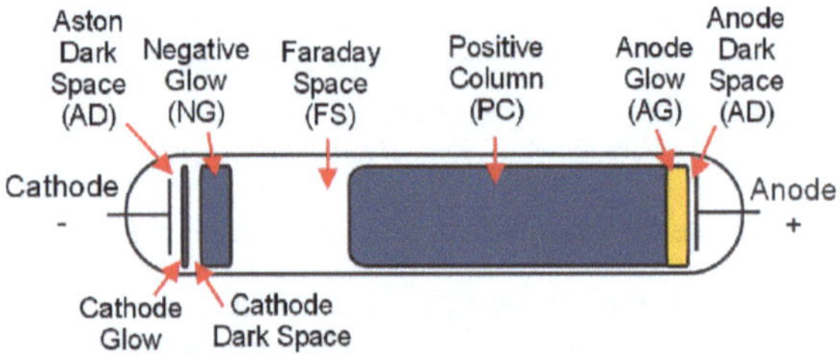

Fig. 22 Gas discharge tube [17]

Fig. 23 Gas discharge in a mercury fluorescent lamp without phosphor

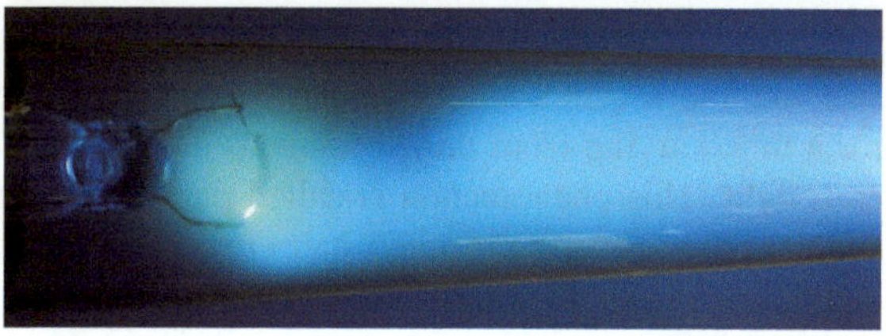

energy of the moving electron, thus raising the valence electron from its normal energy level to a higher one.

2. The conduction electron loses speed in the impact and changes direction, but continues along the tube to excite or ionize one or more additional atoms before losing its energy stepwise and completing its path. It generally ends at the wall of the tube, where it recombines with an ionized atom. A part of the electron current is collected at the anode.
3. Conduction electrons, either from the cathode or formed by collision processes, gain energy from the electric field, thus maintaining the discharge along the length of the tube.
4. After a short delay, the valence electron returns to its normal energy level, either in a single transition or by a series of steps from one excited level to a lower level. At each of these steps, a photon (quantum of radiant energy) is emitted. If the electron returns to its normal energy level in a single transition, the emitted radiation is called resonance radiation.
5. In some cases (as in the high-pressure sodium lamp), a portion of the resonance radiation is self-absorbed by the gas of the discharge before it leaves the discharge envelope. The absorbed energy is then reradiated as a continuum on either side of the resonant wavelength, leaving a depressed or dark region at that point in the spectrum.

Fluorescence

In the fluorescent lamp, UV radiation resulting from luminescence of the mercury vapor due to a gas discharge is converted into light by a phosphor coating on the inside of the tube or outer jacket. If this emission continues only during the excitation, it is called fluorescence.

The phosphors used in fluorescent lamps are crystalline inorganic compounds of exceptionally high chemical purity and of controlled composition to which small quantities of other substances (the activators) have been added to convert them into efficient fluorescent materials. With the right combination of activators and inorganic compounds, the color of the emission can be controlled. For the phosphor to emit light, it must first absorb radiation. In the fluorescent lamp, this is chiefly at 253.7 nm. Stokes' law, stating that the radiation emitted must be of longer wavelength than that absorbed. Because of the oscillation around stable positions, the excitation and emission processes cover ranges of wavelength, commonly referred to as bands [16].

Phosphorescence

In some fluorescent materials, electrons can be trapped in metastable excited states for a time varying from milliseconds to days. After release from these states, they emit light. This phenomenon is called phosphorescence. The metastable states lie slightly below the usual excited states responsible for fluorescence, and energy usually derived from heat is required to transfer the electron from the metastable state to the emitting state. Since the same emitting state is usually involved, the color of fluorescence and phosphorescence is generally the same for a given phosphor. In doubly activated phosphors, the second activator acts longer than the primary activator, so the color changes with time. Short-duration phosphorescence is important in fluorescent lamps in reducing flicker in alternating current (AC) operation [16].

2.2.2 Electroluminescence

The phenomenon of electroluminescence is based on certain phosphors which convert energy directly into light, without using an intermediate step as in a gas discharge.

Light-Emitting Diode (LED)

Light-emitting diodes produce light by electroluminescence when low-voltage direct current is applied to a suitably doped crystal containing a *p–n* junction (Fig. 24). The doping is typically carried out with elements from column III and V of the periodic table of elements. When activated by a forward-biased current, the *p–n* junction emits light at a wavelength defined by the active-region energy gap. When the forward-biased current is applied, minority carrier electrons are injected into the *p*-region, and corresponding minority carrier electrons are injected into the *n*-region. Photon

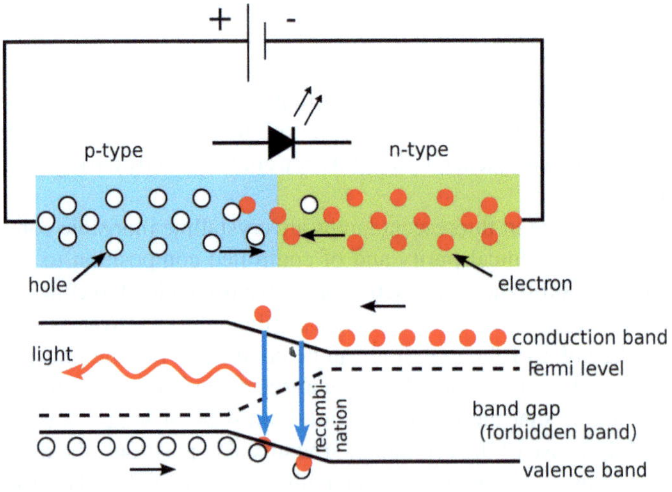

Fig. 24 Schematic diagram of an LED *p–n* junction [18]

emission occurs as a result of electron–hole recombination in the p-region. Electron energy transitions across the energy gap, called radiative recombinations, produce photons (i.e., light), while shunt energy transitions, called nonradiative recombinations, produce phonons (i.e., heat). The energy band gap E_g, shown in Fig. 24, is the separation between the conduction energy band and the valence energy band in the semiconductor crystal. The characteristics of the energy band gap determine the quantum efficiency and the radiative wavelengths of the LED device. For example, the radiative energy wavelength, λ, is given by Formula (6):

$$\lambda = \frac{hc}{E_g} \tag{6}$$

where h is Planck's constant, and c is the speed of light [16].

The efficacy is dependent on the visible energy generated at the junction and losses due to re-absorption when light tries to escape through the crystal.

LEDs are the most modern light sources with a number of advantages. Details about the specific technologies and characteristics of modern LED will be presented in Sect. 2.10 of this chapter.

Electroluminescent Lamp (ac capacitive)

An electroluminescent lamp is composed of a two-dimensional area conductor (transparent or opaque) on which a dielectric-phosphor layer is deposited. The second two-dimensional area conductor of transparent material is deposited over the dielectric-phosphor mixture. An alternating electric field is established between the two conductors with the application of a voltage across the two-dimensional (area) conductors. Under the influence of this field, some electrons in the electroluminescent phosphor are excited. During the return of these electrons to their ground or normal state, the excess energy is radiated as light. The color of the light emitted by an electroluminescent lamp is dependent on frequency, while the luminance is dependent on frequency and voltage. These effects vary from phosphor to phosphor. The efficacy of electroluminescent devices is low compared to incandescent lamps. It is of the order of a few lumens per watt [16].

2.2.3 Miscellaneous Luminescence Phenomena

Galvanoluminescence. Galvanoluminescence is the light that appears at either the anode or the cathode when solutions are electrolyzed.

Crystalloluminescence. Crystalloluminescence (lyoluminescence) is observed when solutions crystallize; it is believed to be due to rapid reformation of molecules from ions. The intensity increases upon stirring, perhaps on account of triboluminescence (see below).

Chemiluminescence. Chemiluminescence (oxyluminescence) is the production of light during a chemical reaction at room temperatures. True chemiluminescences are oxidation reactions involving valence changes.

Thermoluminescence. Thermoluminescence is luminescence exhibited by some materials when slightly heated. In all cases of thermoluminescence, the effect is dependent on some previous illumination or radiation of the crystal.
Diamonds, marble apatite, quartz and fluorspar are thermoluminescent.
Triboluminescence. Triboluminescence (piezoluminescence) is light produced by shaking, rubbing or crushing crystals. Triboluminescent light may result from unstable light centers previously exposed to some source or radiation, such as light, X-rays, radium emissions and cathode rays; centers not exposed to previous radiation but characteristic of the crystal itself; or electrical discharges from fracturing crystals.
Sonoluminescence. Sonoluminescence is the light that is observed when sound waves are passed through fluids. It occurs when the fluids are completely shielded from an electrical field and are always connected with cavitation (the formation of gas or vapor cavities in a liquid). It is believed that the minute gas bubbles of cavitated gas develop a considerable charge as their surface increases. When they collapse, their capacitance decreases and their voltage rises until a discharge takes place in the gas, causing a faint luminescence.
Radioluminescence. Radioluminescence is the light emitted from a material under bombardment from α-rays, β-rays, δ-rays or X-rays.
Bioluminescence. Bioluminescence is a luminous phenomenon which is weakly manifested in nature. It consists of a sparkle emitted by light worms, some classes of fishes, marine algae, rotten wood and similar. This phenomenon is due to the oxidation process of some special chemical or organic substances, like the ones glow worms and photogene bacteria have when in contact with the air or water oxygen. One of the most interesting reactions is the reaction between the naturally occurring chemicals luciferin, luciferase and adenosine triphosphate (ATP) within the firefly, in which luminous efficiency reaches 97%—significantly higher than the artificial light sources created so far [16].

2.3 A Brief History of Light Sources

The earliest man-made light sources were fire, torches and candles. Ancient Egyptians used hollowed-out stones filled with fat, with plant fibers as wicks. These were the first candles, and they date back to about 3000 BC. In the Middle Ages, candles were made of tallow, a type of animal fat; later, they were made of beeswax or paraffin. Modern candles can still be thought of as a type of fat lamp, but their use today is almost entirely decorative.

Ancient Greeks and Romans made lamps from bronze or pottery that burned olive oil or other vegetable oils in their spouts. Many oil lamps appeared during the Middle Ages when reflectors were added to their designs. Early American colonists used fish oil and whale oil in their Betty lamps. Many improvements were made in the design and fabrication of these lamps over the years, but none produced light efficiently until 1784, when a Swiss chemist named Argand invented a lamp that used a hollow wick to allow air to reach the flame, resulting in a bright light. Later, a glass cylinder was added to the Argand lamp allowing the flame to burn better. With the birth of the petroleum industry, kerosene became a widely used fuel in these lamps.

In the 1800s, gas lamps became popular as street lights, originating in London, England. The gas lamp had no wick, but its chief drawback was an open flame that produced considerable flicker. The electric lamp replaced gas lamps in the late 1800s and early 1900s. The first electric lamp was the carbon-arc lamp, demonstrated in 1801 by Sir Humphrey Davy, but electric lights became popular only after the incandescent lamp was developed independently by Sir Joseph Swan in England and Thomas Edison in the USA. The latter patented his invention in 1879 and subsequently made the invention the commercial success that it is today. Figure 25 illustrates the history of different light sources.

Twentieth century has seen a huge increase in the number of available light sources in the marketplace, starting with improvements in the Edison lamp, then the introduction of mercury-vapor lamps in the 1930s, followed closely by fluorescent lamps at the 1939 World Fair. Tungsten-halogen lamps were introduced in the 1950s and metal-halide and high-pressure sodium (HPS) lamps in the 1960s [19].

The introduction of electrodeless lamps and "white" LEDs in the 1990s is an indication that the industry is dynamic, and the introduction of new light sources is expected to continue at least at the present rate well in the next twenty-first century.

2.4 *Incandescent Lamps*

Incandescent lamp technology uses electric current to heat a coiled tungsten filament to incandescence. The main parts of an incandescent lamp are the filament, the filament supports, the glass bulb, the filling gas and the base—Fig. 26.

Early incandescent lamps used carbon, osmium and tantalum filaments. The filament used in modern incandescent lamps is made out of wolfram, which has low vapor pressure and high melting point, 3382 °C, permits high operating temperatures and consequently high efficacies and low evaporation degree. The efficacy of light production depends on the temperature of the filament. The higher the temperature of the filament, the greater the portion of the radiated energy is that falls in the visible region. For this reason, it is important in the design of a lamp to keep the filament temperature as high as is consistent with a satisfactory life. A higher luminous output would be achieved by twisting the filament as a spiral or bi-spiral.

The glass bulb is a cover of sealed glass which encloses the filament and avoids contact with the air outside so it does not corrode and burn. The glass envelope contains filling gas, which is a mixture of inert gases, argon and nitrogen, and in some cases a small amount of krypton or xenon. Filling a bulb with an inert gas slows down the evaporation of the tungsten filament compared to operating it in a vacuum. This allows for greater temperatures and therefore greater efficacy with less reduction in filament life.

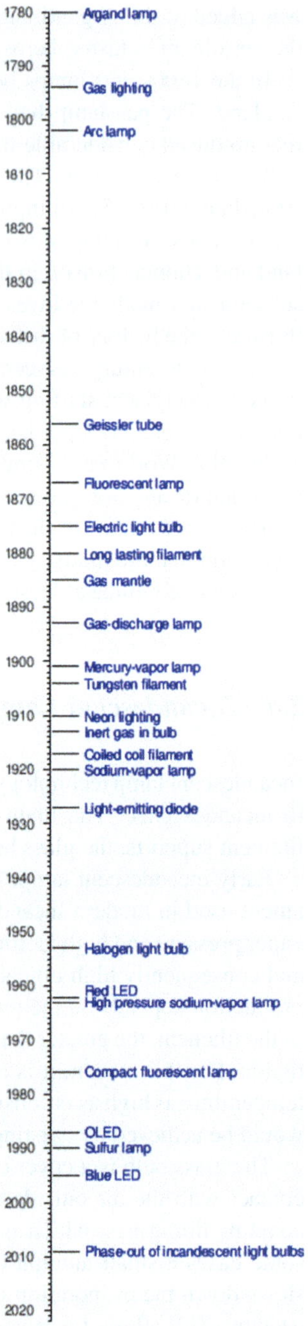

Fig. 25 History of light sources [19]

Fig. 26 Typical incandescent lamp

In incandescent lamps, luminous energy obtained is very little compared to the heat energy irradiated, that is to say, a great amount of the transformed electric energy is lost as heat, and its luminous efficacy is small (it is a "waste-energy" lamp).

The advantage of these lamps is that they are directly connected to the electric current without the need of an auxiliary equipment for their working.

2.5 *Tungsten-Halogen Lamps*

The high temperature of the filament for a normal incandescent lamp makes tungsten particles to evaporate and condense on the wall of the glass bulb, darkening this, as a result. Halogen lamps have a halogen component (iodine, chlorine, bromine) added to the filling gas and work with the halogen regenerative cycle to prevent darkening.

The evaporated tungsten is combined with the halogen to form a halogen wolfram compose. As opposed to wolfram vapor, it is maintained in the form of gas, the glass bulb temperature being high enough as to prevent condensation. When such a gas

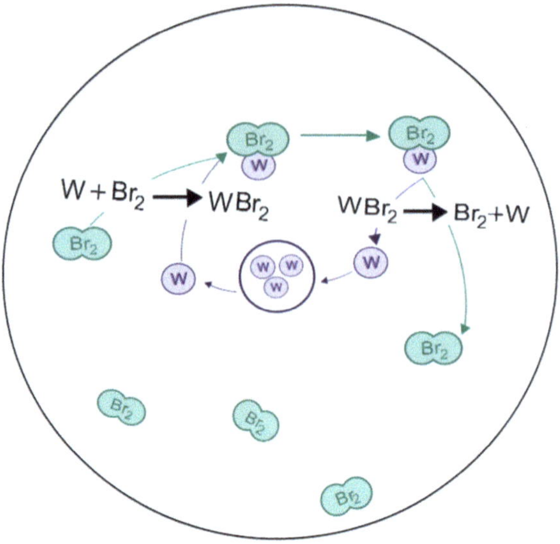

Fig. 27 Regenerative halogen cycle [20]

approaches the incandescent filament, it is decomposed due to the high temperature in tungsten that is again deposited in the filament, and in halogen, which continues with its task within the regenerative cycle—Fig. 27.

The main difference between an incandescent lamp, apart from the halogen additive mentioned before, is in the glass bulb.

Due to the fact that temperature of the glass bulb must be high; halogen lamps are of a smaller size than regular incandescent lamps. Their tubular-shaped glass bulb is made out of a special quartz glass. It should not touch the fingers so as not to alter the distribution of heat on the glass surface, which breaks the regenerative cycle and greatly shortens the life of the lamp.

Since their introduction, tungsten-halogen lamps have entered almost all applications where incandescent lamps were used.

The advantages of tungsten-halogen lamps with regard to regular incandescent lamps are the following: longer duration, 1.5 times greater luminous efficiency, smaller size, higher color temperature and little or no luminous depreciation in time.

2.6 *Low- and High-Pressure Mercury Discharge Lamps*

In this section, discharge lamps, in whose discharge tube mercury is introduced, are going to be presented: fluorescent lamps (FL), compact fluorescent lamps (CFL) and high-pressure mercury lamps.

2.6.1 Fluorescent Lamps

Fluorescent lamp is a low-pressure mercury-vapor gas discharge lamp in which light is produced predominantly by fluorescent powders (luminophores, phosphors) activated by UV energy generated by a mercury arc.

The lamp, generally with a long tubular-shaped glass bulb and a sealed electrode for each terminal, contains low-pressure mercury and a small amount of inert gas for ignition and arc regulation. The glass tube inner surface is covered by a luminophore (commonly called phosphors), whose composition determines the amount of emitted light and the lamp color appearance.

When the proper voltage is applied, an arc is produced by current flowing between the electrodes through the mercury vapor. This discharge generates some visible radiation, but mostly invisible UV radiation, the principal lines being approximately 254, 313, 365, 405, 436, 546 and 578 nm. The UV in turn excites the phosphors to emit light. The phosphors are generally selected and blended to respond most efficiently to 254 nm, the primary wavelength generated in a mercury low-pressure discharge.

The main parts of the fluorescent lamp are the glass tube, the fluorescent layer, the electrodes, the filling gas and the base—Fig. 28.

The glass tube of a regular fluorescent lamp is made out of sodium–calcium glass softened with iron oxide to control short-wave ultraviolet transmission.

The most important factor to determine the characteristics of the light of a fluorescent lamp is the type and composition of the luminophores used. This establishes correlated color temperature T_c (color appearance), color rendering index (R_a) and lamp luminous efficiency, to a great extent. Many different white and colored fluorescent lamps are available, each having its own characteristic spectral power distribution. These types have a combination of continuous and line spectra.

Popular fluorescent lamps use three highly efficient narrow band, rare-earth activated phosphors with emission peaks in the short-, middle- and long-wavelength regions of the visible spectrum. These triphosphor lamps can be obtained with high color rendering, improved lumen maintenance and good efficacy with correlated color temperatures between 2500 and 6000 K relative to halophosphate lamps.

Since the rare-earth phosphors are expensive, the longer triphosphor lamps typically employ a two-coat system consisting of a less expensive halophosphate phosphor applied with the rare-earth type. The rare-earth activated phosphor is closest to the mercury discharge and, as a result, the spectral power distribution of the lamp

Fig. 28 Typical fluorescent lamp electrode and lamps

is more influenced by these phosphors. Common commercial types have correlated color temperatures of 3000, 3500 and 4100 K.

A variety of lamp types is available that radiate in particular wavelength regions for specific purposes, such as plant growth, merchandise enhancement and medical therapy. Various colored lamps, such as blue, green and gold, are obtained by phosphor selection and filtration through pigments.

Electrodes of a lamp which possesses an adequate layer of material emitter serve to drive electric energy to the lamp and provide the necessary electrons to maintain discharge. The majority of fluorescent tubes have electrodes that are preheated by means of an electrical current just before ignition by an independent starter.

Filling gas of a fluorescent lamp consists in a mixture of saturated mercury and an inert gas trimmer (argon and krypton).

Under normal working conditions, mercury is found in the discharge tube both as a liquid and as vapor. The best performance is achieved with a mercury pressure of about 0.8 Pa, combined with a pressure of the trimmer of about 2500 Pa (0.025 atm). Under these conditions, about 90% of the radiated energy is emitted in the ultraviolet wave of 253.7 nm.

Fluorescent lamps require an auxiliary electrical equipment formed by a ballast and an igniter (starter), besides a compensation condenser to improve the power factor.

Working nominal values are reached after five minutes.

2.6.2 High-Pressure Mercury Lamps

In high-pressure mercury lamps, discharge takes place in a quartz discharge tube containing a 20–40 mg of mercury and an inert gas filling, usually argon, to help ignition and to protect the electrodes from destruction during the initial stage of burning—Fig. 29. One part of the discharge radiation occurs in the visible region of the spectrum as light, but some part is also emitted in the ultraviolet one. Within the visible region, the mercury spectrum consists of five principal lines (404.7, 435.8, 546.1, 577 and 579 nm), which result in greenish-blue light at efficacies of 30–65 lm/W, excluding ballast losses. A significant portion of the energy radiated by the mercury arc is in the UV region. Through the use of phosphor coatings on the inside surface of the outer envelope, some of this UV energy is converted to visible radiation. The most widely used lamps of this type are coated with a vanadate phosphor (4000 K) that emits long-wavelength radiation (orange–red). This improves efficacy and color rendering. This phosphor also is blended with others to produce cooler or warmer colors.

Clear mercury lamps generally have a CRI value of approximately 15, while phosphor coated of about 45–60.

Three phases could be distinguished after switching on the lamp to the normal stable operation: ignition, turn-on and stabilization.

The ignition is achieved by means of an auxiliary electrode, placed very close to the main electrode and connected to the other through a high value resistance

Fig. 29 High-pressure mercury lamps

(10–25 kΩ). When the lamp is turned on, a high-voltage gradient takes place between the main and the ignition electrodes, which ionizes the filling gas in this area as a luminescent discharge, the current being limited by a resistance. Luminescent discharge is then expanded through the discharge tube under the influence of the electric field between the two main electrodes. When luminescent discharge reaches the most distant electrode, current increases in a considerable way. As a result, the main electrodes are heated until the emission increases enough to allow the luminescent discharge to change completely to an arch discharge. The auxiliary electrode lacks another function in the process as a consequence of the high resistance connected serially to it. During the ignition stage, the lamp works as a low-pressure discharge (similar to that of a fluorescent lamp). The discharge fills the tube and gives it a bluish appearance. The current is limited by the resistance. The luminescent discharge then expands throughout the discharge tube under the influence of the electric field between the two main electrodes.

During the turn-on phase, the inert gas having been ionized, yet, the lamp does not burn in the desired way and does not offer its maximum production of light, until mercury present in the discharge tube is completely vaporized. This does not happen until a certain amount of time has elapsed, called turn-on time.

As a result of the arch discharge in the inert gas, a heating is generated providing a quick increase of temperature inside the discharge tube. This causes mercury gradual vaporization, increasing vapor pressure and concentrating discharge toward a narrow band along the axis of the tube. With an increase in pressure, radiated energy progressively concentrates along the spectral lines of greater wavelengths, and a small portion of continuous radiation is introduced. This way, light turns whiter. With time, the arc achieves a stabilization point, and it is said that the lamp reaches the total thermodynamic balance point. All mercury is then evaporated, and discharge occurs in non-saturated mercury vapor. The turn-on time, defined as the necessary time for the lamp since the ignition moment to reach 80% of its maximum production of light, is approximately four minutes.

The high-pressure mercury lamp, like most discharge lamps, has a "negative resistance" after emergence of the arc discharge and the electric current would theoretically increase to infinity, thus, it cannot work on its own in a circuit without an adequate ballast—usually electromagnetic choke to limit and stabilize the current through it.

2.7 *Metal-Halide Lamps*

Metal-halide lamps (Fig. 30) are similar in construction to mercury lamps; the major difference being that the metal-halide arc tube contains various metal halides in addition to the mercury and argon. When the lamp attains full operating temperature, the metal halides in the arc tube are partially vaporized. When the halide vapors approach the high-temperature central core of the discharge, they are dissociated into the halogen and the metals, with the metals radiating their spectrum. The use of metal halides inside the arc tube presents two advantages. First, metal halides are more volatile at arc tube operating temperatures than pure metals. This allows the introduction of metals with desirable emission properties into the arc at normal arc tube temperatures. Second, those metals that react chemically with the arc tube can be used in the form of a halide, which does not readily react with fused silica [16].

In order to obtain a desired spectrum, blends of metal halides are used. Two typical combinations of halides used are scandium and sodium iodides, and dysprosium, holmium and thulium rare-earth (RE) iodides. Other metals, such as tin, when introduced as halides, radiate as molecules, providing a continuous band spectra across the visible spectrum. The scandium-sodium system, for example, can produce CCTs between 2500 and 5000 K by varying the blend ratio and arc tube operating temperature. The rare-earth system, on the other hand, has a characteristic CCT of approximately 5400 K, which, when augmented by the inclusion of sodium iodide, may be lowered to 4300 K. A rare-earth system augmented with cesium and sodium iodides can achieve a CCT of 3000 K. The rare-earth system provides a

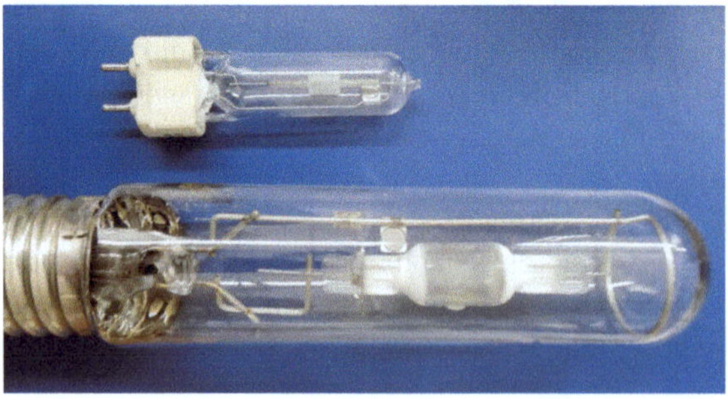

Fig. 30 Metal-halide lamps

somewhat higher general color rendering index than the scandium-sodium system; lithium iodide additions look promising for enhancing the color rendering properties of the scandium-sodium system.

Metal halogen lamps have the highest color rendering index among gas discharge lamps—CRI = 75–95. This allows them to be used in applications requiring high color rendering—football stadiums, sports halls, central city squares and streets, etc.

The efficacy of metal-halide lamps is greatly improved over mercury lamps. Commercially available metal-halide lamps have efficacies of 75–110 lm/W.

Due to metal halides and the absence of auxiliary electrodes, the ignition voltage for these lamps is high. The use of a starter or ignition device with voltage of 0.8–5 kV is needed. After the ignition, a period of about 5 min is required to reach the nominal luminous flux of the lamp.

When the lamp is turned-off, due to a great pressure in the burner, it is necessary to cool down between four and fifteen minutes before it is turned back on. Metal-halide lamps allow for immediate re-ignition with hot lamps (right after being turned-off), by using of specialized ignition device providing shock voltage of 35–60 kV.

2.8 Low- and High-Pressure Sodium Discharge Lamps

2.8.1 Low-Pressure Sodium Lamps

There exists a great similarity between the working of a low-pressure sodium lamp and a low-pressure mercury lamp (or a fluorescent one). However, while light in the latter is produced by transforming ultraviolet radiation of the mercury discharge into visible radiation, using fluorescent powder in the inner surface, visible radiation in the former is produced by direct discharge of sodium.

The discharge tube of a low-pressure sodium lamp is usually U-shaped and is located inside an empty tubular glass cover, with indium oxide coat on the inner surface. The empty part, together with the layer, which behaves as an infrared selective reflector, helps keep the discharge tube wall at an adequate working temperature. Such measurements are necessary for the sodium, which is deposited in slits of the glass when condensed, and it evaporates with a minimum heat loss. Due to this fact, the most luminous efficiency possible is achieved.

The neon gas inside the lamp is used to begin the discharge and to develop enough heat to vaporize the sodium. This responds for the red–orange luminescence during the first few working minutes. The metallic sodium is gradually evaporated, producing the characteristic monochromatic yellow light, with 589 and 589.6 nm lines in the spectrum (Fig. 31). The red color, initially produced by the neon discharge, is energetically suppressed during the working because sodium excitation and ionization potentials are much lower than those of neon.

The lamp reaches its luminous flux established in approximately 10 min. It will re-ignite immediately in case power supply is momentarily interrupted, since vapor pressure is quite low and the voltage applied enough to re-establish the arc.

Fig. 31 Low-pressure sodium lamp (top-down: off, during heating, working)

These lamps need an auxiliary equipment formed by a power supplier with an autotransformer or ballast and igniter with high impulse voltage. A compensation condenser is required.

The lamp has a luminous efficiency up to 200 lm/W and a long life.

Therefore, this lamp is applied to those places where color reproduction is of less importance and mainly where contrast recognition matters, for example: motorways, ports, open warehouses, beaches, etc. [21].

2.8.2 High-Pressure Sodium Lamps

Physically, high-pressure sodium lamps (Fig. 32) are quite different from low-pressure sodium lamps, due to the fact that vapor pressure is higher in them. This pressure also causes many other differences between the two lamps, including emitted light properties. Discharge tube in a high-pressure sodium lamp contains an excess

Fig. 32 High-pressure sodium lamps

of sodium to produce saturated vapor conditions when the lamp is working. Besides, it has an excess of mercury to provide a trimmer gas, xenon excluded, to ease ignition and limit heat conduction from the discharge arc to the tube wall. The discharge tube is housed in an empty glass cover. High-pressure sodium lamps radiate energy in a wider part of the visible spectrum, compared to the low-pressure sodium lamp; therefore, they offer a quite acceptable color reproduction—CRI = 25.

The radiation spectrum close to the maximum sensitivity of the human eye determines the relatively high efficacy of the high-pressure sodium lamp—70 to 130 lm/W. Their application is mainly in street lighting where high-energy efficiency is more important than the color rendering level.

Nominal values of luminous flux and electric power are reached 5 min after ignition. When a lamp is turned-off, due to a great pressure of the burner, it needs to cool down between four and ten minutes before turning it back on.

2.9 *Induction Lamps*

The main feature of induction lamps is that they have no internal electrodes, so they are often called electrodeless lamps. Electrodeless lamps use an electromagnetic (EM) field, instead of an electric current passing through electrodes, to excite the gas in a bulb or tube.

The most vulnerable parts of all the discharge lamps described above are the electrodes. Their wear determines the life of the light source. The absence of electrodes in the induction lamps ensures a significantly longer life of about 60,000 h compared to the lamps with electrodes.

The most common designs of induction lamps are shown in Figs. 33 and 34.

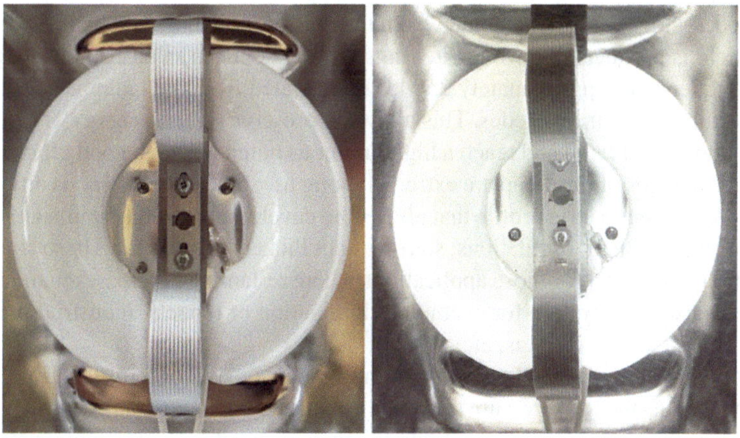

Fig. 33 Electrodeless fluorescent lamp

Fig. 34 Induction lamp

Figure 33 shows the so-called electrodeless fluorescent lamp. The discharge in this lamp does not begin and end in two electrodes like in a conventional fluorescent lamp. The shape of close ring of the glass of the lamp allows to have a discharge without electrodes, since energy is supplied from the outside by a magnetic field. Such magnetic field is produced by two ferrite rings. The system has an electronic equipment (at a frequency of approximately 250 kHz) separated from the lamp besides a fluorescent tube without electrodes. This allows to preserve optimal energy of discharge in the fluorescent lamp and reach a high luminous potency with a good efficacy. The main advantages of this lamp are extremely long life: 60,000 h; luminous efficacy of 80 lm/W; low geometric profile that allows the development of flat luminaires; comfortable light without oscillations; start without flickers or sparkles. These lamps are essentially indicated for those applications where re-lamping increases maintenance expenses excessively, like for example, illumination of tunnels, industrial premises with difficult access ceilings, etc.

Figure 34 shows typical induction lamp which consists of a discharge recipient which contains the low-pressure gas and a voltage coupler (antenna). Such a potency coupler, composed by a ferrite cylindrical nucleus, creates an electromagnetic field within the discharge recipient inducing an electrical current in the gas generating its ionization. Enough energy to begin and maintain discharge is supplied to the antenna by a high-frequency generator (2.65 MHz) by means of a coaxial cable of a determined length, since it forms part of the oscillating circuit.

The main advantages of these lamps are extremely long duration: 60,000 h; luminous efficacy between 65 and 81 lm/W; instantaneous ignition free of flickers and stroboscopic effects; light for a great visual comfort. These lamps are used for many general and special lighting applications, mainly to reduce maintenance expenses, like in public buildings, outdoor public lighting, industrial applications, etc. [21].

2.10 Light-Emitting Diodes (LEDs)

LEDs (Fig. 35) are the most modern light sources with significant advantages over conventional lamps that will allow them to become a major light source for a number of applications in the near future.

The historic beginning of the LED technology starts more than 100 years ago:

1907—Henry Round first describes yellow, green and orange solid diode lighting;

1923—Oleg Losev discovers semiconductor electroluminescence;

1961—Robert Bayard and Harry Pitman patented the infrared LED technology;

1962—Nick Holonyak developed the world's first practical LED with visible red light;

1972—George Kraford invented the first yellow LED and improved the brightness of the red and red–orange LEDs 10 times;

1994—Shuji Nakamura introduces the first high-brightness blue LED, developed jointly with Isama Akasaki and Hiroshi Amano. For the discovery of the cheap blue LED, the three scientists were awarded the Nobel Prize in Physics in 2014.

As can be seen from the historical development, LEDs emitting different, almost monochromatic colors with the absence a blue LED have been consistently created, and therefore it was not possible to obtain a good and continuous spectrum and white

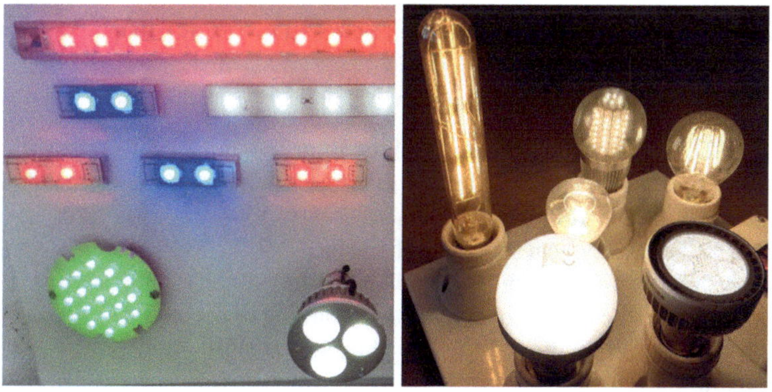

Fig. 35 Light-emitting diodes

Fig. 36 Near monochromatic colors of primary LEDs [22]

Fig. 37 Main light-emitting diode technologies [4]

light. Following the discovery of the powerful blue LED, the possible technical solutions and combinations of monochromatic diodes to produce white light significantly expanded—Fig. 36.

White light LEDs are made using two principal methods: either mixing light from multiple LEDs of various colors or using a phosphor to convert some of the light to other colors. At present time, the following major technical solutions have been established to obtain white light through LEDs (Fig. 37):

- Mixing of the light form red, green and blue LEDs—RGB LED;
- Blue LED chip coated by yellow phosphor;
- Ultraviolet LED covered by red, green and blue/yellow phosphors.

Blue LED + Yellow phosphor technology is most common at this time, due to its highest and continues to increase luminous efficacy achieved by a white light source of 100–230 lm/W and color rendering index—80 to 85.

Figure 38 shows a typical spectrum of Blue LED + Yellow phosphor technology LED. Blue light is directly emitted by the GaN-based LED luminescence (peak at about 465 nm) and the more broadband Stokes-shifted light emitted by the Ce^{3+}:YAG (Yttrium aluminum garnet—$Y_3Al_5O_{12}$) phosphorescence, which emits at roughly 500–700 nm.

RGB technology LEDs use mixing of the light from three color LEDs—Fig. 38, usually embedded in a common housing. The possibility of separately adjusting the

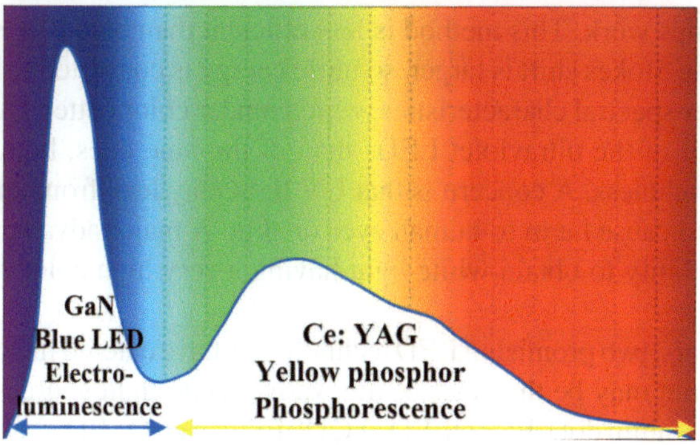

Fig. 38 Spectrum formation of Blue LED + Yellow phosphor technology LEDs

intensity of the individual sub-diodes leads to the main advantage of RGB LEDs—the ability to adjust the color of the light.

Figure 39 shows a typical spectrum of RGB LED. The white light obtained from the relatively narrow light emission spectra of the red, green and blue LEDs has a comparatively lower than of Blue LED + Yellow phosphor technology LEDs color rendering index of 25–65.

In white light in RGB LEDs, except the powerful blue LED, the less efficient green and red are used. This leads to a slightly lower efficacy of RGB LEDs than of the Blue LED + Yellow phosphor technology LEDs of about 70–80 lm/W.

The technology with ultraviolet LED covered by red, green and blue/yellow phosphors (Fig. 36) is not very common in practice and in the market at present. White light is made by coating near-ultraviolet (NUV) LEDs with a mixture of high-efficiency europium-based phosphors that emit red and blue, plus copper and aluminum-doped zinc sulfide (ZnS:Cu, Al) that emits green. This is a method analogous to the way

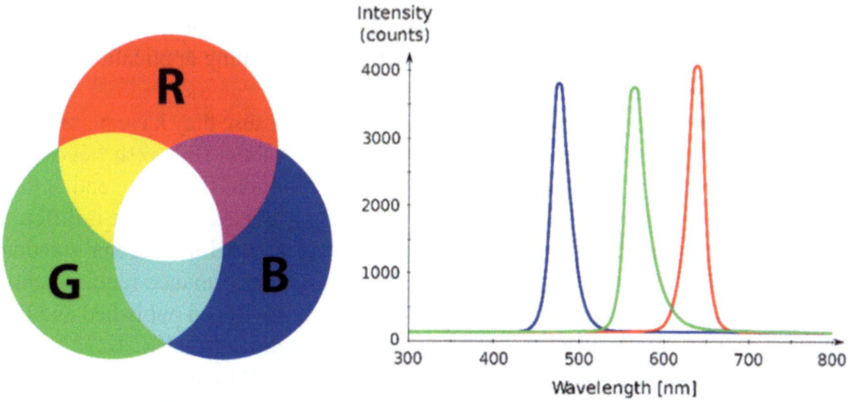

Fig. 39 White color [23] and spectrum [24] formation of RGB technology LEDs

fluorescent lamps work. This method is less efficient than blue LEDs with YAG:Ce phosphor, as the Stokes shift is larger, so more energy is converted to heat, but yields light with better spectral characteristics, which render color better. Due to the higher radiative output of the ultraviolet LEDs than of the blue ones, both methods offer comparable brightness. A concern is that UV light may leak from a malfunctioning light source and cause harm to human eyes or skin. A major advantage of this technology is the ability to obtain white light having a very high color rendering index of 95–99.

In recent years, two promising LED technologies have entered into practice, which in the near future may be the basis of innovative white light sources: organic LED (OLED) and quantum dot LED (QLED or QDLED). In an OLED, the electroluminescent material composing the emissive layer of the diode is an organic compound. The organic material is electrically conductive due to the delocalization of pi electrons caused by conjugation over all or part of the molecule, and the material therefore functions as an organic semiconductor [25]. The organic materials can be small organic molecules in a crystalline phase, or polymers. The potential advantages of OLEDs include thin, low-cost displays with a low driving voltage, wide viewing angle and high contrast and color gamut. Polymer LEDs have the added benefit of printable and flexible displays. OLEDs have been used to make visual displays for portable electronic devices such as cell phones and tablets, while possible future uses include televisions and lighting [26, 27].

Specific parts of the QLEDs are semiconductor nanocrystals with optical properties that let their emission color be tuned from the visible into the infrared spectrum [28]. This allows quantum dot LEDs to create almost any color on the CIE diagram. This provides more color options and better color rendering than white LEDs since the emission spectrum is much narrower, characteristic of quantum confined states. There are two types of schemes for QD excitation. One uses photo excitation with a primary light source LED (typically blue or UV LEDs are used). The other is direct electrical excitation first demonstrated in 1994 [29].

After optimization of still booming LED technology, it is expected to reach the theoretical maximum luminous efficacy of about 270 lm/W of light source with good color rendering (CRI > 80), and even higher, for lighting applications that do not require so good color rendering.

One of the benefits of the LED is its long lifetime, because they have no movable parts or filaments that may break. Technical life of LED is about 100,000 h, but usually they are not fully utilized due to the reduction of the luminous flux at the end of their life. The trend of reduction of the luminous flux of the LED over the time is different depending on various design and operational factors (Fig. 40). International standard IEC 62717:2014 "LED modules for general lighting—Performance requirements" on how lifetime should be declared on LED luminaires have been published and IEC 62722-2-1:2014 "Particular requirements for LED luminaires." IEC 62722 states both test method and minimum required time for testing LED lifetimes. The minimum test time is 6000 h where the luminous flux is recorded every 1000 h. These values are extrapolated using a method stated in IES TM21.

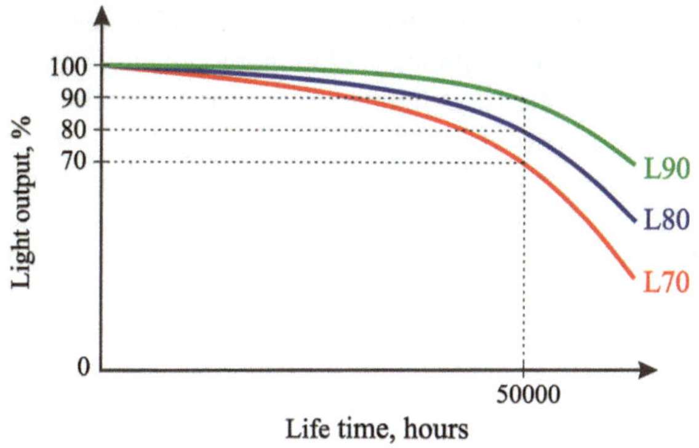

Fig. 40 Reducing the light output of the LEDs over the time

The lifetime for the LED module and the driver should be declared separately. If the lifetime for the driver is shorter than that of the module, a driver replacement may be necessary before the luminaire's lifecycle is completed. This means that there cannot be a single figure declaring the total luminaire lifetime. A helpful metric for "median useful life" has been introduced in IEC 62717. This is the time elapsed until 50% of the LED luminaires in use reaches the stated light output, normally L70, L80 or L90, showing a reduction of 70, 80 and 90%, respectively, relative to the initial luminous flux—Fig. 40.

The operating temperature is the main parameter on which the life and efficiency of the LEDs depend. According to technical data of leading manufacturers and laboratory measurements, when the temperature is increased by 10 °C, the life of the LED is reduced twice, and the light output by 3–8%. The thermal imaging measurements (Fig. 41) of the heating process show that the heating system's cooling duration for the tested LED lighting models varies within 30–120 min. The reached set temperatures are from 35 to 88 °C, with a percentage increase of 30–270% from the initial ambient temperature of 20 °C. Lower temperature overheating is for models with aluminum radiator housings. The relative luminous flux when reaching the set mode

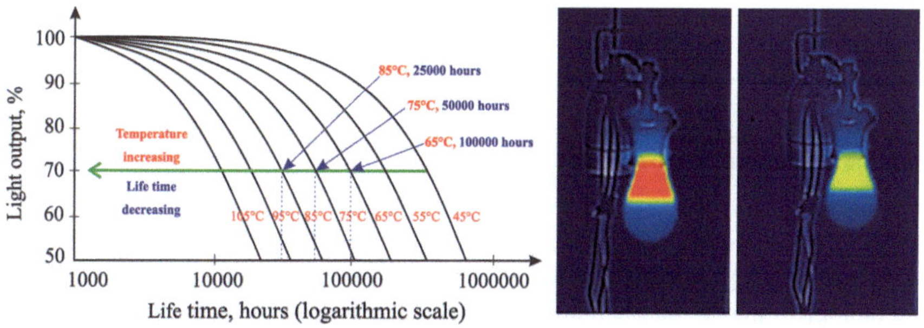

Fig. 41 Temperature dependence of the lifetime of the light output of the LEDs [30]

for the tested LEDs decreases from the original by 4–17%, the largest value being for a relatively powerful LED without an aluminum radiator [30].

The power supply to the LED light sources is not directly carried out by the mains voltage and special drivers need to be developed, based on different electronic circuits. LED luminaires drivers most frequently use the electronic circuit with pulse width modulation (PWM) driver or with bridge rectifier driver with capacitive divider (BRC). With PWM drivers, a stable external current and corresponding light output of the LEDs are guaranteed in a wide range of supply voltage variations. For BRC drivers, the electrical current and the light output of the LEDs are proportional to the supply voltage, which allows dimming the LED light flux [31]. This functionality is also available with PWM drivers by complicating the electronic circuit. The quality of the electronic LED drivers is of great importance for their reliable operation, as the life of the electronic components used in them, especially the capacitors, is commensurate and often less than that of the LEDs themselves. Ensuring a stable low-temperature operating mode for drivers, LEDs are equally important to ensure the long life expectancy of LED luminaires.

2.11 Light Sources Main Parameters Comparison

Table 4 summarizes the main technical and operational parameters of some of the most commonly used in practice light sources. Some of the parameters are shown as value ranges, as they vary considerably for individual constructions of the given light source.

A graphical representation of the increase through technical improvements of the luminous efficacy of the light sources from their invention over the years is given in Fig. 42.

The comparison of the trend of increasing of the luminous efficacy of the different type of lamps and that of the LED show that LEDs represent a kind of innovative revolution in lighting technology that has not yet been completed, but it is now recognized as the most energy-efficient light source invented.

3 Photobiological Safety of Light Sources

3.1 Circadian Cycle

The impact of light and lighting on the human body is still often underestimated. Light is vital for many physiological and psychological processes. Independent studies of the biological impact on the human organism have established that there is a direct connection between daylight and our sense of well-being. Daylight has a positive effect on the human organism and is responsible for important biological processes.

Table 4 Comparison of the main parameters of most used in practice light sources

Parameter	Light source							
	Incandescent	Halogen	High-pressure sodium	Low-pressure sodium	Metal halide	Linear fluorescent	Compact fluorescent	Light-emitting diode
Life (h)	1000	3000	16,000–24,000	10,000–18,000	10,000–16,000	10,000–15,000	6000–12,000	25,000[a]–100,000
Luminous efficacy (lm/W)	9–13	15–25	70–140	140–170	70–110	50–90	40–70	80–180[b]
Color rendering index CRI	98	100	25	0	75–95	80–95	80–90	70–85
Correlated color temperature CCT	2700	3000	2100	1700	2700–6500	2700–6500[c]	2700–6500[c]	2700–6500[c]

[a]It is highly dependent on the temperature mode (cooling) of the LED module
[b]Luminous efficacy of LEDs continues to increase by approximately 10 lm/W per year during the last 10 years
[c]It depends on the chemical composition of the phosphors used

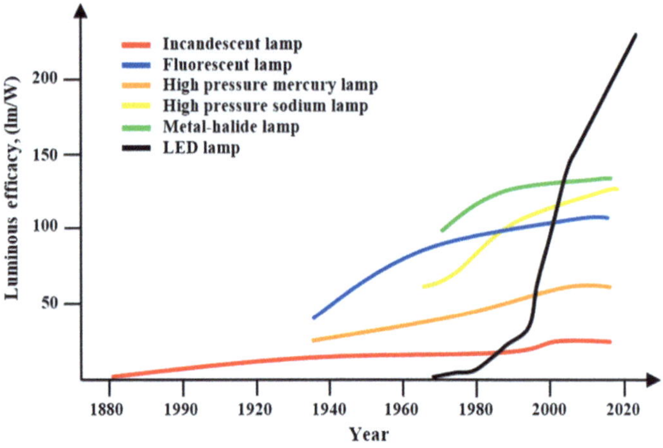

Fig. 42 Increasing of the luminous efficacy of the light sources from their invention over the years

Lack of daylight is often associated with sleep problems. Our sleep/wake rhythm (the circadian system) is controlled by the intensity and spectral composition—especially the blue content—of the light around us. Without daylight, the production of melatonin and cortisol, the hormones that basically regulate sleep/wake rhythm, is thrown out of order.

The circadian rhythm is our 24-h biological clock, which is significantly governed by the hormone melatonin. Produced in the pineal gland in the center of the brain, melatonin regulates many organic processes via signals defined by its concentration in the blood. Our capacity for action is directly related to the amount of melatonin in circulation. If the level is high, we start to feel tired. A high level of cortisol, on the other hand, provides the basis for our waking phases—the phases in which we can be active and productive. Melatonin and cortisol production is directly controlled by the amount of light falling on the retina, regardless of the visual process. A lot of light—especially in the short-wave spectral range—causes the cortisol level to rise and suppresses production of melatonin. That is why we wake up when it is light and feel tired when it gets dark [32].

Light color temperature and illuminance play an important role in the design of biologically effecting lighting indoors. The blue short-wave content of light, with a color temperature greater than 5300 K, has a particularly activating effect on the human body during the day. New fluorescent and LED lamps can extend the short-wave range up to 12,000 K. For biological activation, chronobiologists recommend brief exposure (preferably in the morning) to an illuminance well above that required by standards. In the evening, warm light colors less than 3300 K and a significantly lower lighting level should be used to prepare the body slowly for night.

3.2 Blue Light, Actinic UV and Thermal Hazards

Besides the effects of the circadian cycle, the light in the optical part of the spectrum has a direct effect on the human eye and body, and in some cases it could be very dangerous for them. Recognition of these hazards is of great importance for the correct choice of light source type and its operating mode in a given lighting system. The international standard IEC 62471:2006 “Photobiological Safety of Lamps and Lamp Systems” indicates a new framework for the assessment of the photobiological safety of light sources. A measurement methodology and exposure limit values are given in the consideration of the six dangers to the eye and skin for an exposure duration of up to 8 h, taken as a working day—Table 5. Sources are classified into the following four groups according to hazard, based on permissible exposure time before hazard exceeded—Table 6.

There are four weighting functions used in this standard, three relating to specific photobiological hazards and the fourth, photopic eye response, used in the evaluation of the luminous efficacy of the source in question—Fig. 43.

In all measurements of irradiance, the hazard exposure limit value is divided by the measured irradiance to determine the permissible exposure time, which is then compared with the limits of each risk group. In the case of measurements of radiance, the field of view corresponding to the maximum permissible time to exposure of a

Table 5 Type of hazards according to IEC 62471

Hazard	Wavelength range (nm)	Bioeffect	
		Skin	Eye
Actinic UV	200–400	Erythema elastosis	Cornea—Photokeratitis Lens—Cataractogenesis
Near UV (UVA)	315–400	–	Lens—Cataractogenesis
Retinal blue Light	300–700	–	Retina—Photoretinitis
Retinal thermal	380–1400	–	Retinal burn
Infrared radiation	780–3000	–	Corneal burn Lens—Cataractogenesis
Thermal	380–3000	Skin burn	–

Table 6 Risk groups of light sources according to IEC 62471

Risk group	Philosophical basis
Exempt	No photobiological hazard
RG1	No photobiological hazard under normal behavioral limitation
RG2	Does not pose a hazard due to aversion response to bright light or thermal discomfort
RG3	Hazardous even for momentary exposure

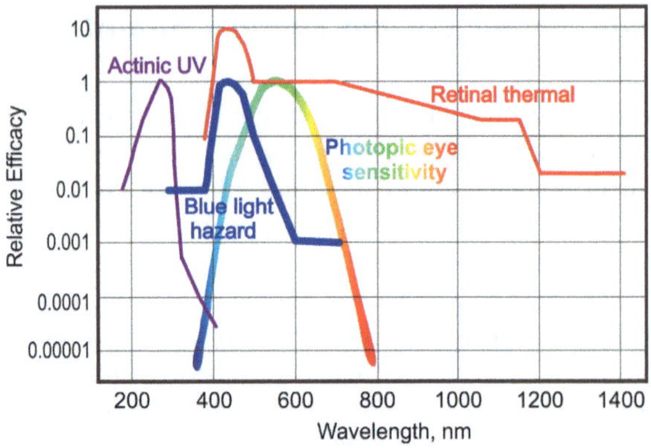

Fig. 43 Photopic eye sensitivity and actinic UV, blue light and thermal hazards spectrums

given class is used as the measurement condition, and pass/fail of the criteria for the given class is determined.

The light source should be evaluated at a distance that produces an illuminance of 500 lx. In practice, the measurement distance may be a number of meters. In all other cases, the measurement should be performed at 200 mm. The measurement of radiance in particular represents a challenge in that within a given field of view, a single device may not pose any hazards, while multiple sources may do so. A further complication is the fact that there is little correspondence between measurements of a bare source and measurements of the same source in a luminaire, making measurements in both instances a potential requirement [33].

In the normative documents, as well as in the technical parameters of the lighting manufacturers, the integrated color characteristics are given: correlated color temperature (CCT) and total color rendering index CRI. Changing the CCT and the CRI in space affects the uniformity of color distribution and quality perception at the various points of the work surface and viewpoints of the observer. At present, there is relatively little information and research on the spatial variation of the color characteristics of light sources and luminaires. Measurements of such characteristics in the Laboratory of Lighting at the Technical University of Gabrovo show the difference of the CCT (Fig. 44) and the CRI (Fig. 45) over the vertical angles of the light distribution curves for several light sources [34].

Table 7 presents the maximum and minimum values of the correlated color temperature and the maximum ΔCCT_{max} and minimum ΔCCT_{min} deviations from the total value of CCT [34].

Table 8 presents the maximum CRI_{max} and minimum CRI_{min} values of the color rendering index and the maximum ΔCRI_{max} and minimum ΔCRI_{min} deviations from the total value of CRI [34].

The analysis of the measurement results of the spatial variation of the color characteristics of LED light sources leads to the conclusion that it can be significant for the higher vertical angles above ±82.5°. Especially for the Blue LED + Yellow

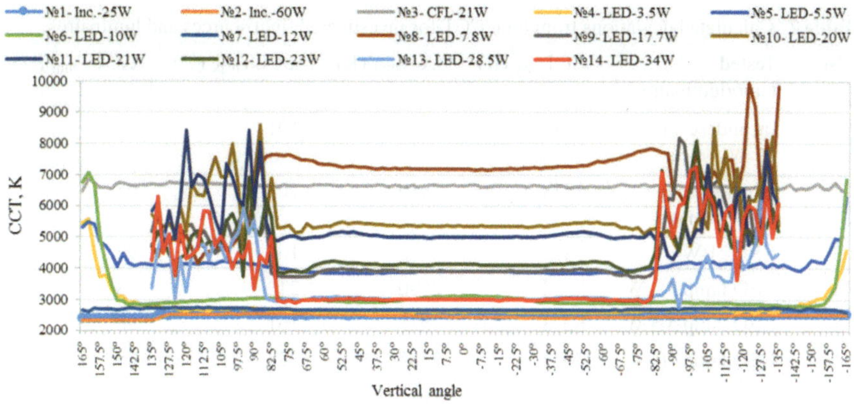

Fig. 44 Spatial variation of the correlated color temperature of the test light sources

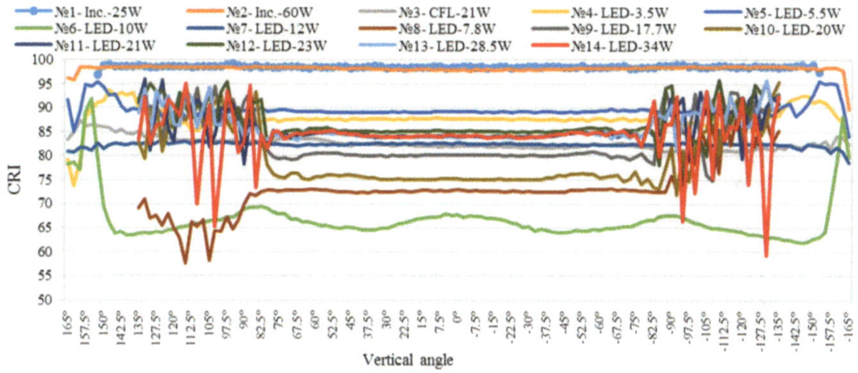

Fig. 45 Spatial variation of the color rendering index of the test light sources

phosphor technology LEDs and transparent lenses, greater differences have been established. This can be explained by an insufficiently thick phosphor coating or even a lack of such coverage in some areas of this spatial range of the LED [34].

The analysis of the relative efficacy and coverage of the spectral curves (Fig. 43) of the blue light hazard ($\lambda_{max} = 435$ nm) and of the human eye photopic sensitivity ($\lambda_{max} = 555$ nm) indicate that the most important attention should be directed to the 380–500 nm range of the visible spectrum. Taking into account the influence of blue light on the circadian cycle, this range of high-energy visible (HEV) light can be divided into two sub-bands: harmful "blue–violet" ($\lambda = 380$–420 nm) and useful "blue–turquoise" ($\lambda = 420$–500 nm) light. Prolonged exposure to "blue–violet" HEV light can cause damage to the retina, the most vulnerable tissue of the eye, accelerating the development of age-related macular degeneration (AMD).

Table 7 Calculated deviations from total CCT for investigated light sources and luminaires

No.	Tested lamp/luminaire	CCT_{max} (K)	CCT_{min} (K)	ΔCCT_{max} (%)	ΔCCT_{min} (%)
1	Incandescent 25 W—clear	2548	2501	101	99
2	Incandescent 60 W—matte	2548	2410	100	94
3	CFL 21 W	6879	6468	104	97
4	LED lamp 3.5 W	5578	2590	212	98
5	LED lamp 5.5 W	6278	3835	161	98
6	LED lamp 10 W	7077	2785	224	88
7	LED lamp12 W	2753	2613	103	98
8	LED plafon 7.8 W	9952	4169	139	58
9	LED luminaire 17.7 W	8195	3720	210	95
10	LED luminaire 20 W	8598	4575	161	86
11	LED luminaire 21 W	8427	4312	169	86
12	LED luminaire 23 W	8096	3742	196	91
13	LED luminaire 28.5 W	6188	2757	206	92
14	LED luminaire 34 W	7256	2907	241	97

Exposure to blue light, hazard light is significantly more dangerous for children and young people due to the greater spectral transmittance of the eye in this range of the spectrum.

Regarding photobiological safety of light sources, it can be concluded that it is most important to choose correctly the spectral composition of their light depending on the application, the time of the day and the duration of illumination.

In almost all recent light sources, the presence of more radiation in the blue light hazard range is characterized by a higher color temperature. An analysis of the spectral power distributions of the most widely used Blue LED + Yellow phosphor technology LED technology having different color temperatures indicates that cool white LEDs have significantly higher emissions in the blue light hazard range than the warm white LEDs (Fig. 46).

A recommendation can be made for most widely used nowadays Blue LED + Yellow phosphor LED technology, in order to reduce blue light hazard for indoor lighting applications with continuous human presence, to be used LED lamps and luminaires with correlated color temperature CCT $\leq$ 4500 K.

Table 8 Calculated deviations from total CCT for investigated light sources and luminaires

No.	Tested lamp/luminaire	CRI_{max}	CRI_{min}	ΔCRI_{max} (%)	ΔCRI_{min} (%)
1	Incandescent 25 W—clear	99.1	98.1	100.5	99
2	Incandescent 60 W—matte	98.7	89.8	100.2	91
3	CFL 21 W	87.3	80.8	106	98
4	LED lamp 3.5 W	93.6	73.7	107	84
5	LED lamp 5.5 W	95.6	84.2	107	95
6	LED lamp 10 W	91.8	61.9	131	88
7	LED lamp12 W	83.4	78.6	101	95
8	LED plafon 7.8 W	90.5	57.6	124	79
9	LED luminaire 17.7 W	94.5	75	118	94
10	LED luminaire 20 W	95.4	71.9	127	96
11	LED luminaire 21 W	96	78.4	114	93
12	LED luminaire 23 W	95.8	74	112	95
13	LED luminaire 28.5 W	95.8	81.2	114	97
14	LED luminaire 34 W	95	59.3	113	70

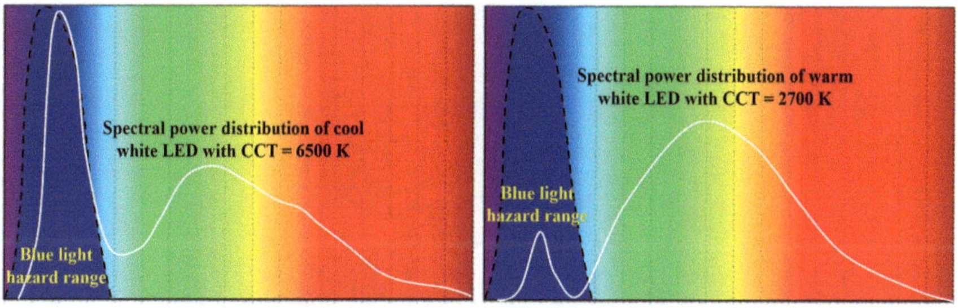

Fig. 46 Spectral power distributions of cool and warm white LED and blue light hazard range in the visible spectrum

4 PV–LED Systems

Structure and principle of work

At the base of the PV–LED systems is the combination of the advantages of two innovative technologies—photovoltaics (PV) and light-emitting diodes (LED) . Both main elements of this type of solar lighting systems are solid state and long life-time, making PV–LED systems very simple and reliable. The application of the two technologies is among the most environmentally friendly options for electric power supply of lighting systems. With the advancement of photovoltaic module manufacturing technology, it is possible to have a variety of sizes, electrical and operational parameters to allow installation in various small and large devices for indoor and outdoor installation.

Among the numerous applications of photovoltaic electricity, the most appealing is certainly for low-power electrical appliances. Among those low-power appliances, domestic lighting is certainly one of the most demanded. Lighting technology has known an important evolution from the carbon arc to the filament light bulb, the fluorescent tubes to the LEDs that are top of the line today. LEDs can cover a number of applications that were not covered before, especially because of the geometry and color range they can reach and their power efficiency and life span which are continuously increasing [35].

Both the PV and LED have low weight, resistance to vibrations and DC voltage operation, enabling the easy creation of portable autonomous PV–LED devices. In recent years, PV–LED lighting systems are widely applied from PV–LED street lighting (Fig. 47) to PV–LED keychain (Fig. 48).

Figure 49 shows a block diagram of the main components of a PV–LED system: PV panel (module), storage battery (accumulator), solar controller and LED luminaire, including itself LED driver and LED lamps (sources). Since both PV modules and LED lamps are low-voltage DC devices, it is more appropriate for PV–LED

Fig. 47 PV–LED outdoor lighting

Fig. 48 LED keychain torch with solar photovoltaic charge

Fig. 49 Block diagram of PV–LED system

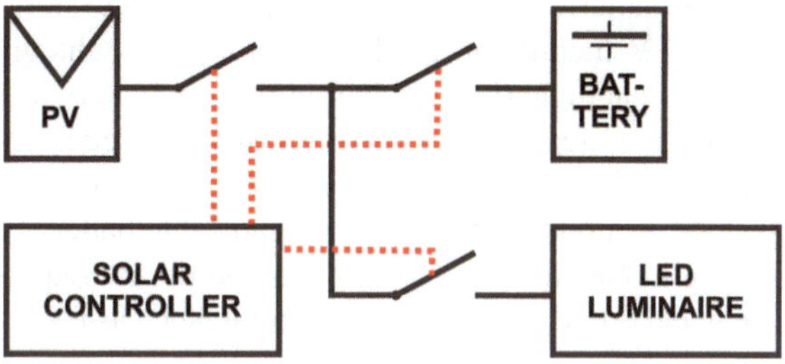

systems to use a DC LED drivers, which are simpler and more energy efficient than the AC LED drivers. Moreover, this way is avoided using of DC/AC inverter, and therefore, the entire PV–LED system is cheaper and more energy efficient.

The main operating modes of the outdoor lighting PV–LED system elements during day and night are shown in Figs. 50 and 51. The generated electric energy from PV modules is stored by rechargeable batteries during the day (Fig. 50), if the

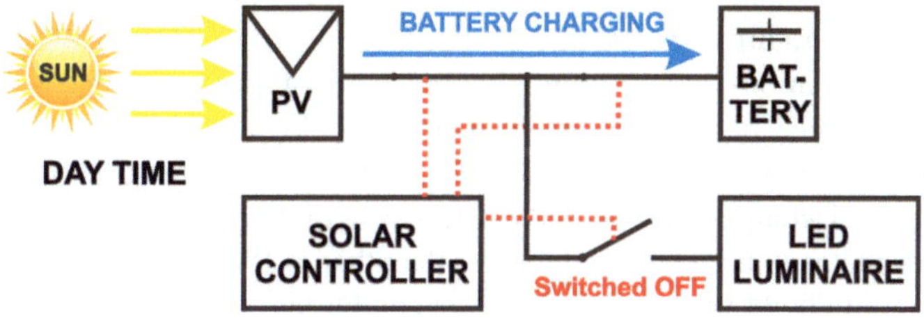

Fig. 50 Operating mode of PV–LED system during the day sunlight time

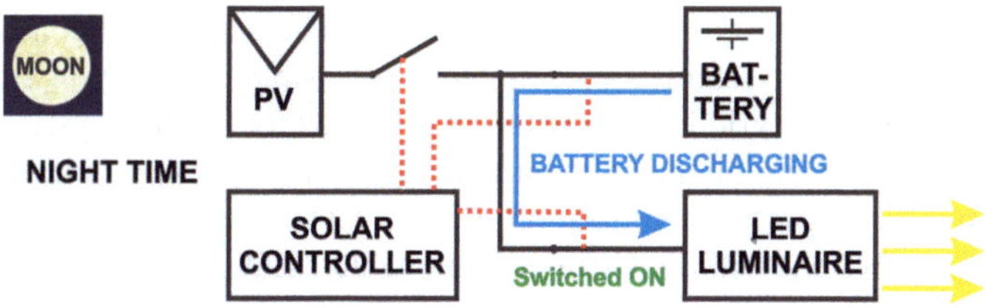

Fig. 51 Operating mode of PV–LED system during the night time

solar irradiation is available, and LED luminaires lighted by the batteries at night (Fig. 51).

The types, parameters and characteristics of PV modules and LED lamps are discussed in detail in the relevant parts of this book, so in this section are presented the other elements of PV–LED system: batteries and solar controllers.

Storage batteries

Storage batteries (accumulators) are designed to store the electrical energy produced by the PV module in the daytime when it is charging, and it gives it to the LED luminaire at night (it is discharging). Each battery technology requires an optimized charging strategy to maximize life span and capacity. A valve-regulated lead–acid battery (VRLA), more commonly known as a sealed battery or maintenance-free battery, is a type of lead–acid rechargeable battery. Due to their construction, they can be mounted in any orientation and do not require constant maintenance. They are widely used in stand-alone systems, such as PV–LED systems, and similar roles, where large amounts of storage are needed at a lower cost than other low-maintenance technologies like lithium–ion. There are two primary types of VRLA batteries, gel cells and absorbed glass mat (AGM). Gel cells add silica dust to the electrolyte, forming a thick putty-like gel. These are sometimes referred to as "silicone batteries," AGM batteries feature fiberglass mesh between the battery plates which serves to contain the electrolyte. Lead–acid batteries with absorption glass mat (AGM) have immobilized electrolyte and offer high capacity at a lower cost per amp-hour compared to other chemistries. All available PV panel energy is stored in the battery until it reaches its maximum charge voltage, at which point trickle charging begins. The battery must be maintained at the proper float voltage, which should be adjusted with a negative temperature coefficient.

For PV applications, most commonly used battery types and with the most appropriate performance parameters are lead–acid, Ni–Cd and lithium–ion. Their life expressed in no of cycles is about 500–1500 [36]. It means that on a normal operated daily charge–discharge cycle of a PV–LED system, they will work for 3–5 years. This life expectancy is significantly lower than that of PV modules and LED lamps, which is about 20–25 years under the same conditions.

The short lifetime, combined with their high cost, turns the batteries into the weakest part of the PV–LED systems. These are the main reasons for the poor economic indicators of stand-alone PV–LED systems in some countries like Bulgaria, where the price of energy from the grid is relatively low. The design and construction of autonomous PV–LED systems are economically feasible primarily in areas where no conventional power grid is built, but there is a need for lighting system.

The feature for dimming the light output of LED luminaires provides an interesting opportunity for further optimization of PV–LED systems. A Photovoltaic-Fed LED Lighting System with SOC-Based Dimmable LED Load allows minimizing the components size. Effective energy management is allowed by battery-driven load dimming, thus avoiding the use of big solar panels and big batteries which, in most cases, provide only marginal benefits in terms of energy productivity. The energy productivity and the quality of the service can be modulated by a dynamic change of the type of dimming criterion with respect to the battery SOC [37].

Depending on the functional capabilities, the solar controllers can be connected to a specialized hardware or PC with software for monitoring the process of operation of a PV–LED system: visualization of the charging and discharging process of the battery, the input voltage, current and power PV modules, amount of charge to the battery, consumed electrical energy from the load.

Continuous improvement in PV–LED system technology causes these systems to be preferred when existing electrical infrastructure is remote or non-existing—for roads, tunnels, car parks, parks and trails. The return on investment to install a PV–LED lighting system can be achieved by avoiding the cost of channeling, buying cables and other electrical and utility costs associated with installing a new lighting system powered by the power grid.

Detailed information on different main components of the PV–LED systems can be found in Chapter "Photovoltaic Solar Energy Conversion," photovoltaic solar energy conversion of the book.

References

1. Alexandre Gondran [CC BY-SA 4.0]. https://commons.wikimedia.org/wiki/File:Interference_electrons_double-slit_at_10cm.png
2. Wolfmankurd [CC BY-SA 3.0]. https://commons.wikimedia.org/wiki/File:Photoelectric_effect.svg
3. Philip Ronan, Gringer [CC BY-SA 3.0]. https://commons.wikimedia.org/wiki/File:EM_spectrumrevised.png
4. Zhu D, Humphreys CJ (2016) Solid-state lighting based on light emitting diode technology. In: Al-Amri M, El-Gomati M, Zubairy M (eds) Optics in our time. Springer, Cham
5. Tsankov P, Platikanov S (2013) Guide for laboratory exercises on lighting and installation equipment. Vasil Aprilov University Publishing House—Gabrovo, ISBN: 978-954-683-506-2 (in Bulgarian)
6. https://commons.wikimedia.org/wiki/File:PlanckianLocus.png,en:User:PAR [Public domain]
7. Adoniscik [CC BY-SA 3.0]. https://commons.wikimedia.org/wiki/File:CIE_CRI_TCS_SPDs.svg
8. Flexfire LEDs, inc., Brenton Patrick Mauriello [CC BY-SA 4.0]. https://commons.wikimedia.org/wiki/File:LED_strip_lighting_Correlated_Color_Temperatures_CCTs.png
9. https://commons.wikimedia.org/wiki/File:Kruithof_curve_2.svg, Hankwang [Public domain]
10. https://www.wikilectures.eu/w/Types_of_Light_Sources. [CC BY-SA 3.0]
11. Vasilev N (1974) Industrial lighting. Technika, Sofia (in Bulgarian)
12. https://commons.wikimedia.org/wiki/File:Planck_law_log_log_scale.png. HighTemplar [Public domain]
13. SiriusB [CC0]. https://commons.wikimedia.org/wiki/File:Blackbody_efficacy_1000-16000K.svg
14. Brighterorange [CC BY-SA 3.0]. https://commons.wikimedia.org/wiki/File:Bohr_atom_model_English.svg
15. Julie Gagnon http://www.umop.net/spctelem (C) 2007, 2013. CC-BY-SA 4.0
16. IESNA Lighting Handbook, Illuminating Engineering Society of North America (2000)
17. http://en.wikipedia.org/wiki/User:Iantresman [CC BY 2.5]. https://commons.wikimedia.org/wiki/File:Electric_glow_discharge_schematic.png

18. User:S-kei [CC BY-SA 2.5]. https://commons.wikimedia.org/wiki/File:PnJunction-LED-E.svg
19. https://en.wikipedia.org/wiki/Timeline_of_lighting_technology
20. Prilepkova [CC BY-SA 3.0], https://commons.wikimedia.org/wiki/File:Hal._proces.png
21. Lighting Engineering, INDALUX (2002)
22. https://cdn.pixabay.com/photo/2012/04/10/16/54/rainbow-26389_1280.png, Pixabay License, Free for commercial use
23. SharkD at English WikipediaLater versions were uploaded by Jacobolus at en.wikipedia. [Public domain]. https://commons.wikimedia.org/wiki/File:AdditiveColor.svg
24. Tijl Schepens [CC BY-SA 4.0]. https://commons.wikimedia.org/wiki/File:RGB_LED_Spectrum.svg
25. Burroughes JH, Bradley DC, Brown AR, Marks RN, MacKay K, Friend RH, Burns PL, Holmes AB (1990) Light-emitting diodes based on conjugated polymers. Nature 347(6293):539–541
26. Kho M-J, Javed T, Mark R, Maier E, David C (2008) OLED solid state lighting. Kodak European Research, Cambridge Science Park, Cambridge
27. Bardsley JN (2004) International OLED technology roadmap. IEEE J Sel Top Quantum Electron 10(1):3–4
28. Neidhardt H, Wilhelm L, Zagrebnov VA (2015) A new model for quantum dot light emitting-absorbing devices: proofs and supplements, nanosystems: physics, Chemistry. Mathematics 6(1):6–45
29. Colvin VL, Schlamp MC, Alivisatos AP (1994) Light-emitting diodes made from cadmium selenide nanocrystals and a semiconducting polymer. Nature 370(6488):354–357
30. Tsankov P, Yovchev M (2015) Thermal imaging of the process of heating up to the steady state mode of LED luminaires. In: Proceedings of Unitech 2015, TU of Gabrovo, pp I 78–83, ISSN 1313-230X (in Bulgarian)
31. Tsankov P, Yovchev M (2018) Study of the electrical characteristics of light- emitting diode luminaires at amendment of the supply voltage. Balkan Light, ISBN 978-1-5386-6730-9/18/$31.00 ©2018 IEEE (2018)
32. licht.wissen 07 Light as a Factor in Health, licht.de, ISBN 978-3-926193-83-4 (2013)
33. http://wonwoosystem.co.kr/pic/catalog/PhotobiologicalSafety.pdf
34. Tsankov P, Yovchev M (2016) Measurement of spatial variation of color characteristics of light sources. In: Youth National Conference with International Participation Lighting 2016, Sofia, 21–23.10.2016. pp 50–53, ISBN: 978-619-160-705-1 (2016) (in Bulgarian)
35. Pode R, Diouf B (2011) Solar lighting. Springer, Berlin, ISBN 978-1-4471-2133-6. https://doi.org/10.1007/978-1-4471-2134-3
36. Manimekalai P, Harikumar R, Raghavan S (2013) An overview of batteries for photovoltaic (PV) systems. Int J Comput Appl (0975–8887) 82(12)
37. Femia N, Zamboni W (2012) Photovoltaic-fed LED lighting system with SOC-based dimmable LED load. In: IECON Proceedings (Industrial Electronics Conference), 2012, pp 1132–1137. https://doi.org/10.1109/iecon.2012. 6388613

Solar Energy and Lighting in Serbia

Tomislav Pavlovic and Nikola Dj. Cekić

Abstract In this chapter, information about geographical position, climate, solar radiation, renewable energy policy, and solar energy research centers in Serbia are given. Also, information about PV plants, early development of lighting, development of electric lighting, modern lighting, public lighting, street lighting, household lighting, lighting in the industry, and PV lighting in Serbia are given.

1 General Information

1.1 Geographical Position

Serbia is located between 41°46′40″ and 46°11′25″ of the north latitude and 18°06′ and 23°01′ of the east longitude.

Republic of Serbia is a continental state situated in the Southeast Europe covering by one part the Balkan Peninsula—region of the Southeast Europe (around 75% of its territory) and by the other part the Pannonian Plain—region of the Mid-Europe (around 25% of its territory).

Serbia borders Hungary to the north; to the northeast Romania, to the east Bulgaria, to the south North Macedonia, to the southwest Albania and Montenegro and to the west Croatia and Bosnia and Herzegovina (entity of the Republic of Srpska) (Fig. 1).

Serbia's total border length with the surrounding countries amounts to 2397 km, 1717 km dry land and 680 km water border. Border length with the surrounding countries is Albania 122 km, Bosnia and Herzegovina 391 km, Bulgaria 371 km, Croatia 315 km, Hungary 166 km, Macedonia 221 km, Montenegro 203 km, and Romania 476 km.

T. Pavlovic (✉)
Faculty of Sciences and Mathematics, University of Niš, Niš, Serbia
e-mail: pavlovic@pmf.ni.ac.rs

N. Dj. Cekić
Faculty of Civil Engineering and Architecture, University of Niš, Niš, Serbia
e-mail: ncekic@yahoo.com

T. Pavlovic (ed.), *The Sun and Photovoltaic Technologies*,
Green Energy and Technology, https://doi.org/10.1007/978-3-030-22403-5_5

Fig. 1 Geographical position of Serbia

The Pannonian Plain covers the northern part of the country while southern parts are covered by hills and mountains. Serbia features more than 30 mountain peaks above 2000 m altitude with the highest peak Djeravica (on the Prokletije Mountain) height of 2656 m. Mountain relief of Serbia explains many canyons, gorges, and caves (Resavska cave, Ceremosnja, and Risovaca). The lowest point is on the border of Romania and Bulgaria, at the estuary of the river Timok into the Danube, at 28–36 m altitude.

Constituent parts of the Republic of Serbia are the autonomous provinces Vojvodina and Kosovo and Metohija. Since the NATO shelling operation, the province of Kosovo and Metohija are under the UN protectorate.

The capital of the Republic of Serbia is Belgrade. With 1,639,121 inhabitants in the wider area and according to the census from 2011, Belgrade is an administrative and economic center of the state. The territory of Serbia is divided into 29 administrative districts and a territory of Belgrade. Administrative districts are a form of the deconcentration of power called detached centers. A district consists of several local government units—municipalities (communes), which, unlike the district, represent a form of the decentralization of power, and as such have their own income and organs of local government.

According to the results of the census in 2011, the Republic of Serbia has 7,186,862 inhabitants. The average population density is 106 inhabitants/km^2. In terms of gender, composition of the population is almost equal. Men account for

3,499,176, and women for 3,687,686 inhabitants in the total population. The average age is 42.2 years. Census in 2011 on the territory of the Republic of Serbia recorded 2,487,886 households. The average number of members per household is 2.88.

The Republic of Serbia is a member of the United Nations, Council of Europe, Organization for Security and Cooperation, Partnership for Peace, Organization of the Black Sea Economic Cooperation, etc. It is also an official candidate for the membership in the European Union and a military neutral country and has the status of the observer in the Organization for Collective Security and Cooperation [1].

1.2 Climate in Serbia

Serbia belongs to the continental climate regions which can be divided into the continental climate in the Panonic lowlands, moderate-continental climate in lower parts of the mountain region, and the mountain climate on high mountains.

Relief substantially influences the climate of Serbia. Parallel to the coast of the Adriatic Sea spreads the range of the Dinars Mountains of the Montenegro which prevents more intensive encroachment of the air masses from the Adriatic Sea toward the areas of Serbia. From the other side, the territory of Serbia is through the Panonic lowlands widely exposed to the climate influences from the north and east. Along the valleys of Kolubara, Velika, and Juzna Morava, the air masses float to the north–south and vice versa. The climate of Serbia is heavily influenced by air masses of certain physical characteristics. The biggest influence is exerted by the air masses formed over Siberia, Arctic, Atlantic Ocean, African land, and the Mediterranean. Over these areas, a field of high air pressure is formed. On the territory of Serbia, often cold air from the Siberia penetrates and rarely from the Arctics (Figs. 2, 3, and 4).

North part of Serbia comprises vast Panonic area which is wide open and exposed to the climate influences coming from the north and the east. The Panonic lowlands show continental climate which encompasses Vojvodina and its edge up to 800 m of height. Continental climate is characterized by excessively hot summers with insufficient humidity. Winters are long and harsh and autumns and springs are mild and short. Average annual air temperatures in the Panonic area are increasing from the west toward the east and from the north to the south. Sombor at the farthest west has average annual temperature of 11.1 °C, and Jasa Tomic on the east has 14.4 °C. Average annual temperature of Palic on the farthest north is 10.6 °C, and of Belgrade on the south is 11.6 °C. The hottest month in the Panonic area is July. However, the whole area of the Panonic lowlands exhibits certain differences. From the west toward the east summer temperatures increase. For example, average July temperature in Sombor (Backa) is 21.2 °C, and in Vrsac (Banat) is 23.3 °C. The highest summer temperatures can reach 35 °C and even 44.3 °C (Stari Becej), and in deserts they rise even up to 60 °C. Winter in Panonic area is extremely cold. The lowest winter temperatures are on the east of the region in Banat and Backa, while it is somewhat hotter on the edge of the Panonic basin. January is the coldest month with the average temperature of −1.9 °C on Palic and 0.3 °C in Smederevo. Precipitations

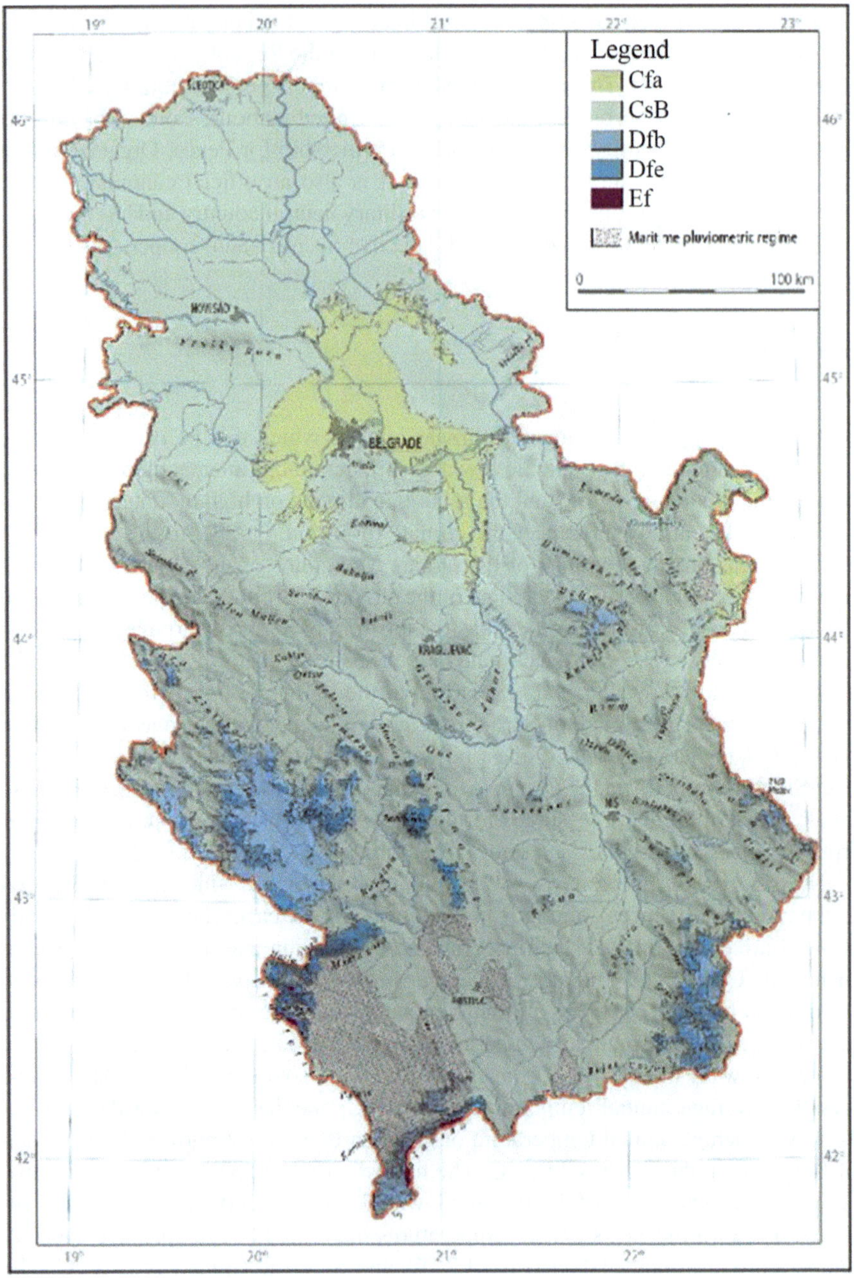

Fig. 2 Climate regions in Serbia Cfa—humid climate; CsB—Mediterranean climate; Dfb—humid continental climate; Dfe—changed Mediterranean climate; and Ef —cold mountain climate [2]

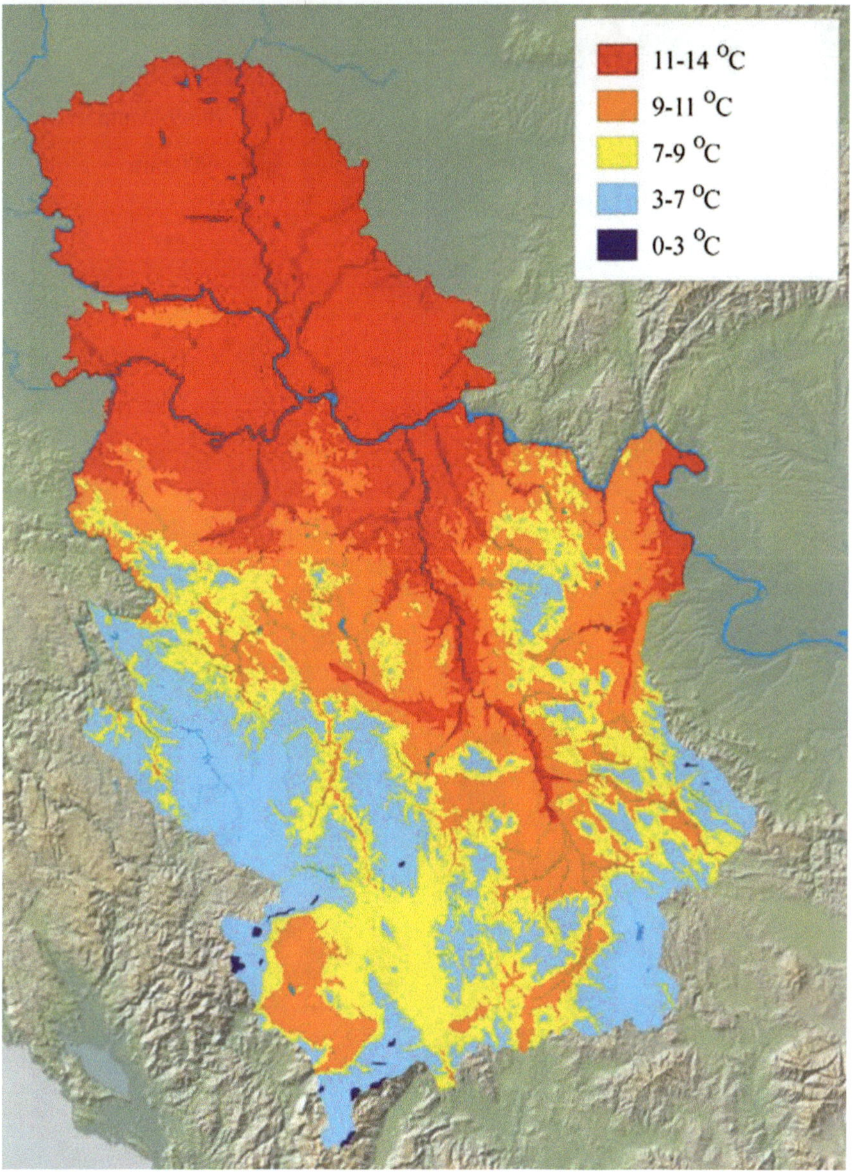

Fig. 3 Mean annual temperature in Serbia [3]

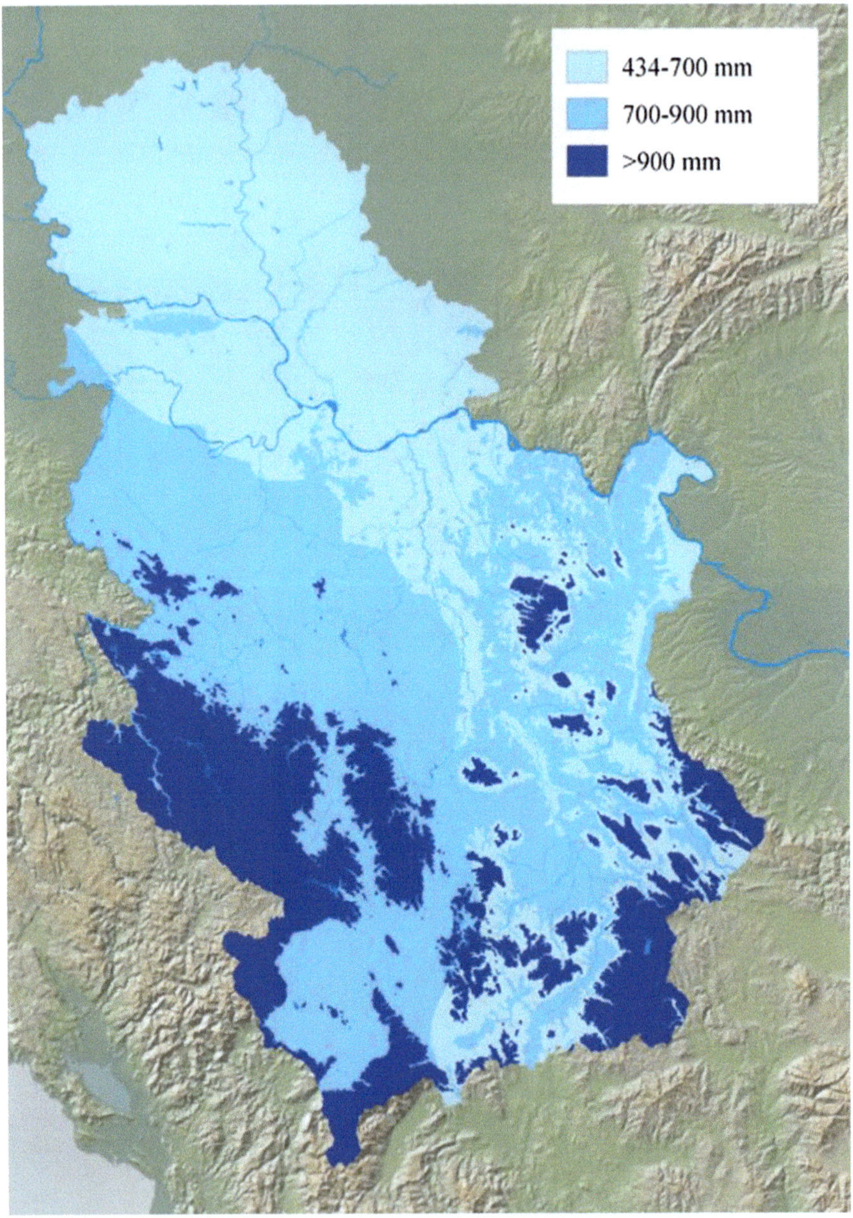

Fig. 4 Mean annual precipitation in Serbia [3]

in the area of the Panonic lowlands are insufficient and unevenly distributed over the year. Also, the territorial distribution of precipitation differs. The lowest annual rate of precipitation is to be found on the whole Panonic area in Vojvodina. On average, Banat and Backa annually have precipitations of around 500–600 mm, and in some years, this is below 400 mm. Therefore, this area is often affected by draught. Starting from the central parts of the Panonic lowlands toward the south, west and east precipitation rises. Area near Vrsac has annual precipitation of 600–800 mm. Toward the south precipitation is slowly increasing so Pozarevac has annual precipitation of 609 mm and Smederevo of 650 mm.

Moderate-continental climate is dominant in the mountain range of Serbia of 800–1400 m altitude. It is characterized by moderate hot summers, autumns longer and hotter than springs and cold winters. Mountain climate prevails in the range over 1400 m of latitude. On the territory of Serbia, it is most present on the mountains Sar planina, Prokletije, Kopaonik, Stara planina, etc. This climate type is characterized by long, cold, and snowy winters and short and chilly summers.

Table 1 is formed on the basis of meteorological data of the Republic Hydrometeorological Institute of Serbia for the period 1961–2010.

In high lime fields and the valleys of the Mountain area of Serbia, climate ranges from the moderate-continental to the mountain one. Due to the temperature inversion, winters are there harsher. Summers in lime fields are pleasant and in higher ones even chilly. Extremely hot weather during summer is rear and lasts short. In confined and windproof valleys in Serbia a real *Zupna* (term derived from the places names) climate prevails. These valleys are in summer and winter hotter than their surroundings. Average monthly and annual air temperatures in mountain region of Serbia are decreasing with the higher latitude and altitude. The lowest average monthly and annual air temperatures in Serbia are on Sar planina, Stara planina, and Kopaonik.

Mountain region of Serbia is characterized by the *Zupna* variant of the moderate-continental climate. This variant of the moderate-continental climate is typical for the Aleksandrovac, Metohija, and Vranje valleys. *Zupna* variant occurs as a consequence of bigger protection of the aforementioned valleys from the penetration of cold air from the north. The mountain area of Serbia is characterized by the temperature inversions. High valleys and lime fields in the mountain region during winter are colder than their surroundings, especially at night when the nearby mountains give

Table 1 Yearly average values of the meteorological data of some cities in Serbia in the period from 1961 to 2010

	Serbian cities				
	Belgrade	Niš	Novi Sad	Kragujevac	Subotica
Sunshine duration (h)	2073.2	1956.3	2062.4	1988.6	2112.9
Temperature (°C)	12.22	11.64	11.18	10.99	10.75
Precipitation (mm)	692.5	591.4	613.3	646.3	550.9
Humidity (%)	68.50	70.15	71.28	73.36	69.73
Overcast (%)	54.00	55.42	52.67	55.50	54.33

away cold air which stays up in the valleys and lime fields and stays there longer. Cloudiness in the mountain region is from 55 to 60% annually. Sunshine duration in the mountain region of Serbia is 1500–2000 h annually. Such a small span of sunshine duration is a consequence of high cloudiness, especially in wintertime. Sunshine duration span is the smallest on the mountains. On Tara, sunshine duration is 1700 h annually or 4.9 h a day. On Kopaonik, annual sunshine duration is 1741 h or 5 h a day. Precipitation in the mountain region is high. On average, mountain region has 1700 mm of precipitation annually [1–4].

1.3 Solar Radiation in Serbia

Average solar irradiation on the territory of the Republic of Serbia ranges from 1.1 kWh/m^2/day in the north to 1.7 kWh/m^2/day in the south during January, and from 5.9 to 6.6 kWh/m^2/day during July. On a yearly basis, average value of the global solar irradiation for the territory of the Republic of Serbia ranges from 1200 kWh/m^2/year in the Northwest Serbia to 1550 kWh/m^2/year in Southeast Serbia, while in the middle part it totals to around 1400 kWh/m^2/year. Due to this fact, Serbia exhibits favorable conditions for the use of solar energy and its conversion into the thermal and electrical energy.

Yearly sum of total solar irradiation incident on optimally inclined south-oriented PV modules in kWh/m^2 for the territory of Serbia obtained by PVGIS is given in Fig. 5.

It is clear from Fig. 5 that average solar irradiation is not dependent on geographical latitude only. There are regional differences in global solar irradiation due to terrain features and climatic conditions.

Results of long-term meteorological measurements have shown that natural potentials of climatic resources of Serbia are very good. In Serbia, the energy potential of the solar irradiation and potential of biomass are around 30% higher than in the Middle Europe (average annual solar irradiation energy in Europe is 1096 kWh/m^2/year, and in Serbia, it is around 1400 kWh/m^2/year). Mean values for January are in the range from 1.1 kWh/m^2 in the north of the country to 1.7 kWh/m^2 in the south. Mean values for July are in the range from 5.9 to 6.6 kWh/m^2 [1, 4–7].

1.4 Renewable Energy Policy in Serbia

Technically, usable energy potential of the renewable sources of energy in the Republic of Serbia is significant and estimated to over 4.3 million of tones of the equivalent oil (tn) annually—out of which around 2.7 million of tonnes of equivalent oil annually is to be used in biomass, 0.6 million of tonnes of equivalent oil annually in unused hydropotential, 0.2 million tonnes of equivalent oil annually is in the existing geothermal sources, 0.2 million of tonnes of equivalent oil annually is in energy of

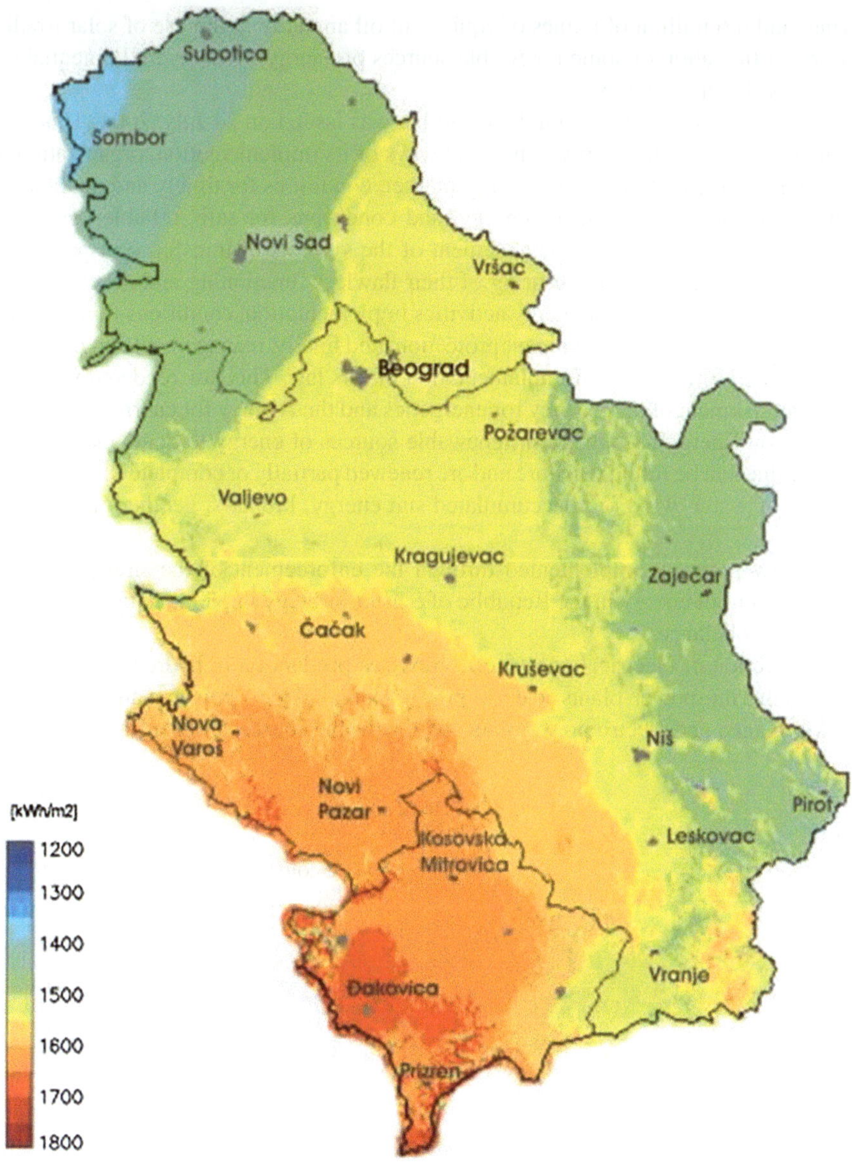

Fig. 5 Yearly sum of total solar irradiation incident on optimally inclined south-oriented PV modules in kWh/m^2 for the territory of Serbia. Adapted for Serbia from PVGIS © European Communities, 2001–2008, http://re.ec.europa.eu/pvgis/

wind, and 0.6 million of tonnes of equivalent oil annually in the use of solar irradiation. Participation of some renewable sources of energy in the overall potential of Serbia is shown in Fig. 6.

Law on Energy of the Republic of Serbia was issued on 24 July, 2004. This law regulates: aims of the energy policy and ways of its implementation, organizational and functioning patterns of the energy market, conditions for timely and qualitative supply of consumers with the energy, and conditions for safe, reliable, and efficient production of energy, management of the systems of transfer, transport, and distribution of energy and ensuring of their flawless functioning and development, conditions and manners of energy activities implementation, conditions for enabling energy efficiency and environment protection, and finally, management and monitoring of the enforcement and implementation of this law. This law rendered possible the establishment of the Agency for energetics and the Agency for energy efficiency.

Law on Energy uses the term renewable sources of energy to denote sources of energy that can be found in nature and are renewed partially or completely, especially energy of water, wind, non-accumulated sun energy, biomass, geothermal energy, etc.

Energy policy is implemented through the enforcement of the strategy of the development of energy in the Republic of Serbia, strategy implementation program, and energy balance.

Production of electrical energy encompasses production in hydroelectric power plants, thermo power plants, electro power plants, electro power plants—thermal power plants, and electro power plants on renewable sources of energy or waste.

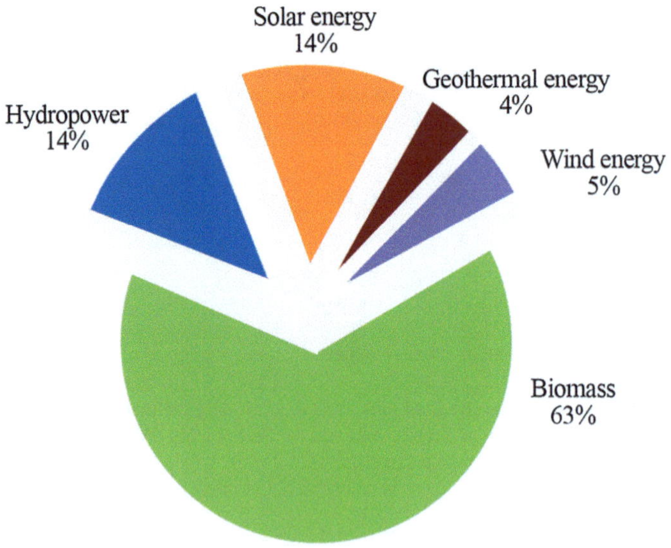

Fig. 6 Participation of some renewable sources of energy in the overall potential of Serbia [1]

Privileged producers of electrical energy are manufacturers who in their production process of electrical energy use renewable sources of energy or waste, manufacturers who produce electrical energy in electro power plants that are according to the Law on energy considered to be small-scale electro power plants, and manufacturers who produce simultaneously electrical and thermal energy if they meet the criteria of the energy efficiency. Preconditions on terms of gaining this special status of the privileged manufacturer of the electrical energy and criteria for the assessment of the fulfillment of these conditions were issued by the Government of the Republic of Serbia on September 3, 2009. Congruent to this Act legal entity or the entrepreneur can attain the privileged manufacturer status if the electro power plant in the production process on a yearly basis uses at least 90% renewable sources of energy and the rest goes to fossil fuel or waste.

Act on incentive measures for the production of electrical energy by use of the renewable sources of energy and by combined production of electrical and thermal energy was issued by the Government of Serbia on November 20, 2009. This Act closely prescribes incentive measures for the production of the electrical energy by use of the renewable sources of energy, its purchase and it defines energy objects to produce electrical energy from the renewable sources. It also determines the content of the agreement on the purchase of the electrical energy under incentive measures, etc.

The right to the incentive measures defined by this Act for the electrical energy produced in plants which use solar energy is limited to the total installed power of (in these plants) up to 5 MW. According to this Act, 1 kWh of the electrical energy produced by solar power plant in the interval of 12 years upon agreement signing is to be paid to the manufacturer at the price of 23 eurocents. Connecting of the solar power plants or small-scale plants installed on private houses to the grid is regulated by the legislative of the EPS of the Republic of Serbia. Law on Energy issued on August 1, 2011 clearly defines incentive measures for the use of the renewable sources in power generation and subsidized power producers which is stated in the Official Gazette of the Republic of Serbia, no. 27/2011.

Republic of Serbia has in 2006 ratified the agreement on the foundation of energy community between EU and Albania, Bulgaria, Bosnia and Herzegovina, Croatia, North Macedonia, Montenegro, Romania, and Interim Mission of United Nations on Kosova.

In September 2008, European Parliament has adopted a set of regulations on the climate changes that aims at ensuring reduction of greenhouse effect gas emission of 20%, improvement of energy efficiency of 20% and participation of the renewable sources of energy of 20% in total energy consumption in EU-until 2020 as compared to 1990. Republic of Serbia has accepted the instructions of the EU on the renewable sources of energy and is putting all efforts to implement it.

Republic of Serbia as on January 26, 2009 becomes a member and the founder of the International Agency for Renewable Energy (IRENA).

Research in the area of renewable sources of energy has solid foundations in the National program of energy efficiency of the Ministry of Science in Serbia.

However, application of attained technological knowledge is legging off, especially the realization of demo projects due to the lack of incentive measures.

The number of installed facilities for the exploitation of renewable sources of energy in Serbia and their current annual energy production is insignificant. Capital invested in up-to-date installed facilities is of small value and is mainly domestic one. From the point of view of the national level, financial results, achieved by running these facilities, are humble.

Serbia nowadays almost does not have manufacturers and maintenance service for the equipment used in the exploitation of renewable sources of energy. However, in the area of the exploitation of water energy, biomass and sun irradiation energy for heating purposes, there are viable possibilities to include domestic equipment manufacturers.

Energy policy defined by the Law on energy among other things envisages taking steps to create conditions for stimulation of the use of renewable sources of energy. Congruent with this initiative, the Law on energy introduces categories of subsidized energy producers who in their production process use renewable sources of energy, and who are entitled to subsidies, tax, customs, and other exemptions in the line with Law and other regulations on taxes, customs, and other subsidies and incentives.

Since production of electrical energy from renewable sources of energy is more expensive than the fossil fuels' energy production, some incentive systems are introduced, that is financial and non-financial incentive measures to invest into facilities using renewable sources of energy.

Most used financial stimulating measure is the increased price of purchased energy produced by renewable sources of energy during the year. Other model deploys application of defined purchase prices for the energy produced by renewable sources of energy, so-called feed-in tariff. Most European countries apply the feed-in tariff model.

One of the significant characteristics of the stimulating measures to increase the use of renewable sources of energy is selective stimulation of the development of chosen technologies. Besides financing research–development projects, it is needed to finance the installation of demonstration projects. Basic criteria for the selection of renewable sources of energy and technologies that are to be stimulated are available energy potential, ability of its own economy, and degree of the international development of technologies and the market (Table 2) [1, 8, 9].

1.5 Solar Energy Research Centers in Serbia

Development of solar energy investigation in Serbia started in 1973 marked by the works of Prof. Dr. Branislav Lalovic (1928–1988) in Belgrade and prof. Zivojin Culum (1911–1991) in Novi Sad.

Table 2 Feed-in tariff in Serbia

Type of power plant	Feed-in tariff (Ec/kWh)
Small hydropower plants	7.8–9.7
Biomass power plants	11.4–13.6
Biogas power plants	12–16
Wind power plants	9.5
PV solar power plants	23
Geothermal power plants	7.5
Cogenerative power plants	7.6–10.4

Now, in Serbia exist solar energy research centers: at the Faculty of Sciences and Mathematics in Niš, Faculty of Mechanical Engineering in Niš, Faculty of Electronic Engineering in Niš, Faculty of Technical Sciences in Novi Sad, and Faculty of Technical Sciences "Mihajlo Pupin" in Zrenjanin.

Faculty of Sciences and Mathematics (FSM) in Niš

Solar energy laboratory at the Faculty of Sciences and Mathematics (FSM) in Niš was established in 2003 and has a sophisticated equipment for investigation physical characteristics spectrally selective absorbers, flat-plate thermal and PV/T collectors with and without concentrators, solar cells, PV solar power plants, etc (Figs. 7, 8, 9, 10 and 11).

A fixed on-grid 2 kW PV solar power plant was installed on the roof of the Faculty of Science and Mathematics (FSM) building in Niš (Republic of Serbia) in October

Fig. 7 Part of Solar Energy Laboratory on the roof of the Faculty of Science and Mathematics: 1. monocrystalline solar module, 2. hybrid collector monocrystalline solar module, 3. hybrid collector with a-Si solar module, 4. thermal collector with spectral selective absorber, 5. thermal collector with non-colored absorber, 6. non-colored absorber, and 7. spectral selective absorber

Fig. 8 Hybrid collector with solar radiation concentrators

Fig. 9 Solar Box

Fig. 10 Rotation PV module

Fig. 11 PV solar plant

2012, functioning from October 15. Until December 31, 2012, trial measurements of the solar PV plant electrical parameters were performed. Experimental determination of the energy efficiency and its performance was conducted from January 1, 2013 until January 1, 2014. The plant consists of 10 monocrystalline silicon solar modules, single power of 200 W (SST-200WM, Shenzhen Sunco Solar Technology Co.). The solar modules based on metal stainless steel with the foundation inclined at 32° toward the South are serial interconnected in a *string*. Using adequate conductors, the mentioned solar modules are connected to a DC distribution box (RO-DC), single-phase inverter (Sunny Boy 2000 HF-30 of 2 kW), AC distribution box (RO-AC), and the city power grid. DC and AC distribution boxes contain protective components providing steady functioning of the solar PV plant. At the output of AC distribution box, there is alternating (AC) voltage 230 V, 50 Hz.

For the monitoring of the solar PV plant, remote diagnostics, data acquisition, and their visualization, *Sunny WEBBox* is used as the central communication interface, which is by *Bluetooth* connected to the inverter and to the sensor *Sunny SensorBox* which is mounted at the angle of 32° in relation to the horizontal surface on the roof of the FSM building in Niš.

Investigation activities

The researchers in *The Solar Energy Laboratory* have implemented following international projects: *Development of Solar Energy, Astronomy and Meteorology Laboratory* (WUS project, 2003–2004), *Physics and Techniques of Solar Energy* (WUS CDP$^+$ project, 2006–2007), *Energetic efficiency and environmental awareness, Experimentation and training for a self sustainable local development* (E.CO.LOC. project, 2007/08) and *Renewable energy sources as a model of sustainable development of the countries of West Balkans* (UNESCO project, 2010–2011).

The researchers in *The Solar Energy Laboratory* have also realized projects funded by the Ministry of Science and Technological Development of the Republic of Serbia: *Development and testing of a hybrid solar flat receiver for thermal and electrical conversion, Development and application of the photovoltaic solar systems as light sources in the individual houses, Atlas of the energy potential of the Sun and wind in Serbia, Development and testing of thermal and hybrid collector with solar radiation concentrators and Testing of the energy efficiency of the photovoltaic solar power plant of 2 kWP.*

The results obtained in the Solar Energy Laboratory at the Faculty of Science and Mathematics University of Niš were published in references [10–33].

Faculty of Mechanical Engineering in Niš

Faculty of Mechanical Engineering in Niš features modern laboratory for thermo-technique, thermo-energy, and processing technique investigating the characteristics of the flat-plate and parabolic solar radiation collectors.

Faculty of Electronic Engineering in Niš

Faculty of Electronic Engineering in Niš features modern laboratory for electronics realizing and investigating rotating PV systems for optimal solar radiation incidence.

Faculty of Technical Sciences in Novi Sad

Faculty of Technical Sciences (FTS) in Novi Sad features modern Renewable & Distributed Energy Sources Laboratory dedicated to research in the field of renewable energy, especially in the wind and solar energy conversion and energy storage. A laboratory setup of a small-scale model of the wind turbine with induction generator connected to public network by back-to-back converter is available. Another setup employs six-phase wind turbine (motor)-generator unit with 6ph-to-3ph grid-connected bidirectional energy conversion system. In addition, there is a complete hybrid wind–solar system with solar power installation of 1.2 kW_P and wind power generator of 0.75 kW, together with battery storage. The laboratory also has a measuring pole for the measurement of wind energy at a height of 12 m and number of instruments for solar energy measurements. The research includes computer simulations and emulation facilities using specialized software for calculation and assessment of wind energy (WaSP—Wind Atlas Analysis and Application Program), solar energy (PVSyst and others), energy converters (Typhoon HIL, Matlab-Simulink, and dSpace), and grid emulator (Cinergia). The laboratory is used for teaching and practical exercises for courses in the field of renewable energy, also (Fig. 12).

Solar power plant FTS1 of 9.6 kW_P is located on the roof of the building of the Faculty of Technical Sciences in Novi Sad (FTSNS). It consists of 40 solar photovoltaic panels arranged in two strings, each of 20 modules. The panels are made of the polycrystalline silicon, individual power of 240 W_P (Jingli Solar) and are directed toward the south under the tilting angle of 30°. The solar power plant uses the inverter of the SMA company type STP8000-TL power of 8 kW with a wireless transmitter. Monitoring and management of the solar power plant is performed via WebBox communication device over the Internet. Solar power plant was designed by the Centre for Renewable Energy and Power Quality (CRESPQ), FTSNS. Installation of the solar power plant was carried out by the CRESPQ experts with the help of the students of the study program—Power Engineering—Renewable Sources of Energy. The solar power plant was put into operation on October 25, 2011.

Fig. 12 PV solar power plant FTS1 on the building of the Faculty of Technical Sciences in Novi Sad

Fig. 13 Flat plate collectors and PV modules

Faculty of Technical Sciences "M. Pupin" in Zrenjanin

Faculty of Technical Sciences *Mihajlo Pupin* in Zrenjanin features a sophisticated solar energy Laboratory which investigates characteristics of flat-plate thermal and PV modules. Solar Energy Laboratory at the Technical Faculty *Mihajlo Pupin* in Zrenjanin, University in Novi Sad, is a part of the Laboratory for machines, appliances and thermo-technics.

Laboratory has the equipment for the experimental research in the field of solar energy, heat exchange, measuring the pressure drop in the pipe installations, etc. Since 2015, the laboratory is equipped with the experimental solar—thermal and photovoltaic systems. Flat-plate collectors and PV modules are located on the south side of the building of the faculty (Fig. 13).

To measure meteorological parameters, the Laboratory uses Davis Vantage Pro 2 meteorological weather station, whose sensor part is located on the roof of the faculty, and the registration part is located in the laboratory. By this meteorological weather station, a continuous measuring of the atmospheric pressure, temperature and humidity, rainfall, wind speed and direction, solar radiation intensity, the intensity of UV radiation, etc., is carried out. By means of the adequate software and the USB port, storing and processing of the measured data in the computer is done [1, 8, 9].

1.6 PV Solar Plants in Serbia

Matarova

In the village of Matarova near Merdare, municipality of Kursumlija, on April 23, 2012 the Italian company *Gascom,* in cooperation with the local company *Solar Matarova* from Novi Sad began the construction of 2 MW_P PV solar power plant. The construction of the PV solar power plant was finished on August 23, 2013, when a contract with EPS (a power distribution company) on the feed-in tariff electrical energy selling was signed. The plant is located on an area of 4 ha, has 8100 polycrystalline solar modules, and the total investment was € 3.6 million. The PV solar power plant was designed by the *Mihajlo Pupin-Automatics* Institute in Belgrade. Technical assessment was performed by the *Jaroslav Cerni* Institute in Belgrade. Construction works at the plant's site were performed by the following companies *Metalac Co.* in Kursumlija, then companies in Niš and Prokuplje. For the purposes of solar power plant, the transmission network Merdare–Degremen a length of 2.3 km was built, which previously did not exist. For the connection, 10.7 km transmission lines Degremen–Kosanicka Raca were reconstructed (Fig. 14).

Kladovo

PV solar park in the village Velesnica near Kladovo consists of two PV solar power plants, *Solaris 1* and *Solaris 2* (Fig. 15).

PV solar power plant *Solaris 1*, of 999 kW_P, began to be constructed in July 2013 and was completed in November 2013, and was work commissioned on December 27, 2013. PV solar power plant *Solaris 1* consists of polycrystalline silicon modules of 245 W_P, manufactured by *Yingli Solar*. PV solar power plant *Solaris 2* was designed by the architectural company *Ceefor Ltd.*, from Belgrade. The contractor was a company *Enertec* from Maribor (Slovenia), and a subcontractor *MT-Komex Ltd.*, from Belgrade.

Fig. 14 PV solar power plant *Matarova* of 2 MW_P

Fig. 15 PV solar park of 2 MW_P in the village Velesnica near Kladovo

PV solar power plant *Solaris 2*, of 999 kW_P, began to be constructed in August 2014, was completed in October 2014, and was work commissioned in October 24, 2014 PV. PV solar power plant *Solaris 2* consists of polycrystalline silicon modules 250 W_P, manufactured by *Yingli Solar*. PV solar power plant *Solaris 2* was designed by the architectural company *Ceefor Ltd.*, from Belgrade. The contractor was a company *MT-Komex Ltd.*, from Belgrade.

Between PV solar power plants *Solaris 1* and *Solaris 2,* there is a transformer station 35/0.4 kV, rated power 2 × 1000 kVA, which allows the distribution of the entire electricity produced in the power distribution system PV. PV solar power plants have following inverters built, TRIO-27.6-TL-OUTD, manufactured by the company *ABB Aurora Power One.*

The PV solar power park is located in an area of 4.5 ha. The total solar modules area is 13,600 m^2. The investment in PV solar power plants *Solaris 1* and *Solaris 2* amounted to 3 million Euros.

Serbia has up to now installed more than 400 small off-grid and on-grid PV solar systems.

PV solar plant Pupin in Belgrade

PV solar power plant *Pupin* of 50 kW_P is installed on the roof of the main building of the *Mihajlo Pupin Institute* in Belgrade. It consists of 180 solar modules installed in 10 strings. Polycrystalline silicon solar modules are of the individual power of 280 W (Schutten Solar STP6-280 W). The solar power plant uses two types of inverters: two Schneider types of 15 kW and Refusol of 20 kW (to compare functioning of the different inverters). The system for the monitoring and management of the solar power plant is fully manufactured in the *Mihajlo Pupin Institute*, consisting of the family *Atlas* hardware and *SCADA software View 4*, which for years have been the Institute's renowned brands in the field of energy, in the country and abroad. In this way, one gathers information on the state of the fuse, performs system protection, manages switches, collects the information on total production, as well as the information on the meteorological measurements (insolation, direction and wind speed, the ambient temperature, and the temperature of the modules). Construction of the PV solar power plant *Pupin* was started in April 2013, and it was connected to the grid network on September 20, 2013. PV solar power plant *Pupin* has received the status of the privileged energy producer on April, 1, 2014 (Fig. 16).

PV solar power plant Energoprojekt in Belgrade

PV solar power plant on the roof of the Energoprojekt Building in Belgrade consists of 492 solar modules of 235 W of polycrystalline silicon and it was installed in 2013. PV solar power plant generated 0.132 GWh of electricity in 2013 (Fig. 17).

PV solar power plant Elektrovat in Čačak

On the roof of the commercial building *Elektrovat Ltd.* in Čačak, a solar power plant power of 54.72 kW_P, consisting of 228 polycrystalline silicon solar modules, individual power of 240 W_P, and two network inverters, individual power of 25 kW, is installed. The solar modules and network inverters were produced by *Schüco*

Fig. 16 PV solar power plant *Pupin* of 50 kWp in Belgrade

Fig. 17 PV solar power plant *Energoprojekt* of 115.62 kWp in Beograde

International KG (Bielefeld, Germany). The solar power plant was designed by Ljubisav Stamenić, Ph.D., from Belgrade and was installed by *Elektrovat Ltd.* from Čačak (Fig. 18) [1, 8, 9].

Residential PV solar plant in Belgrade

See Fig. 19.

2 Lighting in Serbia

2.1 Early Development of Lighting

Lanterns

In 1868, in Belgrade, the Belgrade police issued an order that every nightwalker must carry his own lantern. Wealthier people took their servant with them to carry a

Fig. 18 PV solar power plant *Elektrovat* of 54.72 kWp in Čačak

Fig. 19 PV solar power plant of 20 kWp in Belgrade

lantern or other form of lighting in front of them. At that time, local police prescribed obligation of placement, actual place and the number of street lamps with oil or petroleum lamps, lanterns, etc.

At the beginning of the nineteenth century, first candle lanterns were used in Novi Sad. At that time, the pontoon bridge between Novi Sad and Petrovaradin was illuminated at night with lanterns. In 1846, there were 40, and in 1878, 147 lanterns in Novi Sad.

On February 12, 1828, for the first time, street lanterns were lit in Subotica on the occasion of the City Hall opening ceremony.

In Pančevo, coffee shops started using lanterns in 1833. From 1833 to 1839, only these lanterns lit up the whole city. In 1939, Magistry of Pančevo decided to establish a city-based fund-raising fund, which was assigned to collect money for the procurement and placement of lanterns in the city streets.

With the Turkish Sultan charter issued in 1830, Serbia gained autonomy, and the Turkish government in Belgrade Varoš was replaced by the Serbian one. On this occasion, a big ceremony was held in the town which was lit by lanterns. In 1841, Belgrade was proclaimed the Serbian capital. In 1856, there were only two lanterns in Belgrade, one at the municipal house, and the other at Terazije. In October 1856, Belgrade's town authorities decided to procure new lanterns for the lighting of Belgrade. For the New Year in 1857, Belgrade was "brightly lit" by 336 lanterns. There was also a fireworks display. In 1886, in Belgrade, in addition to 285 restaurant lanterns, there were also 611 municipal lanterns. Also, a service for lighting and extinguishing lanterns was organized. In 1891, the Municipality of Belgrade had 16 paid lantern lighters.

Gas lighting

The first gas lighting in Belgrade was introduced in 1869 in the National Theater, which was a great attraction for Belgrade because it was the only gas lighting in the city.

For the needs of the gas lighting in the theater, a wood-fueled gas factory was built near the theater, on the site of the abandoned Turkish Kara-mosque. The National Theater in Belgrade was officially opened on October 30, 1869 with the "luminous" light of the gas lamps.

Gas lamps—candelabra in front of the National Theater building were the first public lights of the gas lighting in Belgrade (Fig. 20).

In 1877, the city lanterns were replaced with 245 public gas lamps in Pančevo. In 1888, street gas lighting was introduced in Novi Sad and a gas plant for the needs of the city was built. In Subotica, on February 5, 1890, first gas candelabra were lit up. At that time, Donji Milanovac had gas lighting, which in 1924 was replaced by electric lighting.

Fig. 20 Building of the National Theater in Belgrade with gas lamps in 1869 [34]

2.2 Development of Electric Lighting

Cafe "Hamburg" in Belgrade

In the summer of 1880 in Belgrade, Petar Jovanović Šapčanin was the first in Serbia to illuminate the garden of his *Hamburg* cafe with the help of Edison's Bogen electric arc lamp. The tavern was located on the corner of Masaryk and Prince Miloš Street. Electric current was generated by wood-fueled locomotive (mobile plant), which was located in the yard across the street of the tavern. P. J. Sapcanin purchased a locomotive with a dynamo machine in Europe and delivered it to Belgrade. During the work of the locomotive and the arc lamp, they created a noise that did not bother visitors to the taverns and curious people to admire the wonders of the light technology (Fig. 21).

The Locomobile consisted of a vertical boiler with water, a piston, and a dynamo machine that produced enough DC power to power 5–10 arc lamps. People came to the tavern and watched with great interest, with and without tinted glass, the "electric sun" (Fig. 22).

Kragujevac

In August 1884, in honor of the arrival of King Milan and Queen Natalija Obrenović in Kragujevac, an exhibition of products of the Military Technical Institute was

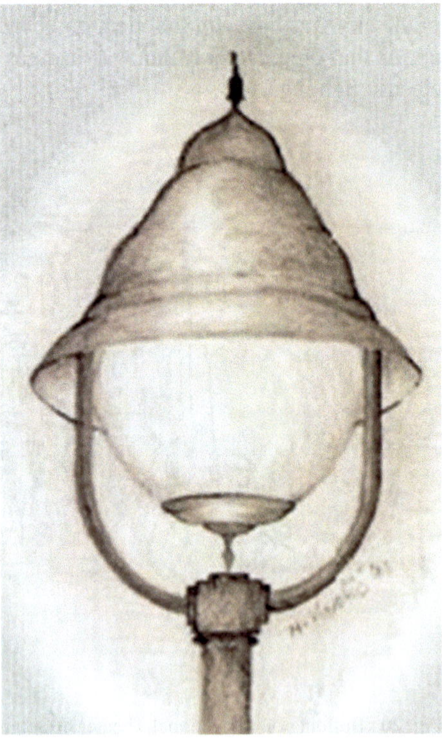

Fig. 21 First lamp, Bogen lamp for street lighting in Belgrade, in 1880

Fig. 22 Locomobile-mobile power station (1900) [34]

organized. On this occasion, the first installation of electric lighting in Serbia was put into operation.

Following the idea and under the leadership of a young engineer of the Todor-Tosa Seleskovic Institute (1856–1901), electric lighting was introduced in the power powder unit of the Military Technical Institute in Kragujevac in 1884. The steam power plant was purchased from *Sigmund Schuckert* from Nuremberg with a 3.67 kW dynamo machine. Along the hall, a conductor was installed in which 30 lamps of 13 W and two arc lamps each of 996 W were connected in series. The first electric lighting in Šumadija created a very strong impression on all visitors of the exhibition in Kragujevac (Fig. 23).

In 1885, the electric power station was expanded with two dynamo machines, one of which supplied eight arc lamps of 996 W, and the other 1000 light bulbs of 13 W, which illuminated the workshops of the Institute.

Paraćin

In November 1879, the Czech industrialist Berthold Minh and his partner Karl Schumpeter contacted the Ministry of Finance of the Principality of Serbia with a request that, with certain privileges, they were authorized to build a factory of cloth and yarns in Paraćin. By the order of the Prince Milan Obrenović of April 16, 1880, they were granted approval and privileges. Applicants were obliged to build the factory by the end of 1881, which they did. The plant's factory had the most modern machines. During the construction of the factory, reconstruction of the dams of the former water mill was carried out, and then the mill of the Prince Milan Obrenović. Water from the Crnica River was used for processing wool and operation of a 36.75 kW turbine, which worked continuously day and night. In the line of major achievements, one can mention the introduction of electric lighting in the factory with the dynamo machine of 4.15 kW of the company *Ganc and Co.*, which provided electricity for 300 bulbs of 8.3 W and 13.28 W. In 1890, all the factory and residential buildings within the factory circle had electric lighting. On the basis of

Fig. 23 Light bulb in the yard of the Military Technical Institute in Kragujevac, 1884 [34]

a list of necessary materials and equipment for the operation of the factory, which the factory submitted to the Ministry of Finance on March 30, 1988 and December 11, 1890, it can be concluded that the electric lighting was introduced in the factory between 1883 and 1890.

The factory was located in the area of today's Serbian glass factory, and the turbine of the central plant (former mill) on the Crnica River was located just next to the factory. The factory was burnt in 1904 in a great fire, "due to a failure on electrical installations."

Subotica

In 1880, an agent propagated electric lighting in Subotica. He brought a small electric power plant which he put on the stairs of the theater building. Upon completion of the theater performance, the agent climbed the theater stairs and gave a speech on the importance of inventing the Edison lamp. Then, all the torches that illuminated the space in front of the theater were extinguished, so there was complete darkness. Then, the agent turned on a large electric lamp and illuminated the space in front of the theater. Then, before the enthusiastic audience, the electric lamp went on and off several times. The crowd whispered "*... they are ignited without matches ...*". In the end, the agent said, "*Look, it's electricity*," and went on to advertise his goods.

Belgrade

In Belgrade, thanks to prof. Djordje Stanojević, in 1891 it was decided to introduce electric lighting. The theater in Belgrade was illuminated with electric light in 1882.

A horse-drawn tram in Belgrade worked from 1892 to 1903. The electric tram in Belgrade was put into operation on June 7, 1894, on the so-called *Topčiderska railway*, length of 5 km (Fig. 24).

Lighting in Kostolac mines

The introduction of electric lighting in Belgrade took place very slowly. Initially, it was introduced only by some intellectuals, senior clerks, and rich craftsmen, and it was also introduced in the building of the *Russian mission*. In many of today's central streets, electric lighting was introduced just before the First World War. In the wider city center, electricity was introduced in the houses only after four decades.

In Kostolac coal mine, electric lighting was introduced in 1903, when a power plant of 45 kW was built. In addition to lighting, electricity was used for the operation of machines in the mining cave, for operation of the fan, for coal loading into wagons, etc. The electric power station consisted of a steam engine of the *Škoda* brand and a DC generator. Thanks to electricity, coal production in Kostolac mines grew year by year.

Lighting in Niš

The development of lighting in Niš began with the installation of the hydropower plant Sveta Petka near Niš, on September 21, 1908. The hydropower plant of Sveta Petka was put into operation by the Crown Prince Djordje Karadjordjević. The lines from the hydroelectric power plant Sveta Petka to Niš, length of 25 km, under the voltage of 8 kV were made on wooden pillars with copper conductors cross section of $3 \times 35\ mm^2$. By the end of August, 1908, all planned work on the electrification

Fig. 24 First electric lighting in Belgrade, in 1891 [35]

of Niš was completed. The first lamps in Niš were lit and enthusiastically received on September 21, 1908. In the first year of electrification, electricity in most households was used exclusively for lighting. Later, it was also used for the operation of various home appliances—electric irons, stoves, boilers, etc., in the houses of the richer Niš inhabitants. Todor Milovanović, president of the Niš municipality, prof. Djordje Stanojević, and an engineer Aćim Stevović take credit for hydropower plant of Sveta Petka and the development of electroenergy in Niš [34–37].

2.3 Modern Lighting

2.3.1 Public Lighting

At the beginning of the twentieth century, attention was paid to the illumination of public spaces in Serbian towns. The light intensity and intensity of the central areas has been significantly improved unlike the peripheral zones where the lighting continued to be of lower intensity. Of particular significance are Belgrade, Niš, Novi Sad, Kragujevac, and other towns where there have been noticeable changes in the way of lighting the city's milestones, and where at night, in certain conditions, some eco-urban architectural physical structures have a visual and iconological, European character. In this regard, very important are public spatial micro-ambient units and their panoramic view, the light-enhanced contrast of certain attractive objects, their physiognomy, contours, and details. It is similar with public buildings: fortresses, religious buildings: churches where dome houses, entrances, crosses, etc. are emphasized, and which appear to be aesthetically better and more representative at night. Then come the buildings of the assemblies and administrative buildings, banks, hotels, etc., in which visual and light effects come to a special, high-quality artistic expression, especially if the color lights on facade surfaces are applied and if it is variable colored periodically, at intervals. Looking back in the near past shows that the illumination of the exterior and interior space in Serbia is in a strong evolutive, progressive phase, and that new technologies of lighting fixtures have significantly suppressed the old conceptions of the lighting of streets, parks, squares, courtyards, factory circles, sports facilities, buildings, airports, religious buildings, museum—gallery objects, historical monuments, hospital-clinical complexes, etc., as well as in other places where high-quality brightness and good visibility are needed. There have been significant inspirational changes in the night appearance of physical structures in the area of Serbian urban agglomerations, to some solutions in the lighting of public buildings that have received iconic cultural character and new aesthetic-visual characteristics. These are new standards and light refurbishment in Serbian cities. Their new identity is revealed slowly, a new conceptualization of hidden urbanism is created, and a new basis for modern development is created (Figs. 25, 26, 27 and 28).

The architecture of these facilities and the overall urban space, using new technological solutions in lighting, live for 24 h. In addition, electricity consumption

Fig. 25 Entrance to the Niš Fortress [38]

Fig. 26 Smederevo Fortress at night [39]

Fig. 27 National Assembly Building in Belgrade [40]

Fig. 28 St. Sava Temple in Vračar, Belgrade [41]

is reduced and visibility in space is improved. Special innovative advantages are observed in LED lamps with solar panels. Electricity is free, there is no digging and laying of cables, and installation is simple and inexpensive. This is a new strategy for discovering urban space in nightlife. It is part of permanent changes in the strategic advancement of cities, using the interpolation of new light content.

In the micro- and macro-ambient parts of the cities of Serbia, in streets and squares, low- and high-pressure sodium lamps of different models and characteristics are predominant, and they are energy efficient, with a predominantly yellow color, so that the colored surfaces of objects in the exterior area, in the night conditions, do not give credible colorful impression.

In general, for public buildings, for example, theaters, congress, and multifunctional halls, halogen lighting with lanterns filled with metal steam is used, a temperature of 5400 K that mimic daylight. In addition, fluo tubes and LED bulbs are used. Particularly important are decorative bulbs, of different colors, which decorate the exterior spaces and facades of architectural objects when they decorate significant places in the open space. Characteristics are decorations for the New Year's holidays where the light effects of LEDs, color flexible plastic tape, LED diode bulbs, advertisements, etc., in the evening give a festive impression and cause a relaxing feeling with the users in the space (Figs. 29 and 30).

For this reason, more and more halogen lamps with metal steam and white light are used in the last few years, which gives a more faithful image and a more accurate impression of the colors of physical structures in the space (Fig. 31).

Fig. 29 New Year's night view of the decorative lighting of Knez Mihailova Street in Belgrade

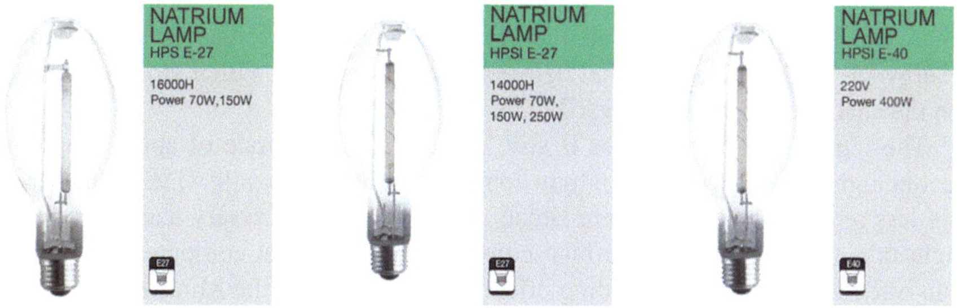

Fig. 30 Sodium bulbs for low- and high-pressure outdoor lighting

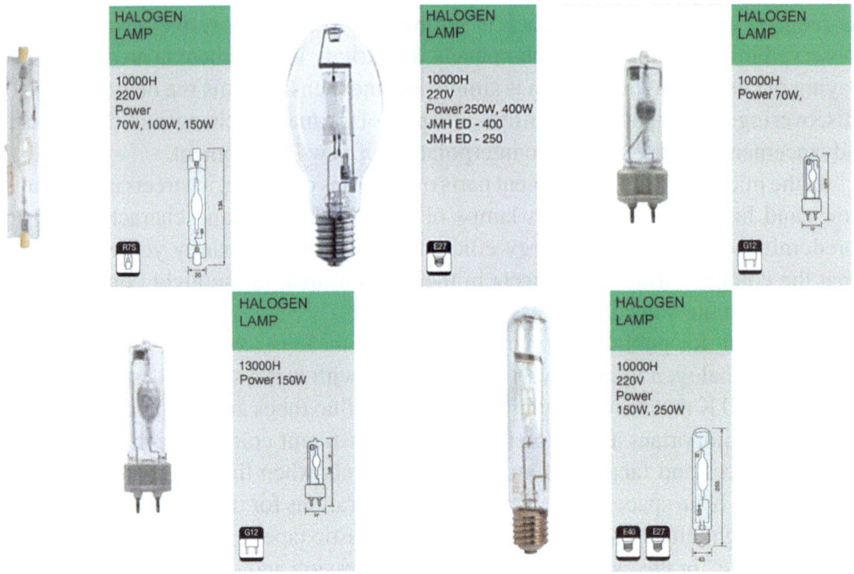

Fig. 31 Halogen lamps with metal steam and white light, of different shapes and characteristics

These lamps were used, for example, in Belgrade, on the plateau near the St. Sava Temple, in the Cyril and Methodius park near the Vuk's monument, in Kalemegdan, and other places. Metal–halogen light sources are especially suitable for the lighting of parks, changeable pedestrian surfaces. In other urban agglomerations in Serbia, significant investments are made for the repair of quality and light comfort in micro-ambient parts. For example, in Novi Sad, Niš and Kragujevac, Zrenjanin, in the central regions, some parts received white light and a good reproduction of colors. Bridges, hotel buildings, museums, theaters, the walls and entrances to the fortress—Petrovaradin and Niš Fortress—are accentuated by a strong white light, so that their appearances in the night conditions has increased the aesthetic and artistic significance. The spirit of the place was changed, the character of the space was changed in the immediate environment, and their historical and cultural value was emphasized. The fortress at Kalemegdan in Belgrade is lit by a bright yellow light that produces too many artistic visual contrasts that are unsuitable for texture and the history of the walls. In Novi Sad, the Varadin Bridge has received the most modern energy-saving LED lighting that delivers a strong contrast with a strong emphasis on the character of the site (Figs. 32, 33, and 34).

The night view of our cities is still in the combined mode of applied lighting technologies. Sodium sources of light dominate the exterior, while white light sources are less represented. The current situation points to the necessary teamwork in the light-urban arrangement of Serbian cities. The point is that computer simulations can often give a wrong, degrading effect, as in the case of the Ministry of Finance

Fig. 32 Night view of the Varadin Bridge in Novi Sad

Fig. 33 Bridge at Ada in Belgrade with LED bulbs [42]

building in Belgrade or the lighting of the City Hall and the Synagogue in Novi Sad (Figs. 35 and 36).

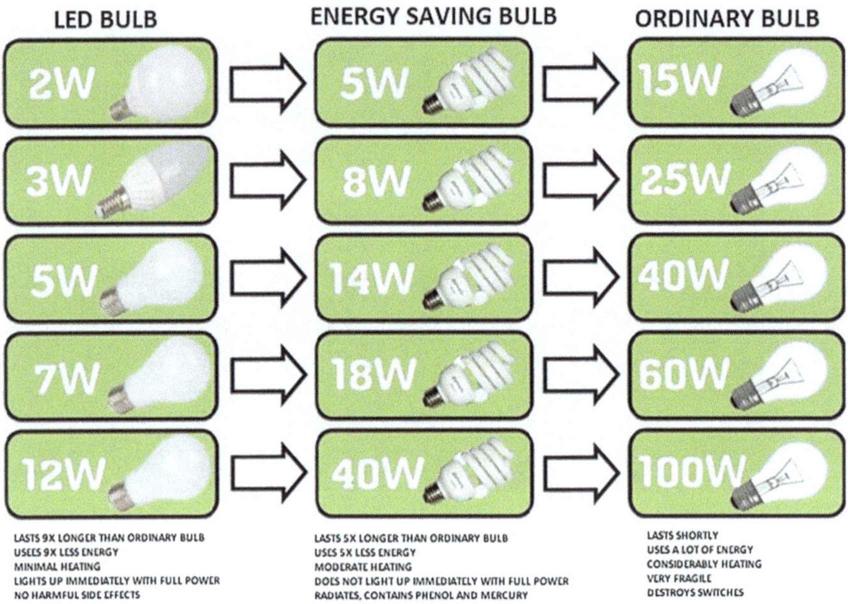

Fig. 34 Tabular display of LED bulbs and comparisons with energy saving and ordinary light bulbs [43]

Fig. 35 Night view of City Hall in Novi Sad

Fig. 36 Night view of the Synagogue in Novi Sad [44]

2.3.2 Street Lighting

Mercury bulbs

In the period from 1996 to 2000, H6 mercury lamps were represented in the street lighting in Serbia. Mercury bulb is a type of lighting body that uses an electric arc through the vaporized mercury to produce light. The arc discharge is generally confined to a small fused quartz arc tube mounted within a larger borosilicate glass bulb. The outer balloon can be transparent or phosphorus-coated. In both cases, the external balloon provides heat insulation and UV protection from the light bulb. Mercury light bulbs are more energy efficient than incandescent bulbs and most fluorescent bulbs, with a light output ranging from 35 to 65 lm/W. Other advantages are: long service life, up to 24,000 h and strong white light (Fig. 37).

As a result, these bulbs are used to illuminate large areas, such as factories with different production programs, warehouses, sports grounds, street lighting, and squares. Translucent mercury bulbs produce white light with a greenish-blue glow due to the mercury combination of spectrum lines. This gives human skin an unsuitable, unrealistic color, and these bulbs are usually not sold in ordinary stores. Pigments with color correction overcome this problem by applying phosphorus on the inside of a balloon that emits white light. This type of bulb gives a better color representation than energy-efficient sodium lamps of high and low pressure. The mercury bulbs function at an internal pressure of about one atmosphere and require special foils as well as electric balusters. Their work also requires a warm-up time of 4–7 min to achieve a

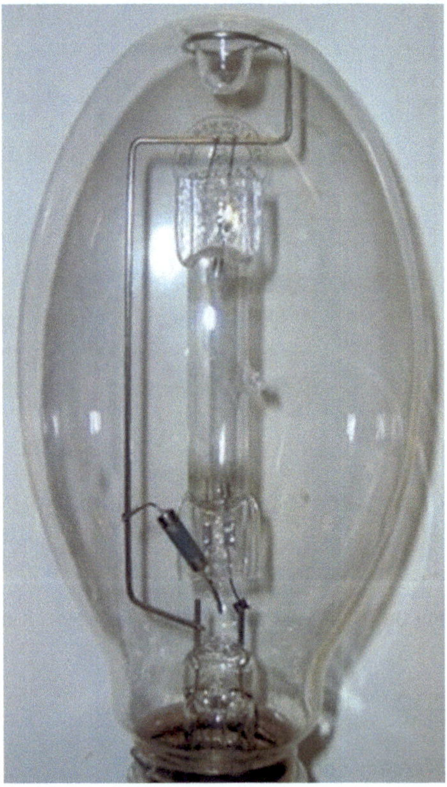

Fig. 37 Appearance of transparent mercury light bulb [45]

full requested, well-distributed light. Mercury light bulbs are becoming more obsolete and non-ecological and are withdrawn from use in favor of metal–halogen bulbs that are more efficient and have a more faithful color display.

Sodium bulbs

In the period 2002–2013 in Serbia, high-pressure sodium lamps were used: "Z1," "Z2," and "Z3" power of 100 W, 250 W, and 400 W, respectively. Thus, the Belgrade highway is illuminated with Z3 lamps with high-pressure sodium lamp. Also, mercury lamps: "ONIX 2"—125 W and "ONIX 3"—250 W are used. These lighting bodies have been used for public lighting in many Serbian cities, especially at petrol stations. Thus, for example, the lighting lamps, such as ONIX, are installed on the Niš–Niška Banja road, at the Zmaj gas station in Belgrade, and other places (Fig. 38).

LED bulbs

Illumination of street space by LED bulbs in Serbian cities is at the beginning phase. In Niš, this type of bulbs is placed in the part of Somborska Street and on the way to the airport building.

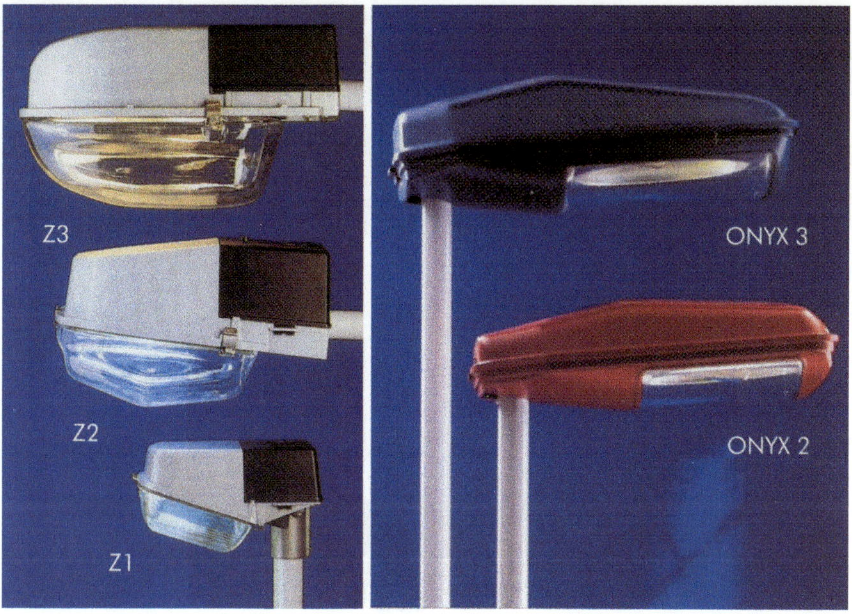

Fig. 38 Display of Z1, Z2, and Z3 lighting bodies as well as ONIX 2 and ONIX 3 lighting bodies

Manufacturers of LED street lights have a number of models in the product range, out of which the most efficient and energy-efficient recommended models are of 60 W, 80 W, 100 W, 120 W, and 150 W, respectively (Figs. 39, 40, 41, 42, and 43).

Fig. 39 LED street lamp, model EL-SL18MA 60 W

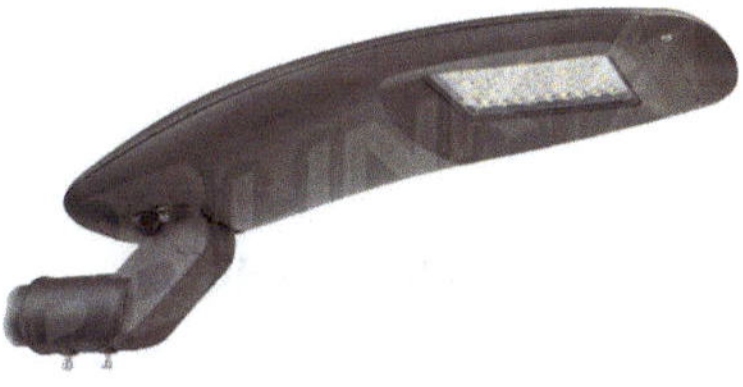

Fig. 40 LED street lamp, model EL-SL18MA 80 W

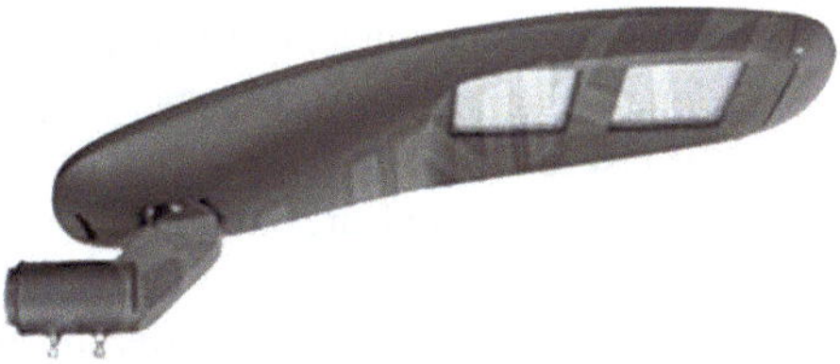

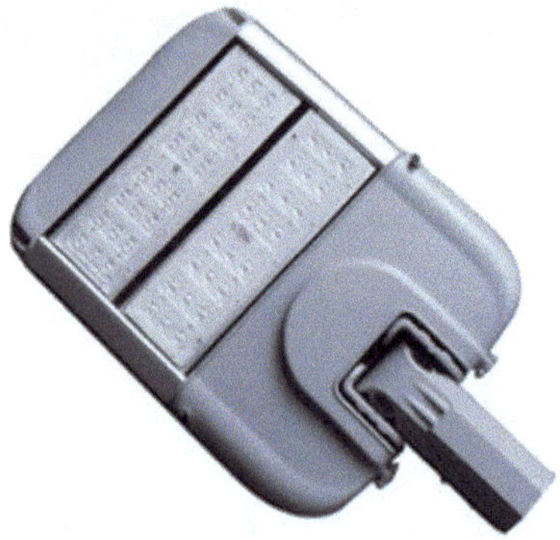

Fig. 41 LED street lamp, model: LED REFLEKTOR LT-T-031 100 W

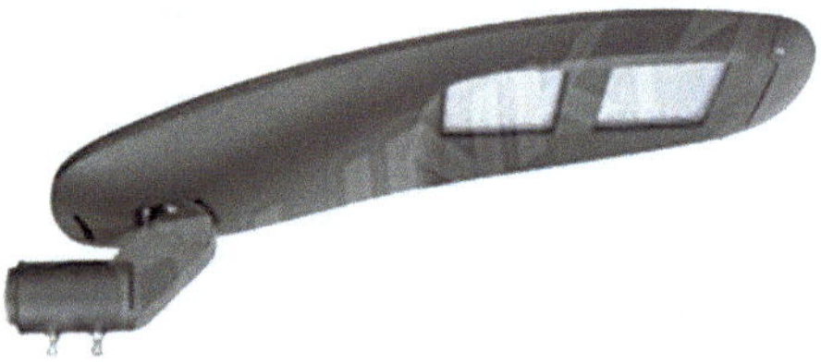

Fig. 42 LED street lamp, model: LED spotlight EL-SL18LA

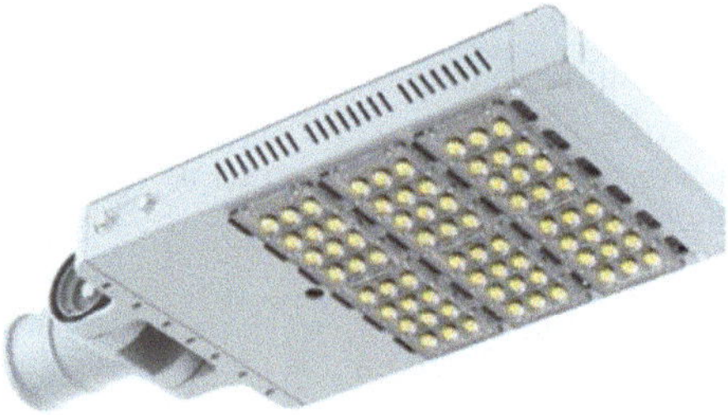

Fig. 43 LED street lamp, LED REFLEKTOR LT-T-036 150 W

2.3.3 Household Lighting

In most cases, households in Serbia use ordinary, classical pear-shaped wolfram filament light bulbs (transparent, amber, and matt), of different designs, sizes and strengths, energy-saving and LED bulbs, from the latest generation of bulbs, with an exceptionally large energy efficiency (Fig. 44).

In the last few years, energy-efficient and lightweight bulbs are used of spiral and standard A-shape which are energy-efficient than conventional bulbs up to four times (Fig. 45).

There are also LED bulbs of different strengths, high efficiency in energy saving, and of a different design. They provide a pleasant light-visual comfort in the interior (Fig. 46).

Fig. 44 Classic bulbs, different shapes with bulbs E27 and E14, clear, amber, and matt

Fig. 45 Spiral and stick energy-saving bulbs

Fig. 46 Different shapes of LED bulbs

2.3.4 Lighting in the Industry

Depending on the type of industrial production, technology in operation and the type of architectural object—dug up, low, floor or high hall, lamps—with different characteristics are used. In principle, halogen lamps of different types are used, white light illumination that does not create the "blind" effect of the users in operation. It provides better concentration in performing precise jobs, especially in workplaces where good visibility is required. In such positions, artificial halogen lighting is of exceptional significance.

From 2007 to 2013 in Serbia, most commonly used were sodium and mercury lamps of the company "Minel-Schreder". Also, in use are rational, fluorescent tubes as well as lamps for special work environments where there are moisture and dust (Fig. 47).

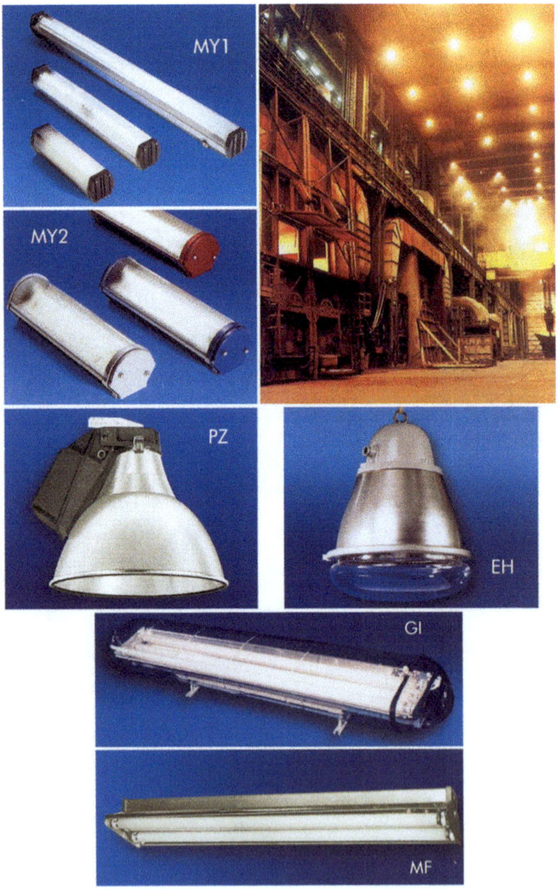

Fig. 47 Bulbs used in industry

Solar street lighting* lamp *from Jagodina

In Serbia, in the period 2002–2017, Feman Co., Ltd. from Jagodina, produced light bulbs similar to the "Minel-Schreder" program. In Jagodina, Feman Co. produces the solar lamp shown in Fig. 48.

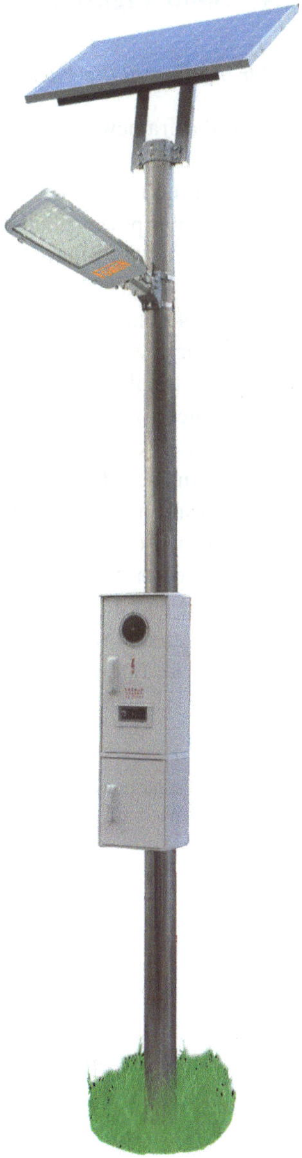

Fig. 48 Solar street lamp lighting Feman company from Jagodina

The solar lamp consists of a support pillar, a 100 W solar module, a battery charging regulator, a 150 battery (Ah) and a 30 W LED lamp. The specified lamp in this system can work continuously for 5–6 h [46–48].

2.4 Solar Lighting

Solar lighting at the Faculty of Sciences and Mathematics in Niš

Within the framework of the project *Ecological lighting*, won by the competition *Run for the Future—1000 young Serbian leaders*, and under the auspices of Philip Moris, on the roof of the Faculty of Sciences and Mathematics in Niš, an off-grid mini PV solar power plant power of 1.05 kW_P was installed. PV solar power plant is intended for the alternative lighting of the faculty and consists of solar modules, control electronics, 24 V DC battery, and inverter, with the output voltage of 220 V AC (Figs. 49 and 50).

PV solar power plant collects solar energy in the daytime; solar modules transform it into electrical and stores it in batteries. In the evening, in accordance with the programmed dynamics, it runs partially the lighting of the yard and parking site of the faculty. The second part of electricity is used to light the interior of the faculty building. In accordance with the dynamics and seasonal change of daylight, lighting of key locations in the building, including stairways, entrance, corridor crossings, etc., is turned on. One group of lamps is in standby mode and is turned on by the motion detector. Such lamps are installed in corridors and rooms where lower traffic is expected and most of the time they are turned off. For winter months, when the lighting is greatly reduced, it is envisaged that the batteries are recharged by the city network grid. In this way, PV system is in operation throughout the entire calendar year (Figs. 51 and 52).

Fig. 49 PV solar power plant, 1.05 kW_P

Fig. 50 Monitoring room

Fig. 51 Solar lighting of the faculty courtyard

PV solar power plant allows lighting of the faculty buildings and its surroundings and in situations where for any reason there is no energy in the electricity network. This is made possible by using the latest generation LEDs, and their total power does not exceed half of the PV solar power plant.

Trim track in Košutnjak in Belgrade

In the national park Košutnjak in Belgrade in January 2012, a 1200 m trim track was lit by electricity generated by solar modules and wind generators. On this occasion, a total of 59 lighting columns height of 450 cm were placed on the trim track. At the 53 lighting columns, one monocrystalline silicon solar module power of 100 W was installed, and on six lighting columns, there are two solar modules of 50 W and one wind generator of 450 W. As light sources, 20 W power LEDs are used, with a lifetime from 50,000 h, with floodlights. In the lighting system, 118 accumulator stationary batteries voltage of 12 V and 70 Ah capacities are installed. The wind

Fig. 52 Solar lighting inside the faculty

turbines have their own charging regulators. The light switches on and turns off using a photocell and a timer. The average daily operation of the system is about 5–6 h, lights are switched off at 1 am, and in the winter at 21.00.

The lighting system on the trim track in Kosutnjak was installed by MiLED Co. in Belgrade (www.ledigps.rs) (Figs. 53, 54, 55 and 56).

The specificity of this system is reflected in the fact that it is located in the forest, with some solar modules occasionally in the shade. For this reason, the solution was used to connect the pillars into one ring, so that the columns with a smaller amount

Fig. 53 Lighting column with solar module

Fig. 54 Lighting columns on trim track in Košutnjak

Fig. 55 Trim track in Košutnjak at night, 2012

of electricity are supplied with energy from the adjacent pillars. This connection facilitates the management of the lighting system (Figs. 57, 58 and 59).

Fig. 56 Lighting column with solar modules and wind turbine, 2012

Ada Huja in Belgrade

Solar lighting at Ada Huja was installed in 2011 and consists of 43 pillars with PV solar modules and 28 W LED lamps (Fig. 60).

Dorćol in Belgrade

In Dorćol, in Belgrade, near the sports center Milan Gale Muškatirović, in 2017, a solar lamp was installed, and it is shown in Fig. 61.

The solar lamp was set up by the Municipality of Stari Grad in cooperation with *Lightinus* and business incubator *Impact Hub Belgrade* [1, 9].

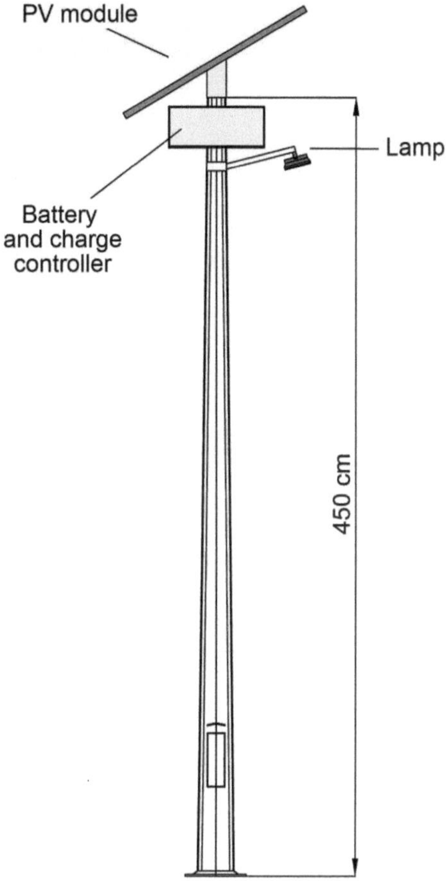

Fig. 57 A schematic diagram of a lighting column with a solar module

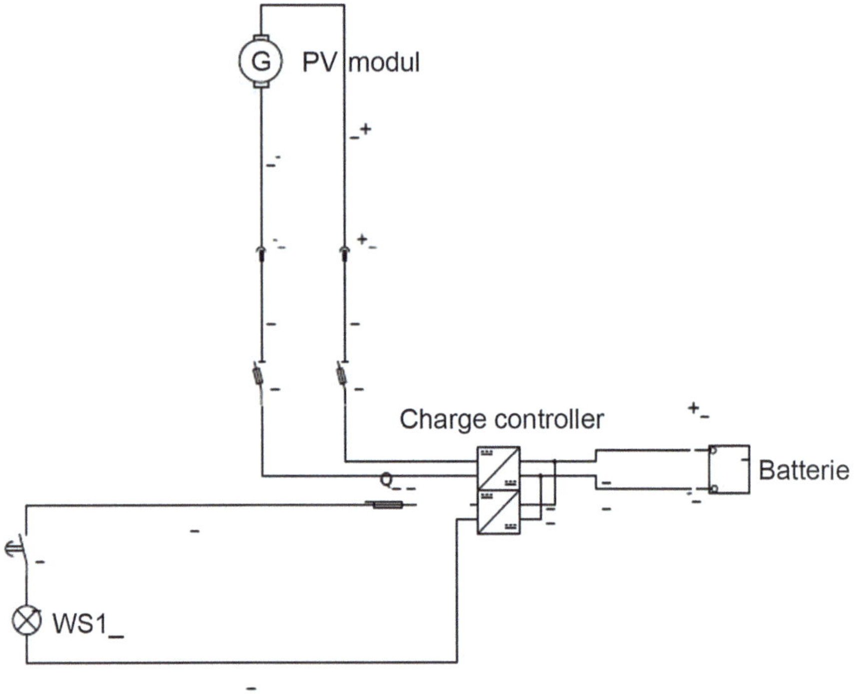

Fig. 58 Schematic view of solar installation for lighting

Fig. 59 LED spotlight

Fig. 60 Solar Lighting at Ada Huja in Belgrade, 2011

Fig. 61 Solar lighting in Dorćol, in Belgrade

References

1. Pavlović MT, Tripanagnostopoulos Y, Mirjanić LjD, Milosavljević, DD (2015) Solar energy in Serbia, Greece and the Republic of Srpska. Academy of Sciences and Arts of the Republic of Srpska, Banja Luka
2. www.hidmet.gov.rs
3. https://www.google.com/search?biw=1366&bih=657&tbm=isch&sxsrf=ACYBGNTF-uqSeEPLSXukfFTOhz822xwMDA%3A1567783920868&sa=1&ei=8HtyXcHSNMri6QS8h4KIAQ&q=raspodele+srednje+godisnje+temperature+u+srbiji&oq=raspodele+srednje+godisnje+temperature+u+srbiji&gs_l=img.3...9016.9016..10046...0.0..0.189.189.0j1......0....1..gws-wiz-img.TXIpIKvnXGg&ved=0ahUKEwjBisCdwrzkAhVKcZoKHbyDABEQ4dUDCAY&uact=5#imgdii=pAhLgwC6Jy7JHM:&imgrc=5NLIYpfqdlTMDM:
4. Ducić V, Radovanović M (2005) Climate in Serbia. Zavod za Udzbenike, Belgrade
5. Gburčik P et al (2004) Study of Serbia's energy potential for the use of solar and wind power. Ministry of Science and Environment, Belgrade (in serbian)
6. Gburčik V (2008) Final report of the technological development project, "Atlas of solar and wind energy potential in Serbia". Institute of Multidisciplinary research of the University of Belgrade (in serbian)
7. Lambić M et al (2011) A study on the assessment of the total solar potential—solar atlas and the possibilities generation and utilization of solar energy on the territory of AP Vojvodina. The Provincial Secretariat for Energy and Mineral Materials, Republic of Serbia, AP Vojvodina, Novi Sad (in serbian)
8. Pavlović MT, Milosavljević DD, Mirjanić LjD (2013) Renewable energy sources. Academy of Sciences and Arts of the Republic of Srpska, Banja Luka
9. Pavlović MT, Mirjanić LjD, Milosavljević DD (2018) Electric power industry in Serbia and the Republic of Srpska. Academy of Sciences and Arts of the Republic of Srpska, Banja Luka (in serbian)
10. Pantić SL et al (2016) The assessment of different models to predict solar module temperature, output power and efficiency for Nis, Serbia. Energy 109:38–48
11. Radonjić SI et al (2017) Investigation of the impact of atmospheric pollutants on solar module energy efficiency. Therm Sci 21(5):21–30
12. Pantić SL et al (2016) Electrical energy generation with differently oriented PV modules as façade elements. Therm Sci 20(4):1377–1386
13. Pantić SL, Pavlović TM (2016) Determination of physical characteristics of horizontally positioned solar module in real climate conditions in Nis, Serbia. Facta Univ 4(1):37–51
14. Radonjić SI et al (2016) Investigation of solar module energy efficiency depending on their surface soiling degree. In: Proceedings of Scientific Conference UNITECH 2016, Gabrovo, vol 1, pp 147–151
15. Cekić N et al (2015) Application of solar cells in contemporary architecture. Contemp Mater VI−2:104–114
16. Radivojević A et al (2015) Influence of climate and air pollution on solar energy development in Serbia. Therm Sci 19:311–322
17. Milosavljević DD et al (2017) Photovoltaic techology: economical framework. In: Proceedings of 7th Scientific Conference "Economics and Management—Based on New Technologies—EMoNT-2017", SaTCIP Publisher Ltd., Vrnjačka Banja, Serbia, pp 136–144
18. Radonjić SI et al (2017) Investigation of the energy efficiency of soiled solar module mounted at the optimal angle. In: Proceedings of International Scientific Conference UNITECH 2017, Gabrovo, vol I, pp 84–88
19. Pavlović T, Milosavljević D, Mirjanić D (2013) Renewable energy sources. Academy of Sciences and Arts of the Republic of Srpska, Banja Luka
20. Lalović B et al (1989) Amorphous silicon solar cells on anodically oxidized aluminum substrate. Solar Cells 26:263–268

21. Kostić Lj, Pavlović T, Pavlović Z (2009) Influence of physical characteristics of flat aluminum concentrators on energy efficiency of PV/Thermal collector. Acta Phys Pol A 115:827–833
22. Kostić Lj, Pavlović T, Pavlović Z (2010) Optimal design of orientation of PV/T collector with reflectors. Appl Energy 87:3023–3029
23. Pavlović T et al (2011) Comparison and assessment of electricity generation capacity for different types of PV solar plants of 1 MW in Soko Banja, Serbia. Therm Sci 15:605–618
24. Pavlović T et al (2013) Possibility of electricity generation using PV solar plants in Serbia. Renew Sustain Energy Rev 20:201–218
25. Pavlović T, Milosavljević D, Piršl D (2013) Simulation of PV systems electricity generation using Homer software in specific locations in Serbia. Therm Sci 17(2):333–347
26. Milosavljević D et al (2014) Assessment of the possibilities of building integrated PV systems of 1 kW electricity generation in some spa resorts in Serbia. SYLWAN 158(6):298–321
27. Milosavljević D, Pavlović T, Piršl D (2015) Performance analysis of a grid-connected solar PV plant in Niš, Republic of Serbia. Renew Sustain Energy Rev 44:423–435
28. Pantić SL, Pavlović MT, Milosavljević DD (2015) A practical field study of performances of solar modules at various positions in Serbia. Therm Sci 19(2):511–523
29. Pavlović T et al (2010) Determining optimum tilt angles and orientations of photovoltaic panels in Nis, Serbia. Contemp Mater I(2):151–156
30. Pavlović MT et al (2013) Experimental determining of energy efficiency of PV solar power plant at the Faculty of Sciences and Mathematics in Niš. Contemp Mater IV(2):112–116
31. Milosavljević D et al (2015) Current state of the renewable sources of energy use in Serbia. Contemp Mater (Renewable energy sources) VI(2):170–180
32. Milosavljević D, Pavlović T (2015) Investigation of the energy efficiency of PV solar power plant installed at the Faculty of Sciences and Mathematics University of Nis. Facta Univ Ser Phys Chem Technol 13(3):141–152
33. Radonjić SI, Pavlović MT (2017) Investigation of the energy efficiency of horizontally mounted solar module soiled with $CaCO_3$. Facta Univ Ser Phys Chem Technol 15(2):57–69
34. Ivanković R (ed) (1993) A century of electricity, 1893–1993. Power Distribution Company of Serbia, Economic Review, Belgrade (in serbian)
35. Roslavcev S (2005) Old thermoelectric power plant in Dorcol-First in Serbia. Electro Power Company, Serbia (in serbian)
36. Stojanović M, Šišković P (eds) (2013) Belgrade from arc lamps and lenterns to LED lighting. Colorgrafx, Belgrade (in serbian)
37. Radojković MŽ et al (1973) 80 years of electrification of Belgrade, 1893–1973. Power Distribution Company, Belgrade (in serbian)
38. http://www.serbia.com/nis-fortress-to-become-a-world-known-attraction/
39. http://www.info026.com/wp-content/uploads/2015/03/tvrdjava-1.jpg
40. http://www.artiscenter.com/wp-content/uploads/2013/06/dom-skupstine-2011-07-25-01.jpg
41. http://forum.pcfoto.biz/?action=dlattach;topic=3857.0;attach=18718;image
42. http://windoworld.ru/serbia/1/13.jpg
43. http://www.unicomled.com/prednosti.html
44. http://www.datourinfo.eu/sites/default/files/imagecache/640/ns2.jpg
45. https://en.wikipedia.org/wiki/Mercury-vapor_lamp#/media/File:MV_Lamp_175_W.JPG
46. Kostić BM (2014) Theory and practice of electrical installation design. Academic Thought, Belgrade (in serbian)
47. Kostić BM (2000) Guide through the world of lighting technology. Minel-Schreder, Belgrade (in serbian)
48. Kostić BM (2006) Lighting of roads. Minel-Schreder, Belgrade (in serbian)

Solar Energy and Lighting in Bulgaria

Plamen Ts. Tsankov

Abstract In this chapter, information about geographical position, climate, solar radiation, renewable energy policy, and solar energy research centers in Bulgaria is given. Also, information about photovoltaic power plants, early development of lighting, development of electrical lighting, modern lighting, public lighting, street lighting, household lighting, industry lighting and solar lighting in Bulgaria is given.

1 General Information

1.1 Geographical Position

Republic of Bulgaria is located in Balkan Peninsula in southeastern Europe. It is bordered by Romania to the north, Serbia and North Macedonia to the west, Greece and Turkey to the south, and the Black Sea to the east (Fig. 1). With a territory of 110,994 km^2, Bulgaria is Europe's 16th largest country.

The borders of Bulgaria have a total length of 2245 km, of them 1181 km is land boundary, 686 km is formed by rivers, and 378 km Black Sea Coast. The northern border with Romania is 609 km. Most of the frontier (470 km) is formed by the river Danube. The northern land border is 139 km long. The eastern border (378 km) is maritime and encompasses the Bulgarian Black Sea Coast. The southern border is 752 km long, of them 259 km is with Turkey, and 493 km is with Greece. The western border is 506 km long, of them 165 km is with the North Macedonia, and 341 km is with Serbia [2].

The territory of Bulgaria is divided into 28 administrative districts and 265 municipalities. The geographic center of Bulgaria (42°45′N 25°30′E) is located in the winter resort Uzana, 22 km away from nearest city Gabrovo.

The relief of Bulgaria is varied. In the relatively small territory of the country, there are extensive lowlands, plains, hills, low and high mountains, many valleys,

P. Ts. Tsankov (✉)
Faculty of Electrical Engineering and Electronics,
Technical University of Gabrovo, Gabrovo, Bulgaria
e-mail: plamen@tugab.bg

T. Pavlovic (ed.), *The Sun and Photovoltaic Technologies*,
Green Energy and Technology, https://doi.org/10.1007/978-3-030-22403-5_6

Fig. 1 Topographic map and geographical position of Bulgaria [1]

and deep gorges (Fig. 1). The main characteristic of Bulgaria's topography is four alternating bands of high and low terrain that extend east to west across the country. From north to south, those bands are the Danubian Plain, the Balkan Mountains, the Transitional region, and the Rilo-Rhodope region. The easternmost sections near the Black Sea are hilly, but they gradually gain height to the west until the westernmost part of the country is entirely high ground. More than two-thirds of the country is plains, plateaus, or hilly land at an altitude less than 600 m. Plains (below 200 m) make up 31% of the land, plateaus and hills (200–600 m) 41%, low mountains (600–1000 m) 10%, medium–high mountains (1000–1500 m) 10%, and high mountains (over 1500 m) 3% [3]. The average altitude of Bulgaria is 470 m.

The population of Bulgaria is 7,101,859 people on December 31, 2016, representing 1.4% of the population of the European Union. Density of the population is 64.3 per km^2. The number of men is 3,449,978 (48.6%) and of women 3,651,881 (51.4%). The average age is 43.5 years. In the cities live 5,204,385 people, making 73.3% of the country's population, and in the villages—1,897,474 people, making 26.7%. The capital Sofia has 1,236,047 inhabitants [2].

1.2 Climate in Bulgaria

Although Bulgaria is not a huge country, it does straddle several climate zones (Fig. 2).

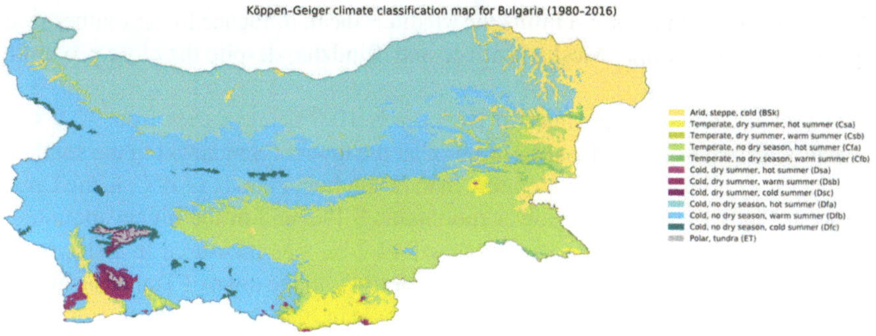

Fig. 2 Köppen–Geiger climate classification of Bulgaria [4]

Bulgaria occupies the southernmost part of the continental climatic zone, with small areas in the south falling within the Mediterranean climatic zone. The continental zone is predominant, because continental air masses flow easily into the unobstructed Danubian Plain. The continental influence, stronger during the winter, produces abundant snowfall; the Mediterranean influence increases during the second half of summer and produces hot and dry weather. Bulgaria is subdivided into five climatic zones: continental zone (Danubian Plain, Pre-Balkan, and the higher valleys of the Transitional geomorphological region); transitional zone (Upper Thracian Plain, most of the Struma and Mesta valleys, the lower Sub-Balkan valleys); continental-Mediterranean zone (the southernmost areas of the Struma and Mesta valleys, the eastern Rhodope Mountains, Sakar, and Strandzha); Black Sea zone along the coastline with an average length of 30–40 km inland; and alpine zone in the mountains above 1000 m altitude (central Balkan Mountains, Rila, Pirin, Vitosha, western Rhodope Mountains, etc.) [3].

Despite the large distance, the most important climate-forming factor is the Atlantic Ocean through the atmospheric circulation of the Icelandic cyclone and the Azores anticyclone, which bring cool and rainy weather in summer and relatively mild weather with abundant snowfall in winter. The influence of the Mediterranean Sea is strongest in the southern parts of Bulgaria, mainly through the Mediterranean cyclones. Due to its small area, the influence of the Black Sea only affects a 30–40 km long strip along the coastline, mainly in summer, when the daily breeze circulation is most pronounced [3].

Another important factor is the relief. The Bulgarian mountains and valleys act as barriers or channels for air masses, causing sharp contrasts in weather over relatively short distances. The Balkan Mountains form a barrier which effectively stops the cool air masses coming from the north and the warm masses from the south. The barrier effect of the Balkan Mountains is felt throughout the country: On the average, northern Bulgaria is about one degree cooler and receives about 192 mm more rain than lowlands of southern Bulgaria. The Rilo-Rhodope Massif bars the warm

Mediterranean air masses and limits the Mediterranean influence to the southern valleys of the rivers Struma, Mesta, Maritsa, and Tundzha, despite the close proximity of the Aegean Sea [3].

The mean annual temperature in Bulgaria is 10.6°C and varies from –2.9°C at the nation's highest peak Musala to 13.9 °C at the town of Sandanski in the southern Struma valley. The average temperature in the Danubian Plain is 11.4°C, in the Upper Thracian Plain 13.9°C, in the lower mountains 8.1°C and in the higher mountains 2.4°C. The highest absolute temperature of 44.2 °C was measured in 1916, in the town of Sadovo, and the lowest of −38.3 °C in 1947, in the town of Tran. The highest temperature in the lowlands and the hilly regions is in June, while in the higher mountains the warmest month is August. The lowest temperature is measured in January and February, respectively. Many valleys experience regular temperature inversions and fogs in winter. The country's lowest absolute temperature was measured during an inversion in the Tran valley [3].

The average precipitation in Bulgaria is about 670 mm. It is uneven in terms of seasons and territory. In northern Bulgaria, the highest precipitation is in May–June, while in southern Bulgaria it is in winter. The average amount of precipitation also varies in terms of altitude—from 450–850 mm in the plains to 850–1200 mm. The lowest mean precipitation is in the eastern part of Dobrudzha and the Burgas Plain (450 mm) and in the area between Plovdiv and Pazardzhik (500 mm); the highest rainfall falls in the mountains—the Petrohan Pass in the western Balkan Mountains and Zlatograd in the Rhodope Mountains. The highest annual rainfall was measured in 1957 in the upper valley of the river Ogosta in the western Balkan Mountains (2293 mm); the highest daily rainfall was recorded at Saints Constantine and Helena resort (342 mm) near Varna in 1951. The total annual amount of the rainfall is 74 billion km^3; of them, 70% evaporate, 20% flow into the rivers, and 10% soak into the soil. Most of the country is affected by droughts in June and August. The snow cover lasts from 20–30 days in the lowlands to 9 months in the highest mountains [3].

Table 1 contains the so-called *Climate normal* data related to precipitation and temperature for the period 1961–1990 for five Bulgarian cities, derived from the

Table 1 Climate data, 1961–1990 normals for Bulgarian cities

	Sofia	Varna	Burgas	Lom	Sandanski
Annual high temperature (°C)	16.3	16.4	17.2	16.2	19.2
Annual low temperature (°C)	5.2	8.5	9.9	7.4	8.8
Average temperature (°C)	10.7	12.0	12.4	11.6	14.0
Average relative humidity (%)	68	76	78	75	65
Sky cloud cover (%)	57	55	54	51	48
Average annual precipitation (mm)	576	464	521	542	483
Days with occurrence of rain (days)	69	90	71	73	56
Annual hours of sunshine (hours)	2043	2033	2202	2031	2575

arithmetic mean of weather measurements from weather station records over three consecutive decades [5].

1.3 Solar Radiation in Bulgaria

The geographical layout of Bulgaria makes 80% of the territory of the country suitable for solar energy utilization. Investigation of the Institute of Hydrology and Meteorology of the Bulgarian Academy of Sciences has sunshine hour data from 45 sites covering 30 years and actual solar radiation measurements from six meteorological stations. The results from the analysis of this data using a correlation relating solar irradiation to sunshine hours, represented as solar energy on the Earth surface that is expressed as the average kilowatt-hours of energy incident on a square meter of horizontal area per year, are shown in Fig. 3 [6].

The territory of Bulgaria could be divided into three zones:

I—with less than 1450 kWh/m^2/year (41% of land area), encompasses mainly the mountainous regions, where sunshine is less than 2000 h/year;
II—with 1450–1500 kWh/m^2/year (52% of land area), encompasses regions in the Danube plain, northwest Bulgaria, the Trace lowland in south Bulgaria and some pre-mountain regions, where the amount of sunshine ranges from 2000 to 2200 h/year;
III—with more than 1500 kWh/m^2/year (7% of land area), encompasses regions in the southeast, part of the southern Black Sea coastal region and the valleys of the rivers Struma, Mesta, and Maritza, where the amount of sunshine is over 2200 h/year.

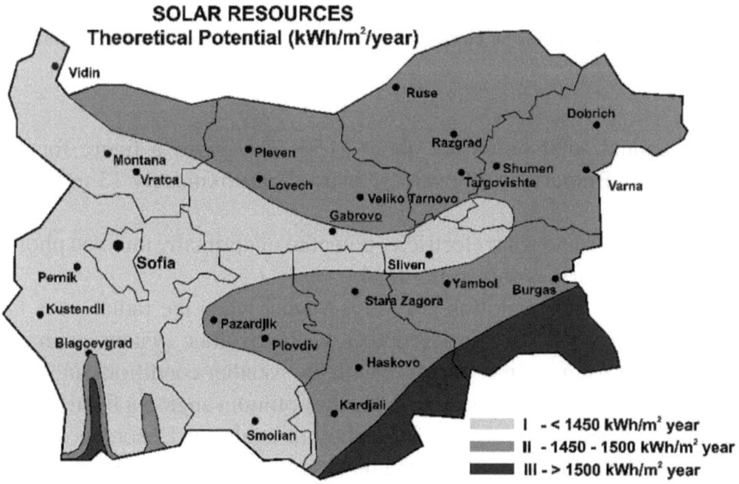

Fig. 3 Solar resource theoretical potential of Bulgaria [6]

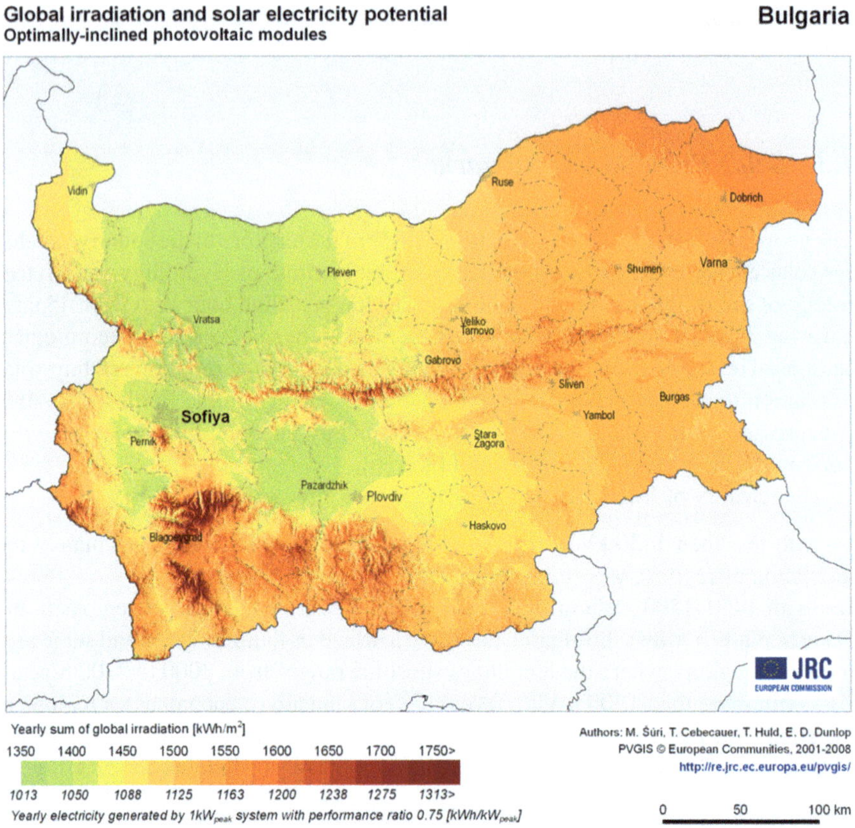

Fig. 4 Global irradiation and solar electricity potential of optimally inclined photovoltaic systems in Bulgaria [7]

Average annual solar hours are about 2150. This gives a figure for the solar energy falling on Bulgarian soil over one year of approximately 13 million tons of oil equivalent.

Global irradiation and solar electricity potential of optimally inclined photovoltaic system map of Bulgaria is shown in Fig. 4.

The best months for operation are May–August when the radiation is far more than the average for the year. In December, a photovoltaic system generates three times less than in August. In connection with the weather conditions and the annual change of the Sun's azimuth, it is proved that the optimum angle for fixed photovoltaic systems lies between 30° and 34°. The optimum angle for December is 63° while for June is 7°. The Linke turbidity in the most southern parts of Bulgaria reach up to 4.8, but the mean value is about 3 on annual basis. The mean annual ratio of diffuse to global irradiation is 0.5.

1.4 Renewable Energy Policy in Bulgaria

One of Bulgaria's priority objectives is bringing the country's social, political, and economic life into line with the European norms and standards.

Bulgaria has taken an active role in the international efforts to help prevent climate change by supporting the concerted actions of the European Union and the wide-ranging package of measures in the energy sector. These measures give a new impetus to Europe's energy security and support the EU's '20-20-20' targets. The widespread use of renewable sources and the implementation of energy efficiency measures are among the priorities of the national energy policy and are in conformity with the objectives of the new energy policy for Europe [8].

The National Renewable Energy Action Plan ("NREAP") is the main instrument developed to ensure the achievement of the national renewableenergy targets. The plan has been drawn up in accordance with the requirements of Directive 2009/28/EC with the template adopted by Commission Decision of June 30, 2009.

Under Directive 2009/28/EC, Bulgaria's mandatory national target for 2020 is a 16% share of energy from renewable sources in the gross final consumption of energy, including a 10% share of energy from renewable sources in the consumption of energy in the transport sector.

The use of energy from renewable sources—in line with the requirements of Directive 2009/28/EC—is analyzed, promoted, and reported separately along three lines:

- consumption of electricity—from wind, solar and geothermal energy, hydropower and biomass;
- consumption of energy for heating and cooling—solar and geothermal energy and biomass;
- consumption of energy from renewable sources in transport—biofuels and electricity produced from renewable sources.

The long-term implementation of the renewableenergy policy is ensured by the national legislation which reflects and fully implements the requirements laid down by the European Parliament and the Council with regard to energy generation from renewable sources.

The national policy on the promotion of energy generated from renewable sources has the following objectives:

- promotion of the development and use of technologies for the production and consumption of energy obtained from renewable and alternative energy sources;
- promotion of the development and use of technologies for the production and consumption of biofuels and other renewable transport fuels;
- diversification of the energy supply;
- environmental protection;
- creation of conditions for sustainable development at local and regional level.

The total technical potential of energy production from renewableenergy sources in Bulgaria is about 4500 ktoe annually. The contribution of different sources is

Table 2 Estimate potential of the renewableenergy sources in Bulgaria as per an updated assessment of 2009

Renewable source according to Regulation 1099/2008 for the energy statistics	Technically available potential, ktoe
Hydropower	1290
Geothermal energy	18 (331 with use of reinjection technologies)
Solar energy	389
Tidal energy	Unspecified
Wind energy	315
Solid biomass	1524
Biogas	280
Liquid fuels	366
Total	4495

uneven (Table 2), the biggest share being hydropower (~31%) and biomass (~36%). Bulgaria's geographical location explains the relatively marginal role of wind energy (~7.5%), tidal energy, and sea wave energy. However, the country has significant forestry resources and a developed agricultural sector as sources of solid biomass and raw material for biogas and liquid fuels. By 2008, Bulgaria chiefly exploited the potential of hydroelectric energy, as well as that of solid biomass, which is used primarily to heat households and public buildings. The production of electricity from wind and solar power plants is under rapid development, as is the use of solar energy for households' hot water needs [9].

According to data for the base year 2005 (in conformity with Eurostat), the energy from renewable sources in the country amounts to 1 mio toe, or 9.4% of the total final energy consumption, of which biomass—70%, hydropower—24%, and other RES—6%.

At present in Bulgaria, the potential of solid biomass is most fully utilized primarily as fuel for heating in households and public buildings, as well as that of hydropower from HPP.

Generation of electricity by wind generators and photovoltaic plants, as well as the use of solar energy for domestic water heating, is developing at high rates.

Table 3 presents installed capacity data, the net generated electricity, and their shares (Fig. 5) by power plant type in 2017 of the enterprises connected to the transmission network provided by Electricity System Operator of Bulgaria (ESO).

Information on the real-time change of the generated electric power by type of the power plants and the electric load of Bulgaria‘s energy system can be viewed on the ESO Web site: www.eso.bg in menu: "Dispatching/Operational Data."

Measures to encourage the use of RES to produce electricity have led to a twofold increase in its share of total electricity generation over the past 10 years, which is already approaching about 20%—Fig. 6.

Table 3 Installed capacity and net generated electricity in Bulgaria for 2017 [10]

Power plant type	Installed capacity		Net electricity generated	
	MW	Share (%)	MWh	Share (%)
1. Nuclear	2000	16.6	14,718,368	36.2
2. Thermal—lignite coal	4119	34.1	17,605,902	43.3
3. Thermal—black and brown coal	362	3.0	246,111	0.6
4. Thermal—gas	563	4.7	1,609,514	4.0
5. Hydro including:	3204	26.5	3,395,131	8.4
5.1. Pumped storage generation	1399	11.6	899,639	2.2
5.2. Pumped storage pumps	933	7.7	647,485	1.6
6. Renewable energy including:	1822	15.1	3,054,993	7.5
6.1. Wind	701	5.8	1,414,564	3.5
6.2. Photovoltaic	1043	8.6	1,325,472	3.3
6.3. Biomass	78	0.6	314,956	0.8
Total	12,070		40,630,018	

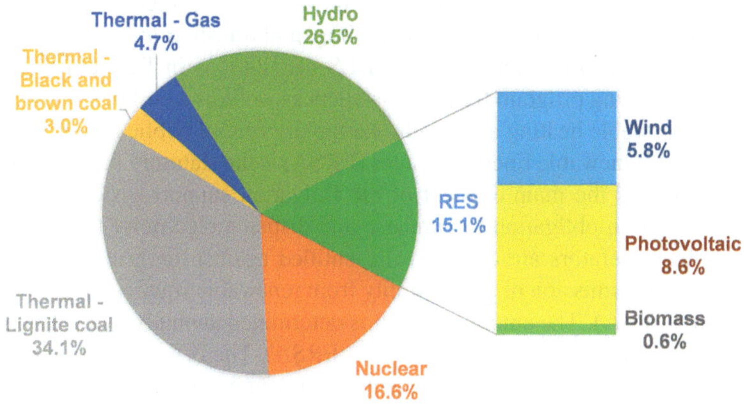

Fig. 5 Installed capacity shares by power plant type in Bulgaria for 2017

Bulgaria ranks 11th in the European Union by specific installed photovoltaic power per capita with 144.8 W/inhabitant and 1390 GWh electricity production from solar photovoltaic power in 2017 [11].

In Bulgaria, electricity from renewable sources is mainly promoted through a feed-in tariff (FiT). Producers of electricity from renewable sources are contractually entitled against the grid operator to the purchase and payment of electricity at a guaranteed price. The connection of renewable energy plants to the grid is subject to the provisions of the general legislation on energy. Renewableenergy is not given priority access. The use of renewableenergy for heating and cooling is promoted through a subsidy from the European Regional Development Fund, several

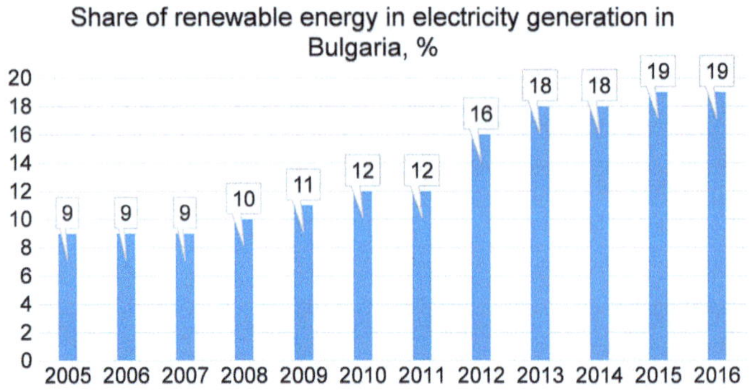

Fig. 6 Change in the annual share of electricity produced from RES in Bulgaria

loan schemes and through an exemption for building owners from property tax. In Bulgaria, the main support scheme for renewableenergy sources used in transport is a quota system. This scheme obliges companies importing or producing petrol or diesel to ensure that biofuels make up a defined percentage of their annual fuel sales. Furthermore, biofuels are supported through a fiscal regulation mechanism. The policies on electricity, heating and cooling, and transport aim at promoting the development, installation, and usage of RES-installations in Bulgaria: There is a professional training program for RES installers as wells as a building obligation for the use of renewable heating and for the exemplary role of public authorities [12].

The Act on Renewable Energy Sources (ERSA) is the statutory basis for the feed-in tariff, which is the main element of the Bulgarian support system. The ERSA also establishes an obligation to purchase and dispatch electricity from renewable sources. Plant operators are contractually entitled against the grid operator to the purchase and transmission of all electricity from renewable sources supplied (art. 18 par. 1 item 2 ERSA). The amount of tariff is determined annually by the Energy and Water Regulatory Commission (art. 32 par. 1 ERSA). The FiT and the purchase obligation apply to power purchase agreements (PPAs) signed for projects implemented before the achievement by the Republic of Bulgaria of the RES end consumption mandatory targets under the National Renewable Energy Action Plan. The Energy and Water Regulatory Commission (EWRC) regulates the electricity selling price at the wholesale market and the FiT at which the RES producers sell electricity to suppliers [12].

EWRC updates the FiT in the middle of each year, based on photovoltaic electricity purchased in the previous year and current market conditions (Table 4). When introduced in 2008, FiT was about five times higher than the average price of electricity for households and industry, for all types of photovoltaic power plants. This high price has led to significant investments being made to design and build a large number of photovoltaic power plants. For only 2 years between 2011 and 2013, about 1 GW_P of photovoltaic power was put into operation (Fig. 7), significantly increasing the share of electricity production from RES. This, in turn, has led to an increase in

Table 4 Change in the years of the feed-in tariff for electricity produced by photovoltaics in Bulgaria, BGN[a]/MWh

Installed power and type of photovoltaic power plants (PPP)		Date												
		01.7.2018	01.7.2017	30.6.2016	30.6.2015	01.7.2014	28.6.2013	29.8.2012	28.6.2012	20.6.2011	30.3.2011	31.3.2010	30.3.2009	31.3.2008
Up to 5 kW_P		–	–	–	–	152.2	195.4	193.4	268.7	576.5	760	793	823	782
Over 5 kW_P											699	728	755	718
Up to 30 kW_P														
Over 30–200 kW_P		–	–	–	–	143.4	191.1	188.1	260.8	567.4				
Over 200–10,000 kW_P		–	–	–	–	134.0	176.3	171.4	237.1	485.6				
Over 10,000 kW_P		–	–	–	–	131.4	160.2	169.9	236.3					
PPP installed on roofs and facades	Up to 5 kW_P	242.1	271.7	255.4	228.0	211.8	354.0	381.2	400.7	605.2				
	5–30 kW_P	206.0	231.2	213.9	211.7	204.0	284.2	290.0						
	30–200 kW_P	–	–	–	–	169.1	211.4	226.9	369.1	596.5				
	200–1000 kW_P	–	–	–	–	144.7	196.6	206.3	316.1	583.8				

[a]Bulgaria uses a fixed exchange rate to the euro: 1 BGN = 0.511292 EUR

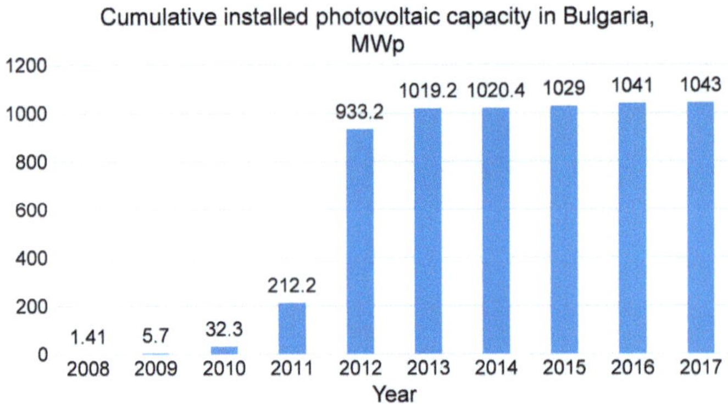

Fig. 7 Amendment during the years of the cumulative installed photovoltaic capacity in Bulgaria, MW_P

the price of electricity as the necessary funding to cover FiT is obtained by increasing the RES penetration in the cost of electricity for the household and the industry. Finally, this resulted in a strong reduction of FiT and its separation according to the type of photovoltaic power plant.

In 2014, the FiT dropped out for all terrestrial photovoltaic power plants and installed on roofs with power over 30 kW_P and FiT currently exists only for small roof and facade photovoltaic systems (Table 4). This explains the comparatively small new installed capacities of photovoltaic systems in Bulgaria in the last 5 years (Fig. 7).

With the lowering of the prices for the construction of photovoltaic power plants in recent years and, at the same time, the large increase in the price of electricity for the industry in Bulgaria, there is a new trend to build PV power plants in industrial enterprises for their own needs without signing contracts for the use of FiT. This trend shows that in Bulgaria comes the moment when the construction of photovoltaic power plants becomes a profitable investment without the need of FiT support, and this is a motivation for the beginning of a new period of increase in the use of photovoltaic systems.

1.5 Solar Energy Research Centers in Bulgaria

Scientific and applied research on solar energy in Bulgaria is distributed in many places, the main ones being Bulgarian Academy of Sciences and a number of universities with a technical profiles in Gabrovo, Sofia, Varna, Ruse, Blagoevgrad, etc., in which specialized scientific research laboratories have been established.

Bulgarian Academy of Sciences—Central Laboratory of Solar Energy and New Energy Sources

Central Laboratory of Solar Energy and New Energy Sources at the Bulgarian Academy of Sciences (CL SENES-BAS) was founded with a decree of the Council of Ministers of Bulgaria dated July 1, 1977. The new scientific institution was organized on the basis and a part of the staff and facilities of the Institute of Solid State Physics (ISSP–BAS) with a decision of the Scientific Council of ISSP from date November 11, 1977.

The main activities of CL SENES are scientific and applied research and development in the field of conversion and utilization of solar energy as well as other renewable sources. In conformity with the law for Bulgarian Academy of Sciences and the decision of the General Assembly of BAS from June 6, 1994, the CL SENES was ratified as permanent scientific institution at the Bulgarian Academy of Sciences with a status of legal person with budgetary support with activities: fundamental and applied scientific research, consulting and expert work, integration of scientific results in practical applications, training of highly qualified specialists in the field of solar energy utilization (photovoltaic and photothermal conversion of solar energy, optical and photoelectric properties of materials, formation of new materials, structures, devices, installations, etc. After the structure reformation of BAS in 2010, CL SENES received (in accordance with article 6 of Statutes of Bulgarian Academy of Sciences), the statute of specialized academy unit (SAU) in the frames of the autonomous research units.

CL SENES is a scientific organization, which has been visible on the national and international aspect from its foundation. The laboratory is one of the founders of the European Center of Solar Energy at UNESCO in 1978. During the period 1978–1989, CL SENES supervised the activities in the field “Nontraditional sources and methods for solar energy conversion” for development of the cooperation among the Academies from Eastern Europe. In 2002, the laboratory has been granted as “Bulgarian Center of Solar Energy” in the framework of program “Center of excellence” of 6 FP of European Commission.

CL SENES organized and implemented the first national conferences on renewableenergy sources (RES) (I National Conference on RES—1994, II National Conference on RES—1999, and III National Conference on RES—2003), and these conferences made popular the renewable energy sources thematic in the Bulgarian society.

The scientists in CL SENES are leaders in solar energy research in Bulgaria. Their efforts are focused in several directions such as

- New materials preparation and study, development of new technological processes for highly effective solar energy converters;
- Photovoltaic and solar thermal system design.

The staff is involved in demonstration and research projects in cooperation with numerous national and international partners in the EU and beyond. The actuality of the thematic and the proved capacity of researchers at CL SENES gave the opportunity of the unit to participate in projects, financed by European scientific programs as follows:

- European projects: 4 FP—4 projects, 5 FP—5 projects, 6 FP—2 projects, and 7 FP—2 projects;
- Program "Intelligent Energy-Europe"—1 project (2009–2011);
- Program "Transnational Cooperation Program South-East Europe"—1 project (2011–2013);
- UNESCO: two demonstrational projects.

Figure 8 shows some of the photovoltaic systems designed and constructed by CL SENES:

- Figure 8a First grid-connected (2006) power plant in Bulgaria. Installed power 10 kW_P monocrystalline silicon modules, CLSENES-BAS, Sofia;
- Figure 8b Grid-connected PV power plant. Installed power 75 kW_P, multicrystalline silicon, Roman;
- Figure 8c First grid-connected (2010) high concentration-level PV system. Installed power 3.3 kW_P concentration ratio 476x. PV cell type, GaInP/GaInAs/Ge, CL SENES-BAS, Sofia;
- Figure 8d Grid-connected PV power plant based on multi and tandem–junction structure (amorphous silicon/microcrystalline silicon); installed power 780 W_P, CLSENES-BAS, Sofia;
- Figure 8e Grid-connected PV system with installed power 750 W_P multicrystalline silicon modules, CL SENES-BAS, Sofia.

Technical University of Gabrovo, Faculty of Electrical Engineering and Electronics, Department of Electric Power Distribution and Electrical Equipment—"Scientific Research Laboratory Photovoltaic Systems"

The laboratory was built gradually in the mid-1990s as a result of participation and work of the department of Electric Power Distribution and Electrical Equipment (EPDEE) (Fig. 9) on several international projects in the field of renewableenergy sources. At present, the laboratory is a participant in a number of university projects in the field of RES, and in the following, international ones with partners form Germany, Italy, Czech Republic, Greece, Romania, Serbia, Bosnia and Herzegovina, etc.

- Project JOU2-CT92-0155 "Development of a Stand Alone PV Power System for Remote Villages Making Use of Pumped Water Energy Storage";
- Project ICOP-DEMO-2145-96 "Demonstration of a Hybrid Powered System for Navigation Lighthouses—Standardised Solution for Remote and Ecologically Sensitive Areas at the Black Sea Coast";
- Project ICOP-DISS-2148-96 "Follow up Activities Concerning the Promotion of Hybrid Renewable Energy Sources in Eastern Europe";
- Project ICOP-DEMO-2154-96 "Development and Application of a Water Pumping System for Remote Areas Consisting of Photovoltaic (PV) Modules with

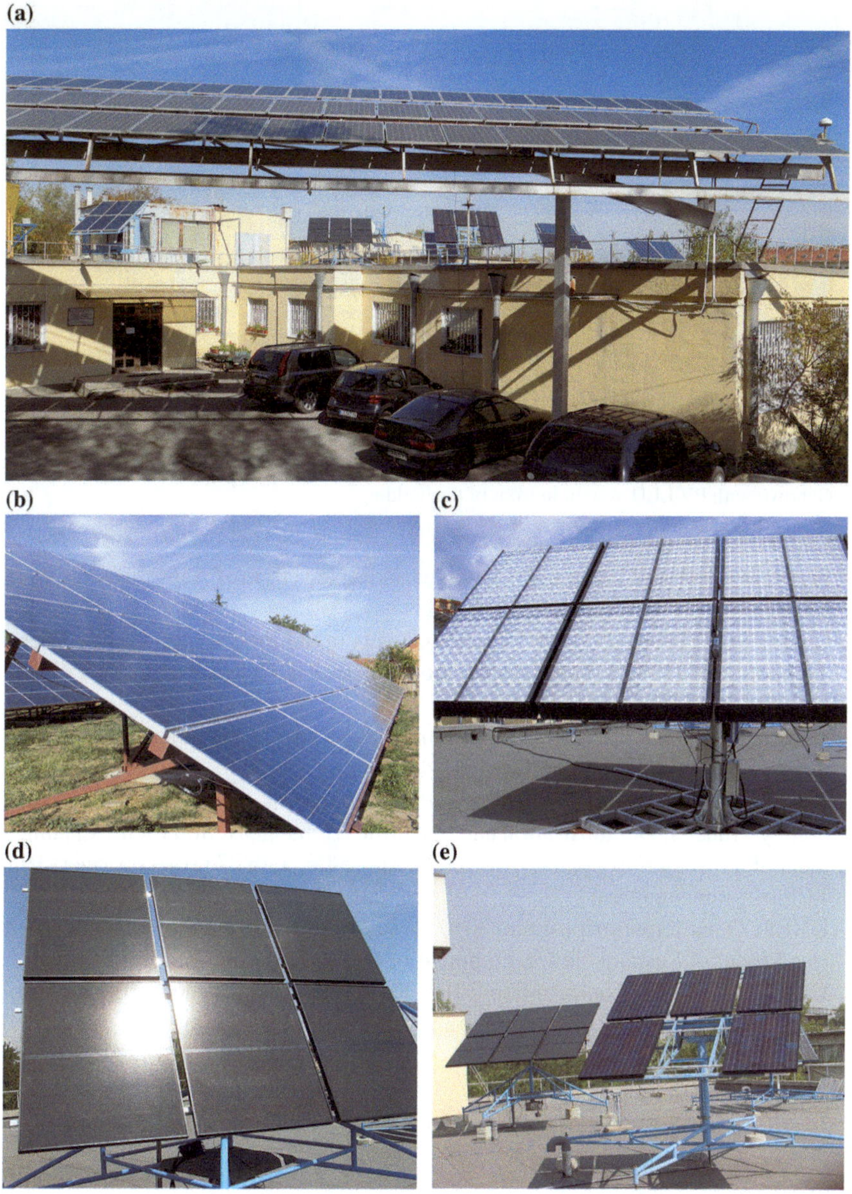

Fig. 8 Photovoltaic systems at Central Laboratory of Solar Energy and New Energy Sources [13]

Fig. 9 Department of Electric Power Distribution and Electrical Equipment at Technical University of Gabrovo with PV-LED system in front of the building

Inverters Integrated into the PV Modules and a New Type of Asynchronous Pump Motor";

- Contract 6.7211-40/I/97-005 "Training, Visits in the Field of Renewable Energy Technologies (Photovoltaic Applications) between Greece/Germany and Bulgaria/Romania";
- UNESCO course "Expert in design, installation and management systems with renewable energy—photovoltaic," 2004, Italy;
- Fifth Framework Program of the EC—PV Enlargement—NNE5/2001/736 "Technology Transfer, Demonstration and Scientific Exchange Action for the Establishment of a Strong European PV Sector," 2002–2006, with 27 project partners from 11 European countries;
- UNESCO participation program 2010–2011, project "Renewable Energy Sources as a Model of Sustainable Development of the Countries of West Balkans";
- UNESCO participation program 2012–2013, project "Influence of Energy Efficiency of Solar Energy on Economic and Sustainable Development for the Western Balkan Region";
- UNESCO participation program 2015, project "The Influence of Renewable Energy Sources to the Protection of the Environment in West Balkan Countries."

At the basis of the solar energy research is the construction of the TU-Gabrovo territory of three different types of photovoltaic systems, equipped with modern measuring and monitoring devices, which will be briefly presented below.

A roof-mounted 10-kW_P grid-connected photovoltaic power plant at the Rectorate of Technical University of Gabrovo was installed in 2005 (Fig. 10). It consists of three subsystems. The first and the second subsystems with single installed power of 3.42 kW_P are identical, each containing 12 polycrystalline silicon PV modules, type ASE-250 DG-FT/MC. The third subsystem with power of 3.22 kW_P is built from a field of 100 amorphous silicon PV modules type ASE-F 32/12. A monitoring

Fig. 10 Roof mounted 10 kW_P grid-connected photovoltaic power plant

system performs regular measurement and storage of the following data: global solar radiation; solar radiation in the plane of PV modules; temperature of the two types of PV modules; temperature of the silicon reference cell; ambient temperature; current, voltage, and power of the DC side of three subsystems; DC and AC energy produced by the three subsystems; total energy produced. Software for real-time monitoring and data storage of meteorological and electrical operational parameters of the photovoltaic system is developed in the laboratory (Fig. 11).

In 2000, a stand-alone photovoltaic system consisting of 5 PV modules, mechanical structure with the ability to modify the inclination of the modules, rechargeable battery, and control panel with solar controller and measuring system was built on the roof of the Department of EPDEE building (Fig. 12). The PV modules are Siemens Sun Power A75 with power 75 W_P—2 modules, Solar Technik—IBC-80 with power 80 W_P—2 modules, and Solara SM 200 PW2 with 50 W_P power—1 module. The rechargeable battery is with immobilized electrolyte in an AGM separator with a capacity of 100 Ah and a nominal voltage of 12 V. The connection of alternating current consumers is done by a sinusoidal inverter EA-SWI-400-12. The main electrical board of the PV system with the measuring system and the computer is located in a laboratory in the department, allowing for various operating modes and experiments to be realized.

In 2010, a PV-LED system was built on two poles in front of the building of the Department of EPDEE (Fig. 13), consisting of 80-W_P monocrystalline silicon photovoltaic module, 100 Ah/12 V storage battery mounted in a metal box, LED luminaire with a power of 18 W and MPPT-type solar controller TRACER 1210 RN. The information on electric processes taking place and on studying the operating

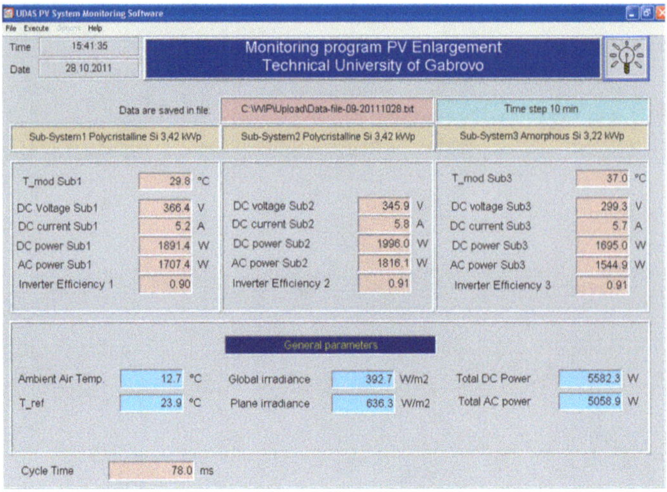

Fig. 11 PV system real-time monitoring and data storage software developed

Fig. 12 Stand-alone photovoltaic system on the roof of the Department of EPDEE

modes of PV-LED system from the solar controller is sent to LCD display meter MT-2 located in laboratory at the Department of EPDEE (Fig. 13) [14].

Except in the field of photovoltaic systems, laboratory "Scientific Research Laboratory Photovoltaic Systems" and Department of EPDEE provide training and research on wind energy, hydrogen fuel cells, Stirling machines, thermal imaging inspection and diagnostics (Figs. 14 and 15), and smart technologies and software for monitoring and management of RES systems.

Technical University of Sofia, Faculty of Electrical Engineering, Department of Electrical Machines, Laboratory on Renewable Energy Sources

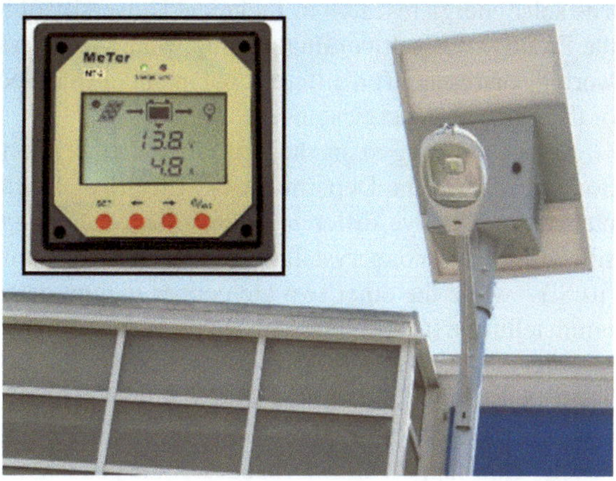

Fig. 13 PV-LED system in front of the building of the Department of EPDEE

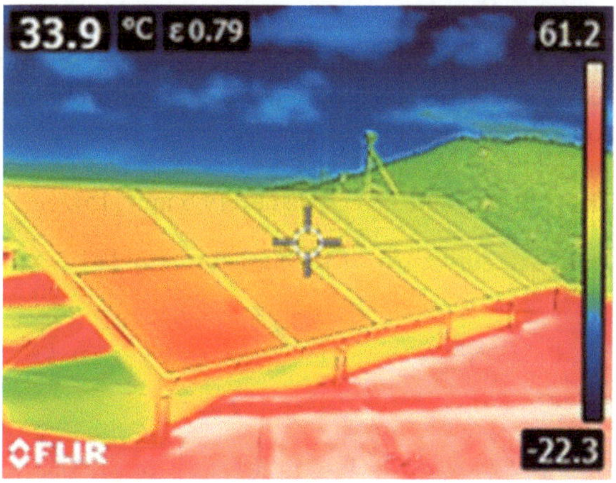

Fig. 14 Thermal imaging inspection of photovoltaic power plant

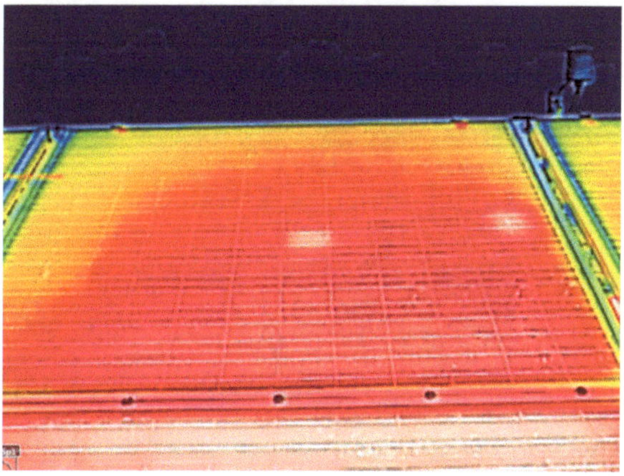

Fig. 15 Thermal image of PV module with probable problems in individual cells

The solar energy research in Technical University of Sofia is concentrated mainly in the Faculty of Electrical Engineering, where in several departments there are teams working on research on different, but mainly in the electrical aspects and applications of the renewableenergy sources.

Among the largest in the institution is the Laboratory on Renewable Energy Sources built in the Department of "Electrical Machines," where an experimental platform using five different photovoltaic technologies is constructed. Three are silicon-based—monocrystalline (mSi), polycrystalline (pSi) and microcrystalline (μcSi)—while the other two are copper indium gallium selenide (CIGS) and cadmium telluride (CdTe). For each technology, a different number of modules are used in order to form five PV generators with similar peak power—around 1200 W_P. The connection to the grid of the PV arrays is made with five single-phase inverters Sunny Boy 1200. Figure 16 shows the three trackers' systems bearing the PV panels. This support allows changes of the panels' orientation by following the Sun or by fixing all systems at the same azimuth and inclination angle. The experimental platform is situated in grounds of the Technical University of Sofia, Bulgaria (42°39′16″N, 23°21′17″E). The experimental platform uses a monitoring and data acquisition system based on the Sunny WebBox data logger to collect information about the meteorological conditions (ambient and cell temperature, solar radiation and wind speed), and about the produced energy (currents, voltages, powers, etc.). The collected data is recorded every 5 min [15].

Technical University of Varna, University Laboratory New Energy Sources

Fig. 16 Experimental platform in the Technical University of Sofia—solar trackers and inverters [15]

Fig. 17 700 and 1000 W_P photovoltaic systems in Technical University of Varna [16]

In 1997, an Inter-University Center "Energy-Nature-Balkan" was established with an affiliate of the University of Chemical Technology and Metallurgy, Sofia, and a section of "Prof. Assen Zlatarov" University, Bourgas, under the TEMPUS JEP S 97258-94 project. Later, the center was transformed into University Laboratory "New Energy Sources". The laboratory has two photovoltaic systems with power 700 and 1000 W_P (Fig. 17). The laboratory is provided with the equipment required for practical student training and scientific research: hybrid solar and wind energy transformation system; apparatus for converting solar energy into heat; laboratory installation for decomposition of water with solar energy; Stirling engine; solar photovoltaic modules 50 and 70 W; aerodynamic test bench; apparatus for exploring the parameters of the sun, wind, and other RES.

In 2002, the SOLTRAIN project from the European program "ALTENER" was launched, which concerns training in solar energy and, in particular, photovoltaic transformation in electricity with the contractor Fraunhofer Institute for Solar Energy, Freiburg, Germany. Partners in the project are universities, research institutes, and companies from the UK, Poland, Hungary, Slovenia, and Romania. On the Bulgarian side, Technical University of Varna participates. As a result of the project in 2004, training materials were developed in three volumes (550 pages) that are used in the learning process. At the same time, there have been educational courses in the field of photovoltaic transformation [16].

"Angel Kanchev" University of Ruse, Faculty of Electrical Engineering, Electronics and Automation, Department of Electric Power Engineering—Laboratory Photovoltaic Systems

(a) pyranometer and PV test modules

(b) 2025 W_P grid-connected PV system

Fig. 18 Laboratory Photovoltaic Systems at "Angel Kanchev" University of Ruse [17]

Research began in 2002, and the first results were published in 2003. Initially, a Kipp & Zonen CM11 pyranometer was purchased and installed on the roof of Building 10 of Rousse University, and solar radiation intensity monitoring was carried out. The first photovoltaic system is autonomous on the roof of the building with a power of 360 W_P (3 PV modules 120 W_P) and an inverter with a power of 400 W (Fig. 18a).

In 2010–2011, a new laboratory under contract BUL/SPG/05P4/Y3/CORE/2010/21 of the Global Environment Facility and Municipal Energy Agency Ruse was built. The 2025 W_P PV system (Fig. 18b) has nine SOLOK P220 polycrystalline photovoltaic modules with a single peak power of 225 W_P connected in one string and single-phase grid-connected inverter SolarMax 2000C. The laboratory performs the following studies: impact of operating factors on the energy produced; quality of electrical energy; exploring the work of the inverters; modeling and research of photovoltaic systems using specialized software; smart grid; communication between devices; cybersecurity in the smart grid [17].

University of Mining and Geology "St. Ivan Rilski," Sofia, Laboratory of Renewable Energy

At the University of Mining and Geology "St. Ivan Rilski," a training laboratory "Renewableenergy sources" was set up under the TEMPUS program 09383-95. The laboratory is equipped with the following facilities: 24 photovoltaic modules with a single power of 120 W_P and a total power of 2880 W_P connected via inverter to the mains; 1 kW wind generator; four monocrystalline modules BP 585F with a single power of 85 W_P and a total output of 340 W_P (Fig. 19), which along with the wind generator, charger, and 400 Ah batteries form an autonomous system; solar collectors for hot water with total area of 3.76 m^2, circulation pump and heat exchanger with a volume of 300 l. Student have been trained and renewable energy sources and their electrical characteristics have been investigated in the laboratory since 1999 [18].

South-West University "Neofit Rilski," Blagoevgrad, Faculty of Mathematics and Natural Sciences, Solar Energy Center

Fig. 19 Roof-mounted photovoltaic system in University of Mining and Geology "St. Ivan Rilski," Sofia [18]

Beginning of the research in Solar Energy Center in Blagoevgrad starts with the participation of specialists of the center who have participated in several research and educational projects in the field of renewableenergy, including: Project "Low technology fabrication of CdS/CdTe solar cells," EU program INCO-COPERNICUS, JOUL II, 1996; Demonstration project "Energy provision using solar systems in isolated locations"; Educational EC project SOCRATES "ICT-Tools in PV-Systems Engineering: Teaching & Learning." On the basis of the project activities, the necessary laboratory facilities and training materials for students' learning and research were developed. Demonstration solar systems as a grid-connected 1.5 kW_P PV generator, active thermo system, and a hot air dry room were installed on the flat roof of University in Blagoevgrad. The integration of solar systems into the roof structure and there combined PV&T work to cover the energy needs of Solar Energy Center [19].

1.6 Photovoltaic Power Plants in Bulgaria

Map of Bulgaria with designated most of the locations of installations with different categories of renewableenergy sources is shown in Fig. 20 [20]. The most widely used renewableenergy technology in the country is solar photovoltaic.

According to the official register of the Agency for Sustainable Energy Development (SEDA) at the Ministry of Energy of the sites for production of energy from renewableenergy sources, 1363 photovoltaic power plants are registered in Bulgaria

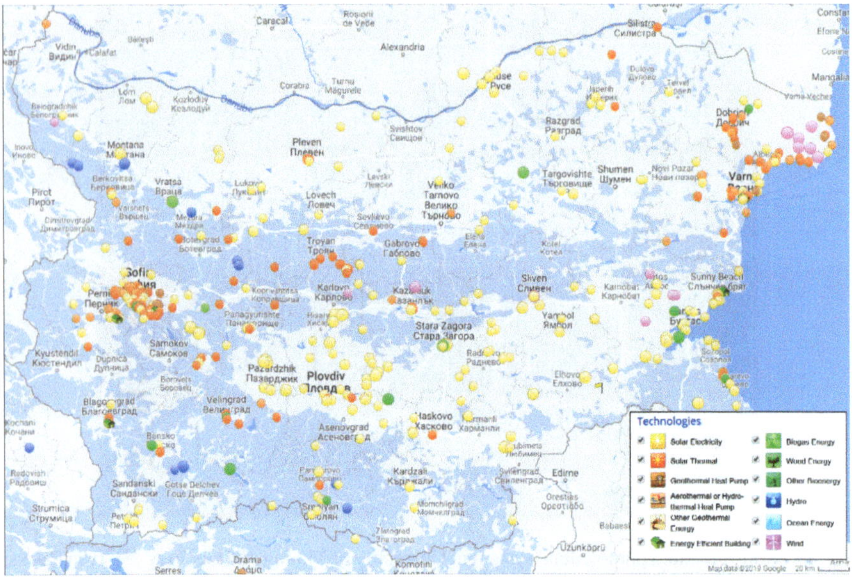

Fig. 20 Map of Bulgaria with designated locations of installations with different categories of renewable energy sources [20]

[21]. An analysis of their number distribution by groups of different installed capacity sizes is shown in Table 5 and Fig. 21.

The largest photovoltaic power plant has a power of 60.4 MW_P, and the smallest officially registered one with a power of 1 kW_P.

Largest photovoltaic power plants in Bulgaria

The *Karadzhalovo Solar Park* (Figs. 22 and 23) is the largest in Bulgaria with its installed power of 60.4 MW_P. It has 214,000 polycrystalline photovoltaic panels and cost 350 million Bulgarian lev (181.4 million in euro). It has been completed in March 2012 after 4 months of construction [22].

Pobeda photovoltaic power plant (Fig. 24), with a total installed capacity of 50 MW_P, is the second largest in Bulgaria. It has 217,632 polycrystalline photovoltaic modules type NA C-Class 3bb with a single power of 225–240 W_P connected to 86

Table 5 Distribution by number and installed power of photovoltaic power plants in Bulgaria

Single installed power (MW_P)	Number	Total power (MW_P)
≤0.03	367	8.2
0.03–0.20	661	69.8
0.2–1	102	64.3
1–10	226	683.1
>10	7	204.8

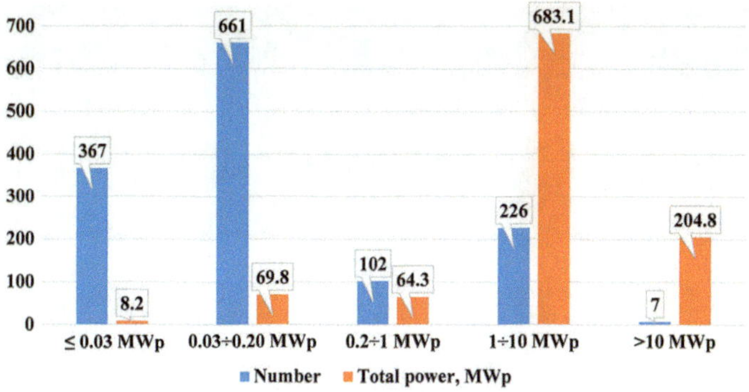

Fig. 21 Comparison of the number and total capacity of the various sized photovoltaic power plants in Bulgaria

Fig. 22 Satellite photograph of biggest photovoltaic power plant in Bulgaria—Karadzhalovo (60.4 MWP) [23]

Fig. 23 Photographs of the biggest photovoltaic power plant in Bulgaria—Karadzhalovo [24]

Fig. 24 Second largest photovoltaic power plant in Bulgaria—Pobeda (50 MW_P) [23]

inverters type PVS800-57-0500 kW-A. The total area of the Pobeda photovoltaic plant is 1016.814 acres. The value of the investment is 118,760 thousand euros. It was put into operation in June 2012 [25].

Cherganovo photovoltaic power plant (Fig. 25) initially is planned to be with total installed capacity of 25 MW_P, now registered as 29.3 MW_P, and is the third largest in Bulgaria. It has 124896 HAREON polycrystalline modules with a single power of 230 W_P to 240 W_P connected to 50 inverters type PVS800-57-0500 kW-A. The

Fig. 25 Third largest photovoltaic power plant in Bulgaria—Cherganovo (29.3 MW_P) [23]

total area of the photovoltaic plant is 600.144 acres. The value of the investment is 61,285,472 euros. It was put into operation in June 2012 [26].

Samovodene photovoltaic power plant (Fig. 26) has a total installed capacity of 21.3 MW_P, and it is the fourth largest in Bulgaria. The total area of the photovoltaic plant is 618,294 m^2. It was put into operation in December 2011.

Zlataritsa photovoltaic power plant (Fig. 27) has a total installed capacity of 20 MW_P, and it is the fifth largest in Bulgaria. It was put into operation in February 2012.

Dobrich photovoltaic power plant (Fig. 28) has a total installed capacity of 14.071 MW_P and it is the sixth largest in Bulgaria. It was put into operation in June 2012.

Examples of roof and facade photovoltaic power plants in Bulgaria

Among the first and largest at that time roof photovoltaic power plants connected to the grid are installed in Central Laboratory of Solar Energy and New Energy Sources and Technical University of Gabrovo as a part of the work on FP5 EC project PV Enlargement 2002–2006, with power of 10 kW_P (Figs. 12a and 14).

132 kW_P roof-mounted solar power plant in the industrial enterprise ZITA in Rousse—Fig. 29. The main components of the DC section of the power plant are monocrystalline PV modules Sinski PV SPV180 M-24, with nominal power of 180 W_P. The power plant consists of 840 modules, forming two subsystems: first subsystem comprising 520 modules with inclination of 5°, and second subsystem of 320 modules with inclination of 11.2°, at azimuth −10.4°. Module surface area is 1072 m^2. The generated electricity is fed to 12 inverters Sunny Mini Central SMA SMC 11000 TL model and to each inverter are connected 5 strings, consisting of 14 serially connected modules. The central monitoring and control device is Sunny WebBox. The 132 kW_P roof-mounted solar power plant in industrial enterprise Zita Rousse is in operation since May 10, 2011 [27].

Fig. 26 Fourth largest photovoltaic power plant in Bulgaria—Samovodene (20.1 MW_P) [23]

Fig. 27 Fifth largest photovoltaic power plant in Bulgaria—Zlataritsa (20 MW_P) [23]

Fig. 28 Sixth largest photovoltaic power plant in Bulgaria—Dobrich (14 MW_P) [23]

Fig. 29 132 kW_P roof-mounted solar power plant in the industrial enterprise ZITA in Rousse [23]

In Gabrovo, three grid-connected photovoltaic roof-mounted systems for kindergartens (Fig. 30) with capacities of 23.76, 21.96, and 20.88 kW_P with monocrystalline PV modules, as well as systems with solar collectors for hot water, have been built. They are equipped with monitoring systems containing SMA Sunny WebBox

Fig. 30 23.76 kW_P grid-connected PV system on the roof of the kindergarten "1 June" in Gabrovo

(including Ethernet interface), RS485 Module, SMA Sunny Sensor Box (including module temp sensor), PT100 Ambient Temperature Sensor, and Wind Sensor.

First public building in Bulgaria with photovoltaic facade is in Gabrovo (Fig. 31). A 416 thin-film photovoltaic panels Schuco MPE 85 AL 01 are mounted on the east, south, and west facades of the building. The total installed power of the photovoltaic system is 35.190 kW_P. The modules are grouped and connected to five single-phase inverters. To assure the autonomy of certain users in the building and to store the unused electrical energy produced by the photovoltaic plant, the inverters are connected to a common three-phase network electric board with a group of three inverters with a built-in charger (bidirectional off-grid inverter/charger). All reserved users are connected to this board. The off-grid inverters group is connected to a VRLA battery array with an installed capacity of 57,600 Wh. A data monitoring and storage system is also implemented, which includes Data Logger with RS485 interface for connection to inverters and LAN port for local or remote access to data via built-in Web interface [28].

Fig. 31 First public building in Bulgaria with facade photovoltaic power plant located in Gabrovo

2 Lighting in Bulgaria

2.1 Early Development of Lighting

The first means of artificial lighting in Bulgaria date from the time of the First Bulgarian Empire (681–1018). Archaeologists have discovered an ancient lighting tool (Fig. 32) for the remains of the palace in the first Bulgarian capital Pliska (681–893). The valuable artifact, serving according to the hypothesis for lighting of large spaces or corridors, is an iron axis about 30 cm long, ending with a bronze part with human hand shape that holds a cylindrical part with a cup with a wick [29].

The first official information on lighting in the present Third Bulgarian State is available after the Liberation of the Ottoman Empire after the Russo-Turkish War of 1877–1878. These events led to the restoration of the Bulgarian State under the Treaty of San Stefano of March 3, 1878. The Treaty forced the Ottoman Empire to return to Bulgaria the vast majority of its territory, conquered in the fourteenth century from Second Bulgarian Empire (1185–1396).

Sofia was chosen as a capital of Bulgaria on April 3, 1879. The street lighting of Sofia was realized with about 200 gas lanterns on wooden posts. They were ignited in the evening and extinguished in the morning by the employees of the municipality (Fig. 33, illustration of a Bulgarian artist, publicist, and writer Prof. Alexander Bojinov from 1900, titled "Every morning the lamplighter travelled his area"). The same was the condition of artificial lighting in other cities of the country [30].

During the Liberation War, some of this lighting was damaged. Therefore, the issue of reconstruction and even improvement of street lighting was put before the newly elected municipal government at the first meeting on February 13, 1878. Initially, 200 oil lamps were purchased from Vienna (Austria) for reconstruction and then 300 lamps, 300 lanterns, and 300 poles for them (Fig. 34). In the next 1879, the municipality decided to improve the living standards of the new Bulgarian capital with 20,000 inhabitants and spends 1/3 of its budget of 153,890 francs for new lighting of the city [31].

Fig. 32 Antique lighting in the first Bulgarian capital Pliska [29]

Fig. 33 Lamplighter

Fig. 34 Eagles' Bridge in Sofia with oil lanterns [32]

2.2 Development of Electrical Lighting

The first electric lamp in Bulgaria was lit on July 1, 1879, on the occasion of the official ascension of the throne of Prince Alexander Battenberg. The Plovdiv Maritsa newspaper (II, 1879, 99, cp. 3–4) publishes the following information about the event (in Bulgarian): "Yesterday it became a superb illumination. The garden against the palace was magnificently decorated. Dr. Dimitar Mollov delivered appliances and other electrical consumables from Vienna to produce electric sun that illuminated all the garden and the city clock... The colorful lanterns were attached to strange wires that connected the top of the pavilion with the surrounding trees and constituted a wonderful visual appearance from which rays spread and joined with those of the electric sun." This is the time when the classic electric lamp was just discovered and the Sun was probably a carbon-electrode lamp, and the "pavilion" is the room where the system was placed. So far, it cannot be proven whether the source was dynamo powered by a steam engine or a rechargeable battery. Probably, the event of 1879 is the first use of electric lighting not only in Bulgaria but also on the Balkan Peninsula [30, 31].

The second presentation of the electric lighting was done in April 1885 when the Palace in Sofia (Fig. 35) was illuminated with electric lamps for the celebration of the 1000th anniversary of the death of St. Methodius—April 6, 885. It was probably filled with incandescent lamps with charcoal wires powered by a battery—galvanic or rechargeable. The ensuing Serbian–Bulgarian war and some political turmoil in

Fig. 35 Palace in 1885. A view from Alexander I Square to the western gate [32]

Bulgaria diverted the attention of the Sofia governance from the problems of street lighting for several years [30].

Far more forward-looking and expeditious in the study and implementation of the latest developments in electrical and lighting engineering were the Bulgarian manufacturers. Some of them quickly appreciated the benefits of electric lighting and invested in it. An example is the “Ferdinand I” manufacturing factory in Gabrovo, which started production in October 1888, illuminated perfectly with incandescent filament lamps. It was owned by the company “Uspeh” (Success) in Gabrovo. Lamps, along with production machines, dynamo, and all other necessary lighting equipment and equipment, have been delivered from England. This is the first factory in Bulgaria with electric lighting. The production machines and the dynamo were driven by a steam engine.

The second factory in Bulgaria with electric lighting was the Kazanlak woolen-spinning factory “Rozova dolina” (Rose Valley), opened in 1889. Its machinery and production facilities and electric lighting were delivered by England.

The third place where electric lighting is being implemented in Bulgaria is again in an industrial site in the town of Gabrovo, in the water mill of Ivan Hadjiberov in Gabrovo, with a dynamo machine, delivered by Germany, driven by the water force of the water wheel of the mill. This happened in 1891, when the first electric power was received from a water force in Bulgaria (Fig. 36). In 1892, the woolen-weaving factory of Ivan Hadjiberov and A. Momerin in Gabrovo was completed with a set of machines and equipment for electric lighting with a carbon fiber lamp delivered by Germany. It is also curious that in the entrance room of the headquarters of the

Fig. 36 Hydroelectric power plant of Ivan Hadjiberov in Gabrovo—1891 [30]

Fig. 37 Painting of the Roman goddess of dawn and the light Aurora [30]

opposite wall was placed a large painting, painted by a specially invited artist from Germany Adolph Selov, the Roman goddess of dawn and the light Aurora, raised in a chariot and holding high in his right hand electric lamp (Fig. 37).

The "First Bulgarian Agricultural and Industrial Exhibition" (now Plovdiv Fair) (Fig. 38), held in Plovdiv in 1892, is the next place with the use of artificial lighting with electric lamps in the open areas and in the fairground pavilions. The exhibition was attended by 24 countries from Europe, Asia, and America and visited by about 170,000 people. A contract for the delivery, installation, and maintenance of electrical lighting equipment was signed with Company "Gantz & Co" from Budapest (Hungary). The power supply of the electric lighting was carried out with two dynamo machines driven by two locomotives with a power of 12 hp. The exterior areas and alleys of the exhibition were illuminated with 24 arc electric lamps, each with lightintensity of 46 candles. The lamps were mounted on wooden posts. In the pavilions, there were installed 60 pieces carbon-filament incandescent lamps.

Until 1900, the following sites with autonomous power supply were illuminated in Bulgaria (Table 6) [30].

Fig. 38 Plovdiv fair—1892 [33]

Initial electrification of cities with electricity grid for lighting—1901 to 1919

Sofia

In 1899–1900 in Sofia the first in Bulgaria and on Balkan Peninsula, large hydroelectric power plant (HPP), Pancharevo, with four generators 430 kVA with a voltage of 8 kV and a new urban electric distribution network was built (Fig. 39). In October 1900, the construction of the following facilities was completed:

- Substation "Yuri Venelin" with two transformers with a voltage of 7/3 kV and power of each of 1500 kVA, as well as a transformer with a voltage of 3/0.15 kV;
- 14 transformer stations on 7 kV with overhead electric power supply;
- 5 transformer stations on 3 kV with electric power supply via underground cables;
- an 8 kV overhead power line of pine-impregnated pillars from Pancharevo to Sofia with a length of 16 km;
- 7 kV overhead distribution power line on pine-impregnated poles for power supply to the 10 km transformer substations;
- 23 km overhead and 9 km cable distribution network with a voltage of 150 V for street lighting and other consumers;
- mounting of 600 luminaires on the pillars of the distribution grid [30].

The official reception of street lighting took place on October 29, 1900, by a special committee with the participation of a specially invited expert from Switzerland

Table 6 Illuminated objects with autonomous power supply in Bulgaria until 1900

No.	Name of the illuminated object	Year of start	Type of power supply
1.	City Garden near the Palace in Sofia	1879	Battery
2.	Princely Palace in Sofia	1885	Battery
3.	"Ferdinand I" manufacturing factory in Gabrovo	1888	Steam machine and dynamo
4.	Woolen-spinning factory "Rose Valley" in Kazanlak	1889	Steam machine and dynamo
5.	Water mill of Ivan Hadjiberov in Gabrovo	1891	Water wheel and dynamo
6.	Weaving factory of Ivan Hadjiberov in Gabrovo	1892	Water wheel and dynamo
7.	First Plovdiv Fair	1892	Locomotive and dynamo
8.	Palace of Evksinograd, Varna	1893	Diesel engine and generator
9.	Palace in Sofia	1895	Locomotive with generator
10.	Palace stables, Sofia	1898	Diesel engine and generator
11.	Mines in the town of Pernik	1899	Steam machine and dynamo
12.	Anti-Plague Institute, Sofia	1900	Diesel engine and dynamo
13.	National Theater, Sofia	1900	Diesel engine and dynamo
14.	National Assembly, Sofia	1900	Diesel engine and dynamo

Fig. 39 Hydroelectric power plant Pancharevo on the river Iskar [34]

Fig. 40 Monument of Vasil Levski in Sofia, the first years of the twentieth century [32]

(University of Lausanne) Prof. A. Pola. From November 1, 1900, Sofia was already illuminated with electric lamps and the Pancharevo HPP was included in regular operation. The street lighting system of Sofia, made of aesthetic steel pillars with a conical shape and cast iron sleeves down to the ground and stylish lighting fixtures, was beautiful at night and during the day (Figs. 40 and 41). Street lighting was manually controlled by the transformer substations. One-third of the total number of lamps lit all night and two-third by midnight. With this, they save only electricity, but lamps and change work, because at that time the lamps had relatively little durability—the arc lamps worked 120–150 h and the filament lamps—about 400 h. Lighting service (changing lamps) was done with single wooden stairs, about 3 m high, which were left in the street, hung on some of the pillars.

The remaining part of the street lighting was accepted and put into operation on May 19, 1901, and the installed lamps in Sofia have already become 1350 [30].

In the coming years in Sofia continued the construction of steam power plants and the gradual expansion of the urban electric grid. The dynamics of this development over the years is shown in Table 7 [30].

Lom

Lom is the second city in Bulgaria, fully illuminated with electric lamps and equipped with a public electricity grid for low voltage, implemented through concession. The tender for the electric lighting of the city through concession was announced in 1907, but for various reasons, its conduct was delayed by several years. A diesel

Fig. 41 Sofia in 1908 [32]

Table 7 Development of electric lighting in Sofia in the period 1901–1918

Indicator	Year				
	1901	1905	1912	1916	1918
Population, number	72,000	83,000	110,000	140,000	160,000
Street lighting lamps, number	600	1580	2030	2400	2900
Subscribers for electricity, number	250	1210	2510	3200	4000
Consumed electricity, MWh/year:					
(a) For lighting	842	1331	2510	3200	4000
(b) Total propulsion power	1430	2685	5820	6550	8200

engine was built with two diesel engines, each with a power of 150 hp, two dynamo machines with a power of 100 kW and a voltage of 440/220 V with a three-wire line. This power supply scheme is unique to Bulgaria with a number of advantages—one line uses two voltages: 220 V for lighting and 440 V for motors. There is another uniqueness in the solution of power supply. A rechargeable battery was fitted to each dynamo machine, which was maintained with a constant supply and was able to withstand a load for 3.5–7.5 h. With this, top loads could be fed. The total length of the electricity distribution network for the town and the village of Golintsi was 20 km. This is the first village in Bulgaria, illuminated by electric lamps. The power distribution network has been cabled for the city center and overhead on wooden impregnated and non-impregnated pillars that have been fitted with street lighting fixtures. The corner posts were steel lattice.

The diesel power station and street lighting were put into operation in November 1912. The number of subscribers by 1915 was over 300. There were four outlets for night and half-night street lighting and 14 outlets for street lighting in the individual neighborhoods of the city. All the lamps were incandescent. The control of street lighting was done manually from the electric panel in the plant [30].

Gabrovo

Gabrovo is the third city in Bulgaria after Sofia and Lom, which began building public street lighting and electrification. For the first time, the initiative is not of the municipality, but of two enthusiastic Gabrovians, devotees of electricity—Ivan Hadjiberov and Hristo Lulev. For a period of 4–5 years, Hadjiberov explored the Yantra River valley, choosing a site for the dump and the plant and finally completed the construction of its "Usteto" Hydro Power Plant (Fig. 42), which is put it into operation in 1906.

This is the second in Bulgaria power plant for public power supply, because it supplied besides Hadjiberov's factory and the street lamps on the street leading to the city. The hydroelectric power station "Usteto" has three water turbines with 80 hp each with 60 kVA generators. To promote the quality of electric lighting, Ivan Hadjiberov has ordered the lighting of one electric lamp to be installed in the houses along the illuminated street of his factory in Gabrovo.

The example of Ivan Hadjiberov was followed by Hristo Lulev in Gabrovo. He decided to build a more powerful power plant for public needs, again on the Yantra River, in the Boaza area. It was designed by the Gabrovian engineer Stefan Manafov

Fig. 42 Power plant of Ivan Hadjiberov—internal view 1912–1940 [35]

in 1908, and construction started in the same year. It was completed in 1911. It had a generating power of 135 kVA and a voltage of 6 kV. A power line to Gabrovo was built with a length of 14 km, and the factories "Georgi Rasheev, Ivan Doinov" and "Buffon" were supplied with electricity. Next to them were transformer stations for voltage 6000/220/127 V. For lighting, a voltage of 127 V is used. The electricity consumed was paid for the installed capacity.

Until 1917, the Boaza Power Plant was banned to sell electric energy for residential lighting in the city. A low-voltage distribution network was built in the city center. This energy has been paid to the manufacturer by the number of lamps installed [30].

Kazanlak

Kazanlak is the fourth city in Bulgaria, where street electric lighting and public electrification has been built. It was initiated by Stoyan and Gencho Stainovi, father and son, who created the Electricity Stock Company "Pobeda," which aimed to build a hydroelectric power plant "Enina" and the necessary facilities for electricity supply and lighting. Work on them started in 2011. Two water turbines were installed in the Enina hydroelectric power plant, each of them propelling a generator with a power of 300 kVA at a voltage of 6 kV. Transformer with voltages 6000/220/120 V was installed for the plant's own needs (lighting and propulsion power) and for other nearby objects. The machine room was illuminated with three arc lamps for a voltage of 120 V. From the "Enina" hydropower station to the town of Kazanlak an 11 km in length 6 kV transmission line was built on oak posts. The street distribution network has been on oak poles with a height of 9 m, with a voltage of 220/120 V, four wires. The street lighting was with a 120-volt carbon incandescent lamps, switched on and off manually from the transformer substations. From January 1, 1914, Kazanlak is considered to have electric lighting and public electrification.

The price of electricity, which was set by the Ministry of Trade and Industry when issuing the permit, was 0.36 BGN/kWh. The price for the housing, determined by Pobeda AD, was BGN 0.60/kWh—at a price in Sofia of 0.70 BGN/kWh. In 1915 has begun replacing burnt bulbs in street lights with incandescent tungsten filament lamps. Tungsten lamps have been imported in Bulgaria since 1908 at a price of 4 BGN, while the carbon lamp price was 0.80 BGN [30].

Varna

Varna became the fifth city in Bulgaria, illuminated by electric lamps, 12 days after Kazanlak on January 12, 1914. In 1911, the municipal commission designated the place of the power plant, which was proposed to be with diesel motors, the street lights to be incandescent lamps, and squares lighting with arc lamps. It was elected engineer Nikola Petkov to draw up the project and the conditions for the tender for street lighting and electrification of the city, which to be financed by and property of the municipality of Varna. The tender was won by "Siemens-Schuckert" company from Vienna, and the contract was signed on December 21, 1911.

A diesel power plant with a machine room was built with the possibility of installing five diesel engines fitted with three-phase generators, each with a power output of 250 kVA, with a voltage of 5 kV. The medium-voltage power grid was with underground cables with a total length of 9 km and powered 14 transformer substations with 50 kVA transformers and a voltage of 5000/210/120 V. The low-voltage distribution network was four-wire with a voltage of 210/120 V. For the lighting, a voltage of 120 V was used. The whole network length was 48 km, of which 33 km was overhead and 15 km cable. The pillars for the overhead grid and the street lighting were with a height of 9 m made by steel with cast iron joints on the ground.

At the opening, there were 1200 incandescent lamps and 60 arc lamps for street lighting. One year later, there were 1590 lamps with a total output of about 160 kW. Street lighting lamps were divided into two groups—for night and half-night lighting—and were automatically controlled by electric watches [30].

Ruse

In 1910, the municipal council decided that the electrification of Ruse was to be built as a municipal enterprise through a tender and to commission the engineer Petar Malchev to make, design, and prepare the tender conditions. The tender was held on May 9, 1911, which was won by the "Siemens-Schuckert" company from Vienna. A contract has been signed that specifies the obligations of both parties in the construction and assembly and the respective deadlines. But the Balkan War and the First World War have led to a number of difficulties that have prolonged the deadlines. The opening of the power plant and the street lighting of Ruse took place a few years later on March 21, 1917. For the electrification of Rousse were built:

- Diesel power plant with three diesel engines with three-phase generators, which have a power of 250 kVA and a voltage of 3 kV;
- Transformers stations—9 pieces with one 50 kVA transformer for voltage 3000/210/120 V;
- Power supply cable line in the form of a ring on voltage of 3 kV;
- Overhead distribution network with four wires for low voltage 210/120 V on wooden impregnated poles, about ¾ of their total number, and ¼ on steel tubular pillars;
- Street lighting was with electric arc lamps for the squares, and for streets—incandescentlamps. The arc lamps were 300 and 500 W, and the incandescent were 100 W for crossroads and 60 W the streets.

The control of street lighting was manual from the transformer stations. The luminaires were grouped on full-night and half-night lighting.

The price of electricity in Rousse was:

- 0.65 BGN/kWh—for street lighting;
- 0.72 BGN/kWh—for lighting of public buildings;
- 0.92 BGN/kWh—for residential buildings [30].

Electrification and electric lighting in Bulgariain the period 1919–1947

In the period 1919–1947, in Bulgaria were used lamps with voltages 110, 120, 127, 150, and 220 V, according to the distribution voltages of the low-voltage networks in the individual cities. At that time, lamps with power of 15, 25, 40, 60, 75, 100, 150, 200, 300, 500, and 1000 W were used. The lamps up to 200 W have and continue to have an Edison socket size E27 and 300–1000 W—socket E40.

By the end of 1942, 93 cities and 539 villages had been electrified and illuminated. By the end of 1944, their number had grown to 758 settlements, and by the end of 1945–871 and by the end of 1946, the number was already 1140, which represents about 14% of all settlements in Bulgaria. All cities were already lit up and 11% of the villages. By the end of 1947, a total of 1350 settlements had been lit and electrified.

During the period 1919–1947 in Bulgaria were built: 62 hydropower plants with capacities from 9 to 7000 kW with a total output of 43,000 kW; 32 TPPs with capacities from 40 to 2000 kW with a total output of 62,000 kW; 86 NPPs with capacities from 4 to 1200 kW with a total power of 14,940 kW—a total of 180 power plants with a capacity of about 120,000 kW. The approximate ratio between the percentages of electricity consumed by type of lighting and consumers and by year is shown in Table 8 and Fig. 43 [30].

Table 8 Electricity consumption in Bulgaria by type of lighting and consumers and by year in the period 1902–1947

Type of consumption	Electricity consumed over the years in %							
	1902	1908	1912	1918	1928	1932	1945	1947
Street lighting	68.2	31.3	19	14	–	10.6	1.9	1.8
Residential lighting	24.2	42.1	38.3	49.1	–	13	12.9	12.7
Industrial lighting	0.9	1.9	2.2	2.5	–	10.7	10	10.2
Total	93.3	75.3	59.5	65.6	41.4	34.3	24.8	24.7
Other electric users	6.7	24.7	40.5	34.4	58.6	65.7	75.2	75.3

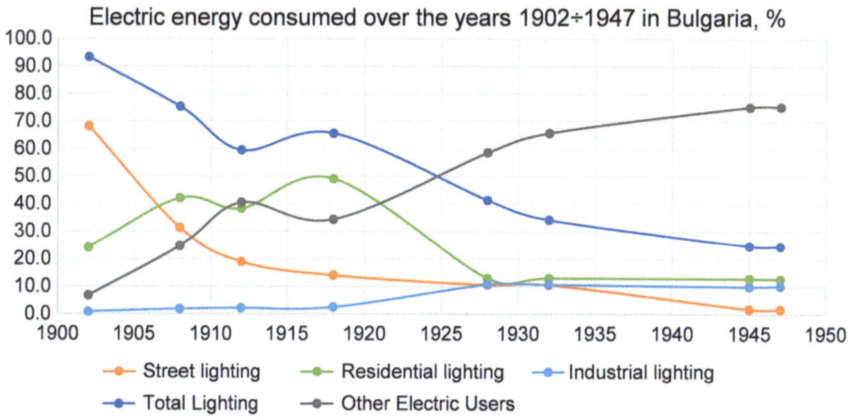

Fig. 43 Electricity consumed over the years 1902–1947 for different kind of lighting and other electric users in Bulgaria

Development of electric lighting in Bulgaria after 1947

Since 1947, the way of energy development in Bulgaria has changed significantly since the end of the year was the "nationalization" in Bulgaria, including the power plants, the substations, the power lines, and the urban electricity networks with the street lighting. The state takes full responsibility of the planning, design, and construction of the electrical system.

In 1949, it was decided for low-voltage networks to adopt a four-wire three-phase system with voltage 380/220 V–380 V for power consumers and 220 V for lighting. Until 1964, the low-voltage residential networks all over Bulgaria were rebuilt into four wires with voltage 380/220 V.

After 1948, many books by Bulgarian authors of lighting and installation equipment for electricians, electrical designers, and people interested in lighting were published. In 1948, lectures on lighting techniques were initiated as a basic discipline for students of Electrical Engineering (high currents) at the Higher Institute of Mechanical and Electrical Engineering in Sofia. In 1953, a textbook on lighting techniques for electrotechnical secondary schools was published and in 1954 for university students.

The development of lamp production in Bulgaria is concentrated in the Svetlina plant in Sliven and can be traced by following the market entry of different types of lamps over the years:

- 1948—incandescent lamps for general lighting;
- 1957—fluorescent lamps, tubular with diameter of 38 mm;
- 1963—automobile lamps for headlights and signaling;
- 1965—infrared and glimmer lamps;
- 1970—miniature and supermini filament lamps;
- 1973—pipeline halogen filament lamps for general lighting;
- 1978—high-pressure mercury discharge lamps;
- 1983—U-shaped fluorescent lamps and car halogen lamps for headlights;
- 1987—high-pressure sodium discharge lamps.

Until 1956, only incandescent lamps for both indoor and outdoor lighting were used as light sources. In 1956, for the first time, some industrial premises equipped with luminaires with fluorescent lamps from abroad were illuminated, and in the same year, the Elprom factory in Stara Zagora started production of fluorescent luminaires with lamps from abroad. In 1957, electric lamp factory in Sliven manufactures and markets the first Bulgarian tubular fluorescent lamps 40 and 20 W. Luminaires with luminescent lamps also come into the street lighting, but they cannot be enforced for a long time, as high-pressure mercury lamps are soon to be used. In 1965, the length of the illuminated streets in Sofia was about 1470 km. The ratio of the lengths of the streets lit by incandescent, luminescent, and mercury lamps in 1965 was 890:190:390 km. In 1970, the luminescent luminaires were almost completely displaced by mercury luminaires.

With their appearance in the market in 1957, the fluorescent lighting fixtures for interior lighting have found a good reception for illuminating industrial workshops

Table 9 Production capacities of the lamp factory in Sliven at the end of the 1980s

No.	Type of lamp group	Maximum annual capacity (number)
1	Incandescent for general lighting	40,000,000
2	Infrared and glimmer	650,000
3	Luminescent	3,000,000
4	Mercury with a high pressure	600,000
5	High-pressure sodium	200,000
6	For vehicles	18,000,000
7	Auto halogen	1,000,000
8	Halogen tubular and special	200,000
9	Glimm miniatures	4,200,000

and premises. Luminescent luminaires suitable for lighting and housing in homes were also produced, but very few people were interested in them. Only after the appearance of compact fluorescent lamps with Edison's socket in the 1990s, this type of lamps became widely used in households in Bulgaria.

In 1970–1971, a general electric lighting plan was developed in Sofia, which gives the general directions for the development of all kinds of external electric lighting necessary for a modern city such as street, park, facade, advertising, showroom, etc.

In 1972, the lighting of "Tsarigradsko shoes" Blvd in Sofia for the first time in Bulgaria was completed with luminaires with high-pressure sodium lamps imported from Belgium. It promotes the advantages of sodium lamps for street lighting, and in 1980, the Lighting Factory in Stara Zagora started regular producing of luminaires with imported high-pressure sodium lamps. The Lamp factory in Sliven started production of Bulgarian sodium lamps 400 and 250 W in 1987, and the lighting fixtures with them illuminated wider streets in all major Bulgarian cities.

At the end of the 1980s, the production capacities of the lamp factory in Sliven for two working shifts for the different types of lamps are shown in Table 9 [30].

2.3 *Modern Lighting*

A real technical revolution in the improvement of electrical lighting (internal and external) occurred in Bulgaria after 1992, when the markets in Bulgaria opened for all types of lighting and installation products. For several years, many Bulgarian manufacturers of high-quality luminaires and relatively low prices also appeared. This has made it possible to build new street lighting systems in cities and villages and households to use energy-efficient lamps and lighting fixtures [30].

2.3.1 Public Lighting

Until the 1989, the ownership of public lighting in the Bulgarian cities and villages belonged to the state electricity supply enterprise—National Electricity Company (NEC) and its units on the whole territory of the country—in which it sells electricity. Under these conditions, NEC was not motivated to invest in modernizing and increasing the energy efficiency of the public lighting, as the same company would then receive less cash revenue from the electricity consumption for this lighting. Since the significant political change in 1989, the privatization of economy transferred the ownership of street lighting from NEC to the municipalities. This burdened additionally the municipal budgets and has created strong economic motivation to increase the energy efficiency and introduce modern control systems of the public lighting. At the same time, municipalities have the opportunity to judge which public objects—historic or modern buildings, monuments, parks, bridges, etc.—can be illuminated in addition to street lighting to emphasize identity and to create a nice and beautiful appearance of the cities.

In the lighting of public objects, lighting fixtures and projectors with different types of light sources and light distributions are used in order to emphasize from a decorative point of view specific features, colors, and relief of the given object. Some examples in the city of Gabrovo are given in Figs. 44, 45 and 46. In recent years, LED light sources of different colors have been used, and in some cases for more impressionable effect, dynamic color control systems have been introduced where appropriate (Figs. 47 and 48).

Fig. 44 National Museum of Education—Aprilov High School in Gabrovo

Fig. 45 House of Culture in Gabrovo [36]

Fig. 46 Church of the Assumption in Gabrovo

2.3.2 Street Lighting

The change in the technologies in the modernization of light sources used for street lighting in the recent 20 years in Bulgaria can be divided into two main periods:

- Transition from mercury to sodium and CFL lamps—1995–2010;
- Transition from sodium lamps and CFL to LED—since 2010.

Fig. 47 District Administration Building in Gabrovo with a visualized through colored facade lighting national flag of Bulgaria

Fig. 48 Dynamic LED color lighting of pedestrian bridge and illumination along the Yantra River in Gabrovo

Transition from mercury to sodium and CFL lamps—1995–2010

In the beginning of the period, almost all outdoor luminaires in Bulgaria were equipped with mercury lamps and electromagnetic ballasts, produced by "Svetlina" factory in the city of Stara Zagora. The bodies of these luminaires were made of steel sheets, their size and weight were huge, and most of them had protruding dispersers made of transparent polycarbonate or plexiglass. During the next approximately 15 years, the outdoor luminaires in the cities, smaller towns, and villages have been consequently modernized. The old luminaires were replaced by new ones with high-pressure sodium lamps. The new luminaires were produced by Bulgarian and several foreign manufacturers. The bodies of outdoor luminaires were made of aluminum alloy; solar UV-resistant polycarbonate; steel sheets, painted with electrostatic powder. Most street luminaires were mounted on poles through consoles. The ballasts of outdoor luminaires were mostly electromagnetic and a small part of them electronic. Ingress protection of the bodies of outdoor luminaires with two parts with different level of ingress protection usually was optical part (IP65 or IP54) and electrical part (IP43) [37].

Summarized results about the modernization of the street lighting systems carried out in the fifteen Bulgarian cities in the period 1997–2005 are given in Figs. 49, 50, and 51. Finally a 2.9-fold decrease in the installed power has been achieved, the quantitative and qualitative requirements of road lighting being satisfied [38].

The approximate number of lamps by different types, installed in outdoor luminaires in Bulgaria in 2010–2011, is given in Table 10 [37].

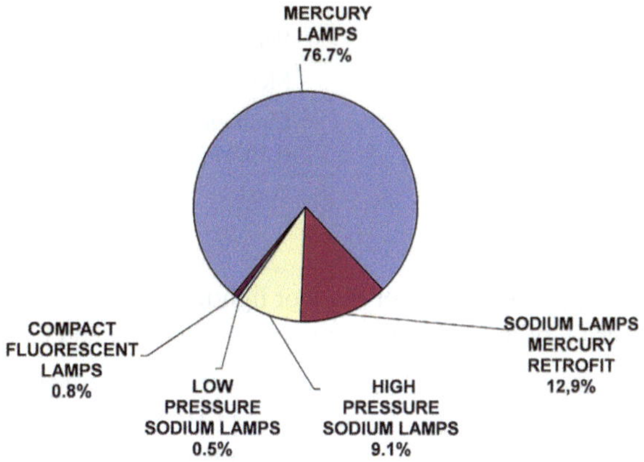

Fig. 49 Lamp types of street luminaries in 15 Bulgarian cities from audits of existing lighting in the period 1997–2005

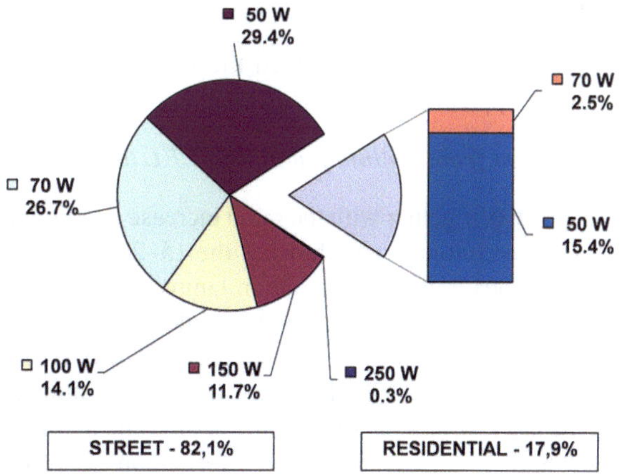

Fig. 50 Distribution of nominal powers of high-pressure sodium lamps of the street luminaries in 15 Bulgarian cities after the realization of the projects for increasing the energy efficiency in the period 1997–2005

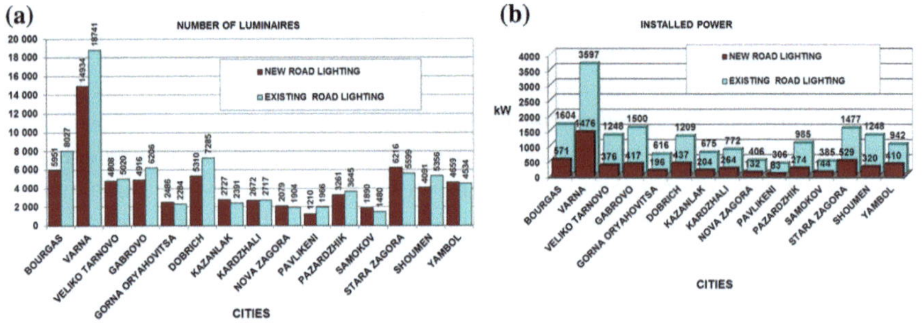

Fig. 51 Number of luminaires (**a**) and installed power (**b**) before and after the modernization of road lighting in the 15 Bulgarian cities in the period 1997–2005

Table 10 Number of luminaires with different lamp types in Bulgaria in 2010–2011

Lamp type	Number	Share (%)
Sodium high pressure	450,000	53.0
Mercury vapor	136,000	16.0
Fluorescent compact	222,000	26.2
Metal halide	36,000	4.3
Light-emitting diodes (LEDs)	4500	0.5
Total	848,500	

The lamps used for outdoor lighting in the Bulgarian settlements at the end of the period of transition from mercury to sodium and CFL as of 2010–2011 include:

- High-pressure sodium lamps of 50, 70, 100, 150, 250, and 400 W;
- Compact fluorescent lamps of 11 W, 13 W, 36 W, and 55 W;
- Metal halide lamps of 35 W, 70 W, 100 W, 150 W, and 250 W;
- Mercury lamps of 80, 125, 250, and 400 W;
- LED modules (20–150) W.

This is the period in which the first demonstration projects for outdoor LED lighting appeared.

Transition from sodium lamps and CFL to LED—since 2010

After 2010, together with the rapid increase in efficiency and the reduction in the cost of LED technology also finishes the 15–20-year lifetime of the light fixtures from the previous transition to sodium lamps. This determines the rapid and widespread penetration of LED lights in the outdoor lighting and initiates the transition from sodium and CFL to LED light sources. The process is not very fast, but all new street lighting systems are implemented with LED luminaires. Expanding the use of LEDs in the last few years is so dynamic in time and in different Bulgarian cities, so it is not easy to fix precise general data. Information can be obtained for specific settlements

Table 11 Audit of the type, power, and number of existing lights in the town of Gabrovo by 2016

Type of luminaires and lamps	Single lamp power (W)	Number of luminaires	Share (%)
Street LED	35–14	200	3.0
Park LED	5–70	375	5.6
Park compact fluorescent	20	108	1.6
Street sodium high pressure	50–150	4816	71.9
Park sodium high pressure	50	1172	17.5
Floodlight metal halide	250	8	0.1
Tunnel sodium high pressure	100	18	0.3
Total		6697	

from the periodical energy audits of the current state and the implementation of new LED lighting projects (Tables 11 and 12).

On the territory of the municipality of Sofia by 2017, there are 89,254 sodium lamps, 4703 LEDs, 1816 mercury lamps, and 2452 others [39].

Phased replacement of sodium lampswith LED starts of the biggest boulevards, intersections, parks, and tunnels in the cities (Figs. 52 and 53).

In 2015, a project for modernization of street lighting system in 128 small settlements in municipality of Gabrovo was implemented. Detailed energy audit of the existing lighting system before implementation of the project is done, the summarized results of which are shown in Table 13 [41].

The modernization includes replacing of 2691 of the ineffective existing luminaires with 35 and 70 W LED luminaires for different light classes of streets (Table 14) and introducing of new centralized GSM-GPRS remote management of the street lighting system in all villages. As a summarized result, the installed electric power of

Table 12 Number, type, and power of the lights in the preliminary design for modernized LED street lighting in the town of Gabrovo announced in 2018

Preserved existing luminaires not subject to modernization		New LED luminaires with power (23–104) W
Light source and power	Number	Number
LED (5–140) W	575	6115
CFL 20 W	108	
Metal halide 250 W	8	
Total 6806 luminaires		

Fig. 52 Street lighting and circular junctions with LED lighting in Gabrovo (2016–2018) [40]

the street lighting system is decreased by 67% and the payback period of the financial investment is 4.6 years, which proves the high energy efficiency and financial profitability of the realized modernization [41].

In 2016, a project was implemented in Varna for the replacement of 3912 sodium lamps on the main streets (about 26% of the city's lights) with 2736 intelligent LED lamps with single powers from 72 to 200 W and a centralized GSM-GPRS system for their individual management and dimming capability (Fig. 54) [42].

2.3.3 Household Lighting

In households in Bulgaria are most commonly used light fixtures with Edison socket E27 and E14 lamps. As light sources inside the light fixtures mainly compact fluorescent lamps and LEDs are used in the last years (Fig. 55).

Fig. 53 Tunnel with LED lighting in Gabrovo (2018)

Table 13 Street lighting luminaires in small settlements in municipality of Gabrovo before 2015

Type of light source	Single lamp power (W)	Number of luminaires	Share (%)
High-pressure mercury lamps	125	888	41.1
	250	613	
High-pressure sodium lamps	50	909	57.5
	100	1190	
Compact fluorescent lamps	20	51	1.4

Table 14 Street lighting luminaires in small settlements in municipality of Gabrovo after modernization in 2015

Type of light source	Single lamp power (W)	Number of luminaires	Share (%)
LED luminaires	35 and 70	2691	73.7
High-pressure sodium lamps	50	909	24.9
Compact fluorescent lamps	20	51	1.4

Yet rarely can be seen filament incandescent and halogen lamps, they are already banned for production and sale in Bulgaria and in the European Union, which is why after the end of their life, they are being replaced with new, more energy-efficient technology light sources.

Knyaz Boris I Boulevard Hemus Motorway to Varna Airport

Fig. 54 Intelligent LED street lighting of Varna (2016) [42] Knyaz Boris I Boulevard Hemus Motorway to Varna Airport

Compact fluorescent lamps LEDs

Fig. 55 Most common types and forms of household lamps in Bulgaria Compact fluorescent lamps LEDs

2.3.4 Industry Lighting

The lighting of high-ceiling industrial workshops is mostly done with high-bay "bell"-type luminaires with 250 W or 400 W metal halide lamps (Fig. 56) and in recent years with 100–200 W LEDs (Figs. 57 and 58).

Relatively rarely in industrial sites with lower color rendering requirements, the same type of luminaires with mercury and sodium high-pressure lamps is used.

In industrial workshops with lower ceiling height, luminaires with lamps fluorescent type T5 (14, 28, 49, 54 or 80 W) or T8 (18, 36, or 58 W), having a degree of protection IP54 or higher (Fig. 59).

In recent years, luminescent industrial lighting has been gradually replaced by LED technology luminaires with relatively two times higher luminous efficacy (Fig. 60).

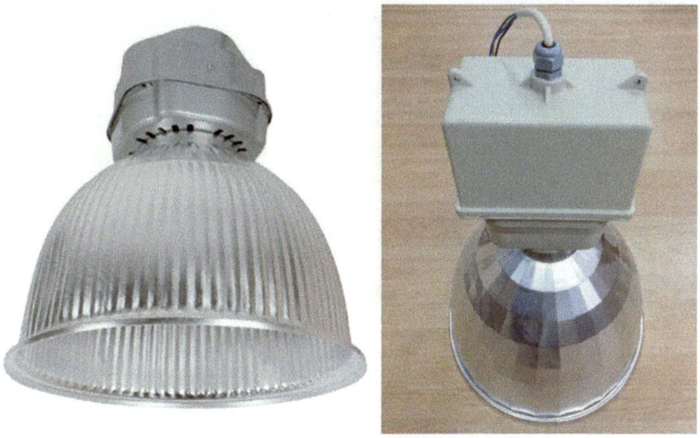

Fig. 56 High-bay "bell"-type luminaires with metal halide lamps

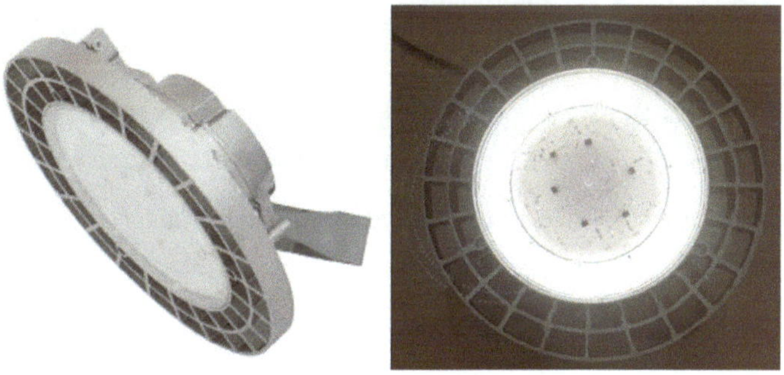

Fig. 57 Industrial high-bay luminaires with LEDs produced in Bulgaria

Fig. 58 High-bay industrial LED lighting produced and used in workshops of ATRA Ltd—Plovdiv

Fig. 59 Industrial lighting with fluorescent luminaires

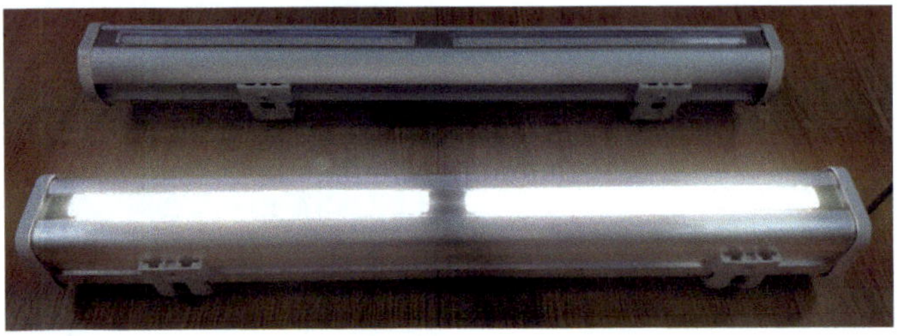

Fig. 60 Industrial 40 W LED luminaires produced and used in Bulgaria

2.4 *Solar Lighting*

The first solar lighting systems in Bulgaria are installed about 20 years ago in some of solar energy research centers in Bulgaria, shown in Sect. 1.5 of this book.

Significant entry into practice of solar lighting systems occurs simultaneously with the boom of the installation of photovoltaic power plants in Bulgaria in 2012–2013. During this period, with funding from European programs, autonomous solar street lighting systems were built in several smaller Bulgarian municipalities and settlements: Municipality of Nikolaevo—506 pillars, Kaynardja Municipality—486 (Fig. 61), Municipality of Straldja—220, and municipalities of Antonovo, Balchik, Sozopol, Pavlikeni, and others.

This period of installation of relatively small number of solar street lighting systems has been short, since, after the removal of additional grant funding, these systems are financially disadvantageous in Bulgarian market conditions, due to the low cost and the availability of electricity in all settlements. For example, the following calculations can be made from the operation of the PV-LED system for external lighting at the Technical University of Gabrovo (Figs. 9 and 13): From the solar power supply of a 20 W LED, luminaire for 1 year is saved 20 W $\times$ 4100 h = 82 kWh $\times$ 0.2 BGN/kWh = 16.40 BGN/year. For the entire 5 years, battery life are saved 5 $\times$ 16.40 = 82 BGN, which is not enough to buy a new battery, which is priced at 400 BGN, or five times higher than savings.

Fig. 61 Solar lighting in villages in Kaynardja municipality [43]

Due to the reasons described, solar lighting systems have been installed in Bulgaria in recent years only in places where the power supply would be too complicated or expensive (Figs. 62, 63, 64 and 65).

In recent years, Bulgarian researchers have had interesting research on the big potential of photovoltaic lighting projects for road tunnels [46–50], for which the maximum necessary adaptation luminance of the lighting system coincides with the maximum sunshine in the middle of the day and maximum power of photovoltaic modules, respectively.

Fig. 62 PV-LED systems for additional illumination of road signs and pedestrian walkways

Fig. 63 Additional solar lighting for fence security of an industrial site

Fig. 64 Autonomous PV system for power supply of Vezhen hut—high-altitude mountain information center in National Park "Central Balkan" [44]

Fig. 65 PV-LED lighting of a park in the city of Kardzhali [45]

References

1. Ikonact [CC BY-SA 3.0]. https://commons.wikimedia.org/wiki/File:Bulgaria-geographic_map-en.svg
2. Statistical Reference Book 2018 (2018) National Statistical Institute of Bulgaria, Sofia
3. Donchev D, Karakashev H (2004) Topics on physical and social-economic geography of Bulgaria, Ciela, Sofia (in Bulgarian)
4. Beck HE, Zimmermann NE, McVicar TR, Vergopolan N, Berg A, Wood EF [CC BY 4.0]. https://commons.wikimedia.org/wiki/File:Koppen-Geiger_Map_BGR_present.svg
5. http://gisdata.lib.ncsu.edu/fedgov/noaa/clino/TABLES/REG_VI/BU/
6. Markova D, Platikanov S, Konstantinoff M, Tsankov P (2011) Opportunities for using renewable energy sources in Bulgaria. J Contemp Mater (II-2 Dec 2011, Academy of Sciences and Arts of Republic of Srpska)
7. Šúri M, Huld TA, Dunlop ED, Ossenbrink HA (2007) Potential of solar electricity generation in the European Union member states and candidate countries. Sol. Energy

81: 1295–1305. http://re.jrc.ec.europa.eu/pvgis/, https://commons.wikimedia.org/wiki/File:Pvgis_solar_optimum_BG.png
8. National Renewable Energy Action Plan (2011) Ministry of Economy, Energy and Tourism, Republic of Bulgaria
9. Forecast Document in Accordance with Directive 2009/28/EC
10. Energy and Water Regulatory Commission of Bulgaria, Annual Report to the European Commission (2018)
11. Photovoltaic Barometer, EurObserv'ER (2018)
12. http://www.res-legal.eu/search-by-country/bulgaria
13. http://www.senes.bas.bg, photos by Plamen Ivanov
14. Platikanov S, Yovchev M (2013) Study of operating modes of a stand-alone photovoltaic system for outdoor lighting. J Contemp Mater (Renew Energy Sour) (IV-2 Academy of Sciences and Arts of the Republic of Srpska)
15. Zarkov Z, Stoyanov L, Milenov V, Voynova H, Lazarov V (2016) Modeling of PV generators from different technologies—case study. In: IEEE 17th international power electronics and motion control conference
16. http://www.tu-varna.bg, photo by Iliya Hadzhidimov
17. Stoyanov I, Iliev T, Evstatiev B, Mihaylov G (2019) Harmonic distortion by single-phase photovoltaic inverter. In: The 11th international symposium on advanced topics in electrical engineering (ATEE), Bucharest, Romania, 28–30 Mar 2019. 978-1-7281-0101-9/19/$31.00 ©2019 IEEE
18. Istalianov R, Lakov N, Spasov V (2016) Evaluation efficiency photovoltaic modules in continuous service. Annual of the University of Mining and Geology "St. Ivan Rilski", vol 59, Part III (in Bulgarian)
19. Stoev M, Stoilov A, Shtrakov S (2005) Study on combined solar systems. In: Proceedings of international conference on "The Integration of the renewable energy systems into the buildings structures", Patra, Greece, pp 205–211
20. www.repowermap.org
21. www.seea.government.bg/bg/registers
22. https://en.wikipedia.org/wiki/Karadzhalovo_Solar_Park
23. Satellite photos from www.google.com/maps
24. https://photos.wikimapia.org/p/00/02/30/60/52_big.jpg, https://photos.wikimapia.org/p/00/02/30/60/54_big.jpg (CC-BY-SA)
25. http://old.dker.bg/KAPDOCS/rep_Helios_12.pdf
26. http://old.dker.bg/KAPDOCS/rep-BCI-Cherganovo.pdf
27. http://solar.sts.bg/en/reference/223
28. http://kab-sofia.bg/novini/2979-parvata-obshtestvena-sgrada-v-balgariya-s-fotovoltaichna-fasada-e-v-gabrovo
29. https://trud.bg›откриха-стенно-осветление-в-плиска
30. Deyanov D (2004) History of the lighting in Bulgaria, Vasil Aprilov University Publishing House, Gabrovo (in Bulgarian)
31. https://www.sandacite.bg›история-на-осветлението-в-софия
32. http://www.stara-sofia.com
33. Ivan Karastoyanov [Public domain]. https://commons.wikimedia.org/wiki/File:BASA-3K-7-327-4-International_Fair_Plovdiv,_1892.jpg
34. Bulgarian Archives State Agency [Public domain]. https://commons.wikimedia.org/wiki/File:BASA-3K-7-511-7.jpg
35. Regional History Museum of Gabrovo, RIMG6019 479NI Sp.f
36. https://mapio.net/images-p/5452272.jpg
37. Walraven H (2012) ESOLI project, WP2 1.7 Assessment of framework conditions
38. Platikanov S, Tsankov P (2005) Design and modernization of road lighting in Bulgarian Cities. In: 7 Internationales Forum Lux junior 2005, Ilmenau, Germany
39. https://vizia.sofia.bg/2018/05/31/lighting_map
40. https://www.facebook.com/Dronchepurnoy

41. Yovchev M, Tsankov P, Bardarski N (2016) Modernization of the street lighting system in small settlements in the Municipality of Gabrovo. In: Conference energy forum 2016, Varna, St. St. Constantine and Elena (in Bulgarian)
42. Matev D, Gyurov V, Kirov R, Nikitasov S (2017) Intelligent street lighting of Varna 2015—concept, technologies, realization. In: National lighting conference for young scientists with international participation Lighting 2017, Varna (in Bulgarian)
43. http://www.briagnews.bg/solarnite-lampi-na-kainardja-osvetiha-i-evrostolicata-bryuksel, photos by Aleksey Minev
44. www.centralbalkan.bg, photos by Gencho Iliev
45. SAMEL-90 Ltd. ©2012 [CC BY-NC-ND 2.5 BG]. https://led-lighting-from-bulgaria.blogspot.com/2013/02/fotovoltaichno-svetodiodno-osvetlenie-kardjali.html
46. Pachamanov A, Zhelev S, Pavlov D, Ivanov D (2017) Powering the Vittinja tunnel lighting systems from photovoltaic systems on both sides of the tunnel. In: National lighting conference for young scientists lighting 2017, Varna (in Bulgarian)
47. Velinov, K., Stefanov, R.: Potential for energy saving lighting in road tunnels in the republic of Bulgaria. In: Annual of the University of Mining and Geology Geology "St. Ivan Rilski", vol 58, Part III, Mechanization, Electrification and Automation in Mines (2015) (in Bulgarian)
48. Vassilev H, Ganchev G, Manoilov P, Sgurev A (2014) Power supply system and tunnel lighting, Bulgarian National Committee on Illumination. In: XV National conference with international participation BulLight 2014, Sozopol (in Bulgarian)
49. Boychev B, Hristov K, Valova G (2018) Power supply of tunnel lighting with photovoltaic systems. In: Third national lighting conference for young scientists lighting 2018, Varna (in Bulgarian)
50. Pachamanov A, Hristov K (2018) Photovoltaic systems to reduce road tunnel lighting costs. In: Third national lighting conference for young scientists lighting 2018, Varna (in Bulgarian)

Solar Energy and Lighting in the Republic of Srpska

Tomislav Pavlovic and Dragoljub Lj. Mirjanić

Abstract In this chapter, information about geographical position, climate, solar radiation, and renewable energy policy in the Republic of Srpska are given. Also, information about Banja Luka, Academy of Sciences and Arts, solar energy laboratory at the Academy of Sciences and Arts and lighting in the Republic of Srpska are given.

1 General Information

1.1 Geographical Position

Bosnia and Herzegovina (B&H) consist of two entities: the Republic of Srpska (RS) and the Federation of Bosnia and Herzegovina (FB&H). The Republic of Srpska was founded on January 9, 1992.

The Republic of Srpska is located in the central part of the Balkan Peninsula, between 42°33′ and 45°16′ north latitude and 16°11′ and 19°37′ east longitude and occupies the northern and eastern parts of the geospatial area of Bosnia and Herzegovina (Fig. 1).

The borders of the Republic of Srpska are determined by an internationally recognized border to Serbia, Montenegro, and Croatia and an inter-entity line toward the Federation of Bosnia and Herzegovina. Regarding its territory surface, the Republic of Srpska has disproportionately long and irregular borders. They are very elongated and broken down forming in certain areas narrow belts (so-called "pockets") that bind Serbian spaces. The narrowest and most sensitive belt, until it was allocated to a separate district, was the one around the city of Brčko whose width was only

T. Pavlovic (✉)
Faculty of Sciences and Mathematics, University of Niš, Niš, Serbia
e-mail: pavlovic@pmf.ni.ac.rs

D. Lj. Mirjanić
Academy of Sciences and Arts of the Republic of Srpska, Banja Luka, Republika Srpska, Bosnia and Herzegovina
e-mail: mirjanicd@gmail.com

T. Pavlovic (ed.), *The Sun and Photovoltaic Technologies*,
Green Energy and Technology, https://doi.org/10.1007/978-3-030-22403-5_7

Fig. 1 Map of the Republic of Srpska [1]

5 km. The total length of the border of the Republic of Srpska is about 2170 km, out of which 1080 km belongs to the inter-entity line. The coefficient of boundary breaking is 3.6, which is the rarity in the world and the only to be compared with Chile. The Republic of Srpska has atypical form of state territory whose northern part is elongated west–east, and eastern in the north–south direction. This peculiar form is an aggravating circumstance of internal communication and economic integration of the interdependent western and southern parts of the Republic of Srpska.

The Republic of Srpska occupies a smaller part of the unique Serbian ethnic space, west of the Drina, i.e., it occupies the northern and eastern part of the geospatials of Bosnia and Herzegovina. The Republic of Srpska covers an area of 25,053 km^2 or about 49% of the territory of Bosnia and Herzegovina. RS according to the 2013 census has 1,170,342 inhabitants.

The Republic of Srpska belongs to a group of continental countries, i.e., there is no exit to the sea. The Republic of Srpska is located in contact with two large natural-geographic and socio-economic regional units—Pannonian and Mediterranean.

The natural features of the Republic of Srpska are very complex, which is the result of its belonging to different natural-geographical units and their geomorphological evolution. In the geomorphological appearance of the Republic of Srpska, different forms are represented. In the northern Peripanese part, the hilly terrain built of the Cenozoic deposits gradually descends into the plain areas with alluvial ravine and river terraces, which at the same time make the most fertile part of the Republic of Srpska. In this area, only a few lonely mountains—Kozara, Prosara, Motajica, Vučijak, Ozren, and Trebava, and the northeastern slopes of Majevica—rise. In the south, the plain area over the hilly terrain passes into the mountainous region that occupies the largest part of the territory of the Republic of Srpska.

Brčko District of Bosnia and Herzegovina

The Brčko District of B&H was established by the decision of the Chairman of the Arbitral Tribunal for Dispute over the Inter-Entity Boundary in the Brčko District on March 5, 1999. The state-legal order of B&H is included in the Amendment 1 of the Constitution of B&H, dated March 26, 2009. That Amendment stipulates that the Brčko District is under the sovereignty of B&H and under the jurisdiction of its institutions, in the same way, that those competencies derive from the Constitution and whose territory is in the joint ownership (entity condominium) of the entity. It represents the unit of local self-government with its own institutions, laws and regulations, and with the powers and status, finally, prescribed by the Arbitral Tribunal for Dispute Award in relation to the inter-entity demarcation line in the Brčko District. The relationship between the Brčko District of B&H and the institutions of B&H, and the entities, according to the Constitution, can be further regulated by the law passed by the Parliamentary Assembly of B&H. The Constitutional Court of B&H is competent to decide on disputes concerning the protection of the established status and authority of the Brčko District of B&H, which may occur between the entities and the District or between B&H and the District, under the Constitution and decisions of the Arbitral Tribunal. Such disputes can be initiated by a majority of deputies in the Brčko District Assembly, which includes at least one-fifth of the elected deputies from the ranks of each of the constituent peoples (Fig. 2).

The territory of the Brčko District covers an area of 458 km^2 and the city of Brčko, the administrative seat of the District, 183 km^2. The city is about 95 m above sea level, and the greater part of the territory of the District is below 200 m sea level. Brčko is located on the right bank of the Sava River and on both sides of the River Brka, which flows into the Sava River near the city center.

According to the preliminary results of the 2013 census, the city of Brcko had 43,859 inhabitants [1, 2].

1.2 *Climate in the Republic of Srpska*

The climate of the Republic of Srpska is determined by the main climatic factors: the geographical position, the geological background, the relief, the proximity of the

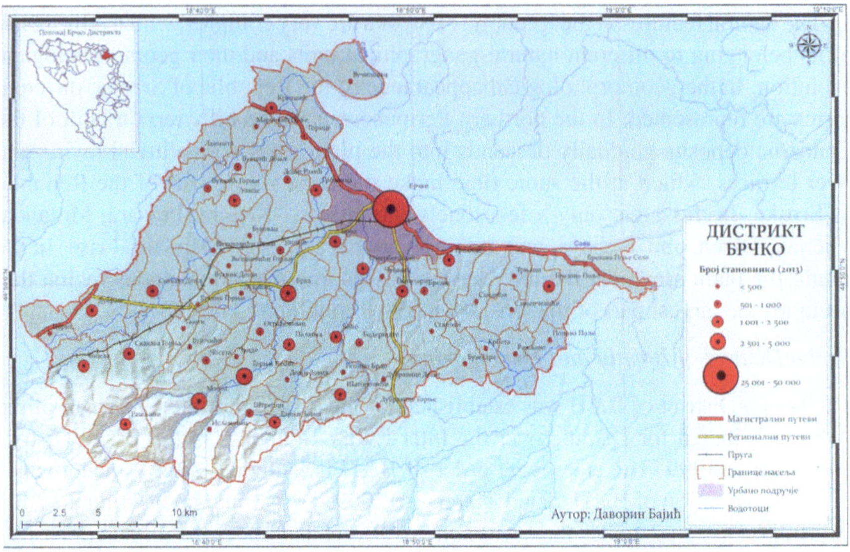

Fig. 2 Geographical position of the Brčko District of B&H [2]

Adriatic Sea, and the coverage of the terrain by the plant world. In addition to these basic factors, there are additional extreme factors that significantly affect the climate of the Republic of Srpska. These are the currents of the subtropical belt, high air pressure, sub-polar belt, and low air pressure, all of which results in the exchange of polar and tropical air masses, and then, streams from the Atlantic, cyclones from the Mediterranean and the Adriatic Sea, and anti-cyclones from the continental part of Asia. All these processes are greatly disturbed by the relief that appears as the main modifier. That is why, on the territory of the Republic of Srpska, there are three basic types of climate: moderate continental, mountain and mountain valley, and Mediterranean. Moderate continental climate occurs in the north, Mediterranean in the south, and the line dividing these two regions is the area of high mountains, plateaus, and ravines in which, depending on the altitude, the mountain climate dominates.

Moderate continental climate is present in the north of the Republic of Srpska. It includes Krajina, Posavina, and Semberija. Measuring stations located in this climate type are Banja Luka, Bijeljina, Derventa, Doboj, Novi Grad, Gradiška, Prijedor, Srbac, Višegrad, Srebrenica, Zvornik. In Semberija, Pannonian (steppe) climatic influence is felt due to the proximity of the Pannonian plain. The main features of this type of climate are warm summers and cold winters. Summer temperatures can rise above 40 °C, and the absolute maximum was measured in 2007. In Bijeljina and Višegrad, it tops at 43 °C. The average air temperature in the warmest part of the year (in July) ranges between 20 °C and 23 °C, while the average temperature in the coldest part of the year (January) is about zero degrees celsius. Absolute minimum can reach up to −30 °C. The average annual temperature is above 10 °C. The amount of rainfall in the RS is affected by humid air masses coming from the west (from the Atlantic) and from the south (from the Adriatic). Precipitation is the most viable climate parameter in terms of space and time. In the area where temperate continental climate is present, the highest precipitation occurs in the warm part of the year and the maximum occurs in June. Precipitation amounts to around 750 l/m^2 per year in the north along the Sava River and 1500 l/m in the west of the Krajina. Moderate continental climate part is also present in the mountainous valley areas that are up to 1000 m above sea level. Following stations can be found there: Mrkonjić Grad, Šipovo, Ribnik, Rudo, Foča, and Drinić with the rise of altitude and climatic changes.

The mountainous climate is present in the mountainous regions of the Republic of Srpska. At an altitude of 1000–1400 m, there is a submountainous (premountainous) climate, and with an elevation above sea level of 1400 m, it turns into a true mountainous climate. Stations, where this type of climate occurs are Sokolac, Čemerno, and Han Pijesak. The climatic characteristics of the mountain climate are short and fresh, and long and cold winters with abundant snowfall. The transition seasons (spring and autumn) are poorly expressed. The average January temperatures range from −3.5 °C to 6.5 °C, and in July, from 14.5 °C to 17 °C. Absolute minimum temperatures range from −25 °C to 35 °C, and absolute maximums are from 30 °C to 35 °C. The amount of precipitation is about 1200 l/m^2, and the snow cover often lasts for a long time. The submountainous climate is slightly milder with moderately warm summers and cold winters. The amount of precipitation in these areas is slightly lower, up to 1000 l/m^2. In mountainous areas, there are places, mostly basins, known as frosts where frequent thermal inversion is common. In these places, the lowest minimum temperatures are measured.

Mediterranean (Mediterranean, Adriatic, and subtropical) climate occurs in the southwest of the Republic of Srpska, i.e., in Herzegovina. As Herzegovina can be geographically divided into the low Herzegovina-Humine and High-Rudine, different climates occur in these areas. Humine has Mediterranean and modified Mediterranean climate, and Rudine is alternating between Mediterranean and mountainous, depending on the altitude (moderate mountainous—Mediterranean and mountainous climates).

The Humine climate is directly affected by the Adriatic Sea, thus winters are mild with the average temperature in January ranging from 3 °C to 6 °C. The summers are very warm in mid-July with temperatures ranging from 22 °C to 25 °C. Extreme winter temperatures depend on altitude and range from −8 °C in lower areas to −15 °C. In summer, maximum temperatures reach and often exceed 40 °C. The main feature of this region is precipitation. This is dominated by the maritime pluviometric regime under the influence of the Mediterranean, so the highest rainfall occurs late in autumn and early winter with frequent precipitation, while in summer, there is the minimum precipitation with frequent drought. The annual amount of precipitation is about 2000 l/m^2 up to 3000 l/m^2 as recorded in Grab, the rainiest place in the Republic of Srpska.

A moderate mountainous-Mediterranean climate occurs in the Rudine area. Here, the influence of the Mediterranean, as well as the mountains, is still felt. Stations that describe this type of time are Gacko, Bileca, and Nevesinje. The temperature of the air decreases with the rise of altitude and distance from the sea. For every 10 km from the sea, the temperature drops from 0.6 °C to 0.8 °C. Winters here are sharper with mid January temperatures ranging from −3 °C to 0 °C. Absolute minimums range from −15 °C to 20 °C. The summers are slightly milder, but here too, but they can reach extreme temperatures up to 40 °C.

1.3 Solar Radiation in the Republic of Srpska

In terms of solar potential, Bosnia and Herzegovina belong to more favorable locations in Europe, with the intensity of solar radiation falling on a horizontal surface of 1240 kWh/m^2 in the north, and up to 1600 kWh/m^2 in the south of the country.

The total annual amount of solar radiation energy per square meter of horizontal surface and surface area set at optimum angle in relation to the horizontal plane in the territory of Bosnia and Herzegovina are given in Figs. 3 and 4, respectively.

The results of calculating the mean annual values of the optimum angle of solar modules mounting, energy of solar radiation that falls on one square meter of an area set at an angle of 0°, 90° and below the optimum angle in relation to the horizontal plane, the relationship between diffuse and total solar radiation, and the turbidity of the atmosphere, obtained by the PVGIS software, for 13 towns of the Republic of Srpska are given in Table 1.

Based on data given in Table 1, it can be seen that in the Republic of Srpska, values of mean annual solar radiation energy falling on 1 m^2 of the horizontal surface range from 3450 Wh/m^2 (Derventa) to 4220 Wh/m^2 (Trebinje), values of mean annual solar radiation energy falling on 1 m^2 of surface tilted at the optimal angle in relation to the horizontal plane range from 3930 Wh/m^2 (Zvornik) to 4890 Wh/m^2 (Trebinje), and values of mean annual solar radiation energy falling on 1 m^2 of the vertical surface range from 2570 Wh/m^2 (Zvornik) to 3240 Wh/m^2 (Trebinje). Mean annual values of the optimal angle of solar modules mounting range from 33° do 35° from the north to the south of the country. Mean annual values of the direct solar radiation intensity obtained by SWERA software for some towns in the Republic of Srpska are given in Table 2.

Based on data given in Table 2, it can be seen that in the Republic of Srpska daily values of the direct solar radiation intensity range from 3640 Wh/m^2 on the west (Banja Luka) to 5250 Wh/m^2 on the south of country (Trebinje). In the Republic of Srpska, the highest daily values of the direct solar radiation intensity of 5250 Wh/m^2 are to be found in the area with the following towns: Trebinje, Lastva, Ljubinje and Plana.

The results of calculating the mean annual energy values of the total solar radiation energy, which during the year falls on one square meter of solar modules set at an optimum angle in relation to the horizontal plane in the fixed, single axis and two axis tracking PV solar power plant, using the PVGIS software, for 13 towns of the Republic of Srpska, are given in Table 3.

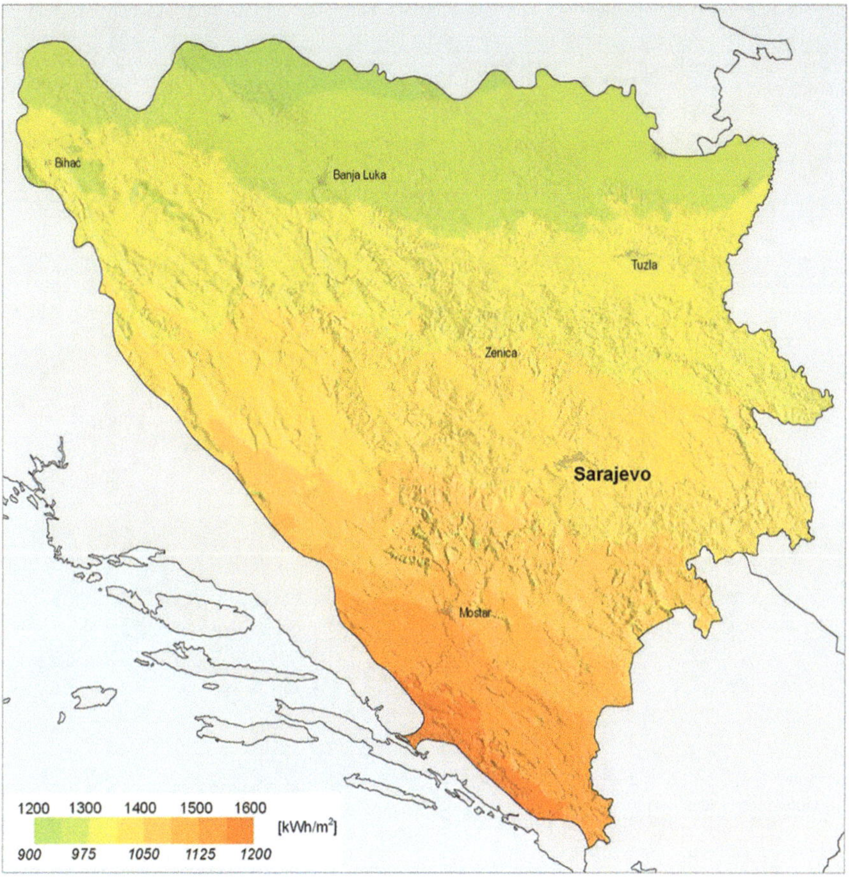

Fig. 3 Total annual amount of solar radiation energy that falls on one square meter of horizontal surface on the territory of Bosnia and Herzegovina [3]

In the light of all previously said, it can be concluded that there are favorable conditions in the Republic of Srpska for the use of solar radiation for the generation of heat and electricity [1, 3].

1.4 Renewable Energy Policy in the Republic of Srpska

In the Republic of Srpska, in 2009, the Energy law (Off. Gazette of the Republic of Srpska No. 49/09) was passed which further regulates the use of the renewable energy sources and efficient cogeneration. This law regulates the basis of the energy policy of the Republic of Srpska, adopting energy development strategy, plans, programs, and other acts for its implementation, the basic questions of the regulation and performing

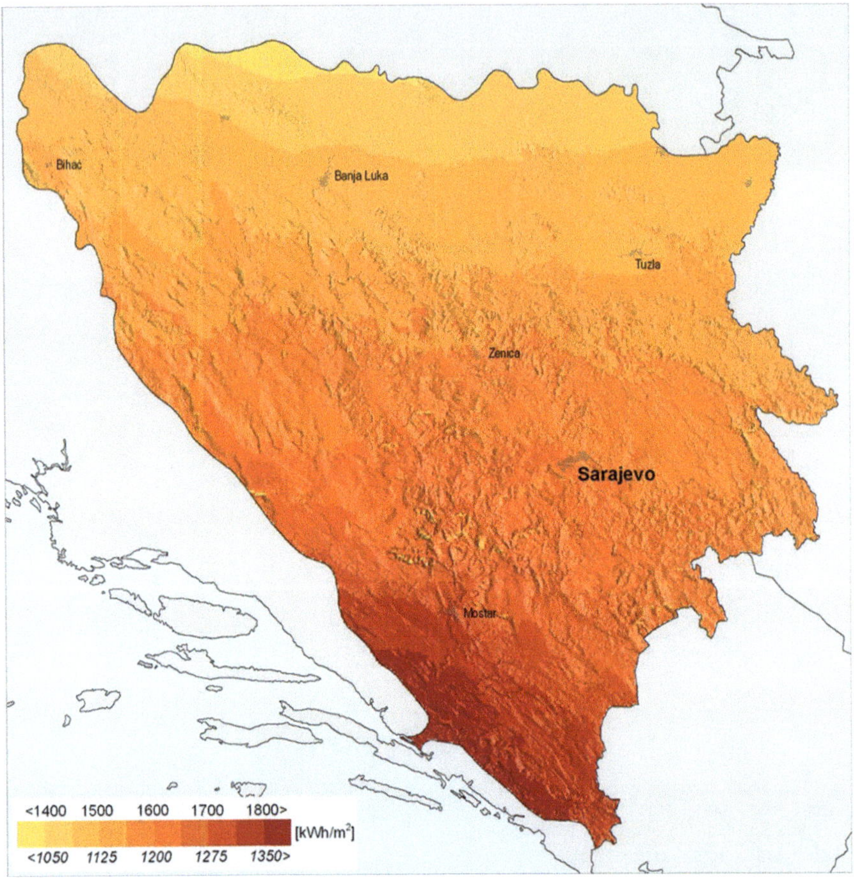

Fig. 4 Total annual amount of energy of solar radiation, which falls on one square meter of the surface, set at an optimal angle in relation to the horizontal plane, on the territory of Bosnia and Herzegovina [3]

energy activities, the use of the renewable energy sources, and ensuring the energy efficiency. In addition, the law's enactment has transferred by-laws in this area to the authority partly of the Regulatory Commission, and partly to the Government of the Republic of Srpska.

Moreover, in 2012, the *strategy of energy development of the Republic of Srpska to 2030* was adopted. This strategy focuses on the development of energy sector in the Republic of Srpska on using domestic resources, inclusion of renewable energy sources for meeting energy demands, inclusion and stimulation of energy efficiency measures, and use of modern energy technologies. At the same time, the strategy requires preservation of the environment and reducing harmful impacts of the energy sector to a minimum. Development of the energy sector in Republic of Srpska is considered in terms of gradual market opening, introducing competition, and setting

Table 1 Results of calculating the mean annual values of the optimum angle of solar modules mounting, energy of solar radiation that falls on one square meter of an area set at an angle of 0°, 90° and below the optimum angle in relation to the horizontal plane, the relationship between diffuse and total solar radiation, and the turbidity of the atmosphere, obtained by the PVGIS software, for 13 towns of the Republic of Srpska

Towns of the Republic of Srpska	Optimum angle of solar modules mounting (°)	Solar radiation energy falling on 1 m^2 of horizontal surface (Wh/m^2)	Solar radiation energy falling on 1 m^2 set at optimum angle in relation to horizontal plane (Wh/m^2)	Solar radiation energy falling on 1 m^2 of vertical surface (Wh/m^2)
Novi Grad	34	3500	3960	2620
Derventa	33	3450	3900	2580
Prijedor	34	3500	3980	2640
Brčko	34	3520	3990	2650
Bijeljina	34	3560	4040	2690
Banja Luka	34	3540	4010	2660
Doboj	34	3530	4000	2650
Zvornik	33	3480	3930	2570
Pale	35	3770	4350	2920
Sarajevo	35	3800	4380	2930
Višegrad	34	3740	4270	2820
Foča	35	3850	4430	2950
Trebinje	35	4220	4890	3240

Table 2 Mean annual values of direct solar radiation intensity obtained by SWERA software for some towns in the Republic of Srpska

Towns in the Republic of Srpska	Geographical position	DNI (Wh/m^2/day)
Trebinje	42°42′40.32″ N, 18°20′33″ E	5250
Pale	43°49′0″ N, 18°34′10″ E	4240
Liješće	45°4′59″ N, 18°5′2″ E	4100
Gornji Podgradci	45°3′21″ N, 17°2′48″ E	4070
Višegrad	43°46′58″ N, 19°17′28″ E	3980
Bijeljina	44°45′0″ N, 19°13′0″ E	3880
Omarska	44°53′30″ N, 16°53′57″ E	3830
Doboj	44°43′48″ N, 18°5′24″ E	3750
Vlasenica	44°11′0″ N, 18°56′0″ E	3750
Banja Luka	44°46′0″ N, 17°11′0″ E	3640

Table 3 Results of calculating the mean annual energy values of the total solar radiation energy, which during the year falls on one square meter of solar modules set at an optimum angle in relation to the horizontal plane in the fixed, single axis and two axis tracking PV solar power plant, using the PVGIS software, for 13 towns of the Republic of Srpska

	The results of calculating the mean annual energy values of the total solar radiation energy, which during the year falls on one square meter of solar modules set at an optimum angle in relation to the horizontal plane in the fixed PV solar power plant (kWh/m^2)	The results of calculating the mean annual energy values of the total solar radiation energy, which during the year falls on one square meter of solar modules set at an optimum angle in relation to the horizontal plane in the single-axis tracking PV solar power plant (kWh/m^2)	The results of calculating the mean annual energy values of the total solar radiation energy, which during the year falls on one square meter of solar modules set at an optimum angle in relation to the horizontal plane in the two-axis tracking PV solar power plant (kWh/m^2)
Novi Grad	1440	1790	1840
Derventa	1420	1780	1830
Prijedor	1450	1820	1870
Brčko	1450	1850	1890
Bijeljina	1480	1890	1930
Banja Luka	1460	1830	1880
Doboj	1460	1810	1860
Zvornik	1430	1670	1720
Pale	1590	2010	2070
Sarajevo	1600	2040	2100
Višegrad	1560	1940	1990
Foča	1620	2000	2060
Trebinje	1780	2290	2360

energy prices to the economically sustainable level. Economic possibilities of the Republic of Srpska and its citizens are also considered because this has primary impact on possibilities of energy sector development. The strategy encompasses all elements of the energy sector, from individual sectors (coal, oil, gas, electricity, etc.) to the legal, organizational, and institutional moments that are important for successful operation and development of energy sector in the observed period (up to 2030). The Republic of Srpska has a significant potential to generate electricity using photovoltaic solar systems.

The Republic of Srpska adopted a new law on Renewable Energy in May 2013. That, together with the decision of the Regulatory Commission for Energy of the Republic of Srpska on the tariff level and premium prices, governs the promotion of renewable energy. The Republic of Srpska prioritizes grid connection for renewable energy source operators and also offers other incentives for foreign investors, such as corporate tax exemption in the Republic of Srpska. In the Republic of Srpska, energy generation is licensed. Licenses are issued by the Regulatory Commission for Energy of the Republic of Srpska (REERS). To be eligible for the feed-in tariffs, renewable energy plant developers must have qualified producer status, which is obtained from the REERS. The new renewable energy legislation significantly improves capacity authorization and access to distribution networks, which is likely to increase effectiveness of renewable energy promotion.

The purchase price for 1 kWh of electricity produced by renewable energy sources in the Republic of Srpska was determined by the government decision from 17.01.2017 on the amount of the guaranteed purchase prices and premiums.

In Table 4, purchase prices for electricity generated by solar power plants with photovoltaic cells according to the place of installation are given [1, 4–6].

Table 4 Prices for kWh of electricity generated by solar power plants in the RS

Solar power plants with photovoltaic cells according to the place of installation	Sale in compulsory purchase at guaranteed purchase prices			Sales on the market and consumption for own needs	
	Guaranteed purchase prices (KM/kWh)	Reference prices (KM/kWh)	Premium (in guaranteed prices) (KM/kWh)	Reference price (KM/kWh)	Premium (KM/kWh)
On installations up to incl. 50 kW	0.2734	0.0570	0.2164	0.0667	0.2067
On installations over 50 kW up to incl. 250 kW	0.2341	0.0570	0.1771	0.0667	0.1674
On installations over 250 kW up to incl. 1 MW	0.1856	0.0570	0.1286	0.0667	0.1189
On the ground up to incl. 250 kW	0.2169	0.0570	0.1599	0.0667	0.1502

1 km = 0.5113 EUR = 0.5882 USD

1.5 Banja Luka and Academy of Sciences and Arts of the Republic of Srpska

Banja Luka

Banja Luka is the administrative, political, economic, cultural, and university center of the Republic of Srpska. In 1997, Banja Luka gained the status of the city.

The narrower city area is 160 m above sea level and covers about 80 km^2. The wider urban area extends from 150–500 m above sea level and occupies an area of 1239 km^2. Banja Luka is mostly located in the basin of the Vrbas Mountain, which from the south and southeast are closed by the branches of the Dinaric system. Valley is spreading eastward toward the valley of the River Vrbas. In 2013, Banja Luka had 65,010 households and 180,053 inhabitants (Fig. 5).

Academy of Sciences and Arts of the Republic of Srpska

The Academy of Sciences and Arts of the Republic of Srpska (ASARS) was founded on 29.12.1993. ASARS represents the most important scientific and cultural institution in the Republic of Srpska. At the beginning of 2019, ASARS has 84 members—29 regulars, 13 correspondings, and 42 foreign members. The working bodies of the academy are departments (social sciences, literature and arts, natural sciences and mathematics and technical sciences, and medical sciences), institutes (for history, social research, natural and technical sciences, Serbian language, and literature), and committees (Fig. 6).

The ASARS realizes its tasks through various forms of work: publishing activity, scientific research, scientific meetings, and inter-academic cooperation. Library and gallery are working within the academy.

Fig. 5 Banja Luka

Fig. 6 Academy of Sciences and Arts of the Republic of Srpska [4]

The ASARS is located in its own, modern-equipped building covering 2500 m^2 in the center of Banja Luka since 2011. The building was built in 1940 for the needs of the Chamber of Commerce and Industry and later on, it was upgraded several times. It houses the Chamber of Commerce, the Pedagogical Academy, and the faculty of Philosophy. Part of the building's interior (columns) is protected as a cultural-historical monument [1, 2, 5, 6].

1.6 Solar Energy Laboratory

The Solar Energy Laboratory of the Academy of Sciences and Arts of the Republic of Srpska, Banja Luka was formed in October 2012 as a result of work on scientific research projects in the field of renewable energy sources—especially solar energy. The projects were financed by the Ministry of Science and Technology of the Republic of Srpska and UNESCO.

Devices

Solar Energy Laboratory in ASARS has following devices:

- *Davis Vantage PRO* (*Davis*, USA) meteorological station for measuring the intensity of solar radiation, wind speed, temperature, pressure and humidity, UV index, etc.;
- *KLA and Mini-KLA* (*Ingenieubüro Mencke & Tegtmeyer*, Germany) for measuring the electrical characteristics of solar modules;

- *SolarUsbSW* (*Metering Solution*, Republic of Srpska) for simultaneous measurement of electrical characteristics of five solar modules;
- *KIPP & ZONEN CMP22* pyranometer (*KIPP & ZONEN*, Holland) for measuring the solar radiation intensity;
- Inverter dc in alternating current *Sunny Boy 2000HF* (*SMA Solar Technology AG*, Germany);
- *Sunny WebBox with Bluetooth* (*SMA Solar Technology AG*, Germany) for monitoring the operation and measurement of electrical parameters of PV solar power plant;
- *HW–SW* device (*Metering Solutions*, Banja Luka) for consecutive recording of physical characteristics of solar modules every 15 min;
- Digital temperature sensors *DS18B20* are used to measure the temperature of the solar module;
- computers, printers, etc.

PV solar power plant

In October 2012, on the roof of the ASARS, the *IRC Alfatec Co.* company from Niš, installed a fixed on-grid PV solar power plant power of 2.08 kW_P with monocrystalline silicon solar cells. The inverter and accompanying equipment for monitoring, acquisition, and data processing were obtained as a donation from the German company SMA. Pyranometer *KIPP & ZONEN CMP22* (the Netherlands) was purchased to measure the intensity of the solar radiation. In addition, the automatic weather station *DAVIS Vantage PRO (USA)* was purchased. In order to determine the physical characteristics of solar modules, *Mini-KLA and KLA Ingenieurburo Mencke & Tegtmeyer* devices from Germany were purchased. Thanks to these devices, the influence of the intensity of solar radiation, air temperature, wind speed, and air humidity on the energy efficiency of the photovoltaic solar power plant in Banja Luka region can be continuously monitored (Fig. 7).

The automatic weather station *DAVIS Vantage PRO*, the *Sunny Boy 2000HF* inverter, and the accompanying equipment for monitoring and acquisition of PV solar power on the roof of the Academy of Sciences and Arts of the Republic of Srpska are shown in Fig. 8.

PV solar power plant on the roof of the ASARS is used for the scientific and educational purposes.

SolarBox

In order to determine the energy efficiency of solar modules, depending on the geographical orientation, angle of inclination and their temperature in the *ASARS Solar Energy Laboratory* in 2014 a solar system was developed—*SolarBox*.

Solar system *SolarBox* consists of a metal base with five photovoltaic solar modules made of polycrystalline silicon, with individual power of 50 W (*SOLE SL-50P*). Three solar modules are positioned vertically and oriented toward the east, south, and west, respectively. The fourth solar module is positioned horizontally, and the fifth

Fig. 7 PV solar power plant on the roof of the ASARS

Fig. 8 Automatic weather station *DAVIS Vantage PRO* (left), inverter and the accompanying equipment for monitoring and acquisition of PV solar power plant on the roof ASARS (right)

is mounted at an optimal angle of 33° (the annual optimal angle for the fixed solar module for Banja Luka) and is oriented to the south. The solar system is installed on the ANURS roof and is raised from the base, which ensures natural air circulation and ventilation of the system (Figs. 9 and 10).

PV-KLA (Ingenieurburo Mencke & Tegtmeyer, Germany) was used to measure the I–V characteristics of the solar panels and the device for simultaneous

Fig. 9 *SolarBox*

Fig. 10 Interface for KLA

automatic measurement of the electrical characteristics of five solar modules *SolarUsbSW (Mettering Solutions*, B&H). Using a *SolarUsbSW*, computer-controlled device enables the connection and synchronization of the measurement cycle. *SolarControlM* computer software (*Mettering Solutions,* B&H) is used for automatic control of this device, monitoring of working conditions, reliability, synchronization, and control and data acquisition.

For consecutive recording of the physical characteristics of solar modules every 15 min between solar modules and PV-KLA devices, *HW–SW* device (*Metering Solutions*, Banja Luka) is used.

To measure solar module temperature, a digital temperature sensor *DS18B20* is used, which is placed in the center of the back side of each solar module.

The TempLogger measuring device was used for continuous monitoring and acquisition of solar module temperature during the day. This device transmits measured temperature values to a computer or other microcomputer system compatible with the RS232 digital communication standard (Fig. 11).

Two-axis tracking PV system

In mid-October 2017, a two-axis tracking PV system was installed on the roof of ASARS in Banja Luka. Two-axis tracking PV system consists of electronic, mechanical, and measuring subsystem.

The electronic part of the PV system contains sensors (two per axis of rotation), differential amplifier, output degree for motor control, and degree for determining the minimum amount of light. Mechanical subsystem is a special mechanical assembly with steel structure and gears that allow automatic optimal orientation of the solar module with regulation by step motors. The total angle of clearance along the vertical axis is about 50° (from 20 to 70°), and on the horizontal axis 150° (± 75°). The

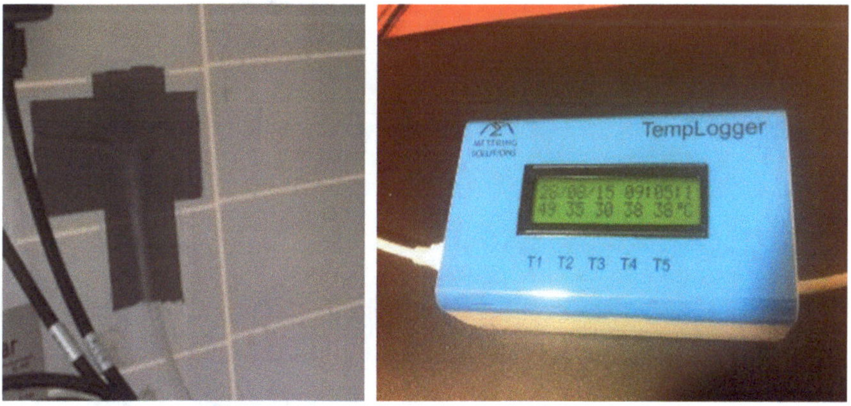

Fig. 11 Sensor for temperature measurement of solar modules *One Wire Digital Temperature Sensors DS18B20* (left) and *TempLogger* measuring device for continuous monitoring and acquisition of solar module temperature (right)

measuring part of this system consists of measuring devices (PV-KLA analyzers) located in the Solar Energy Laboratory at the ASARS (Figs. 12, 13 and 14).

Research activities

In *Solar Energy Laboratory* at the ASARS, following research activities are conducted:

Fig. 12 Experimental two-axis tracking PV system

Fig. 13 50 W solar panel with measuring sensor for solar radiation intensity determination

Fig. 14 Sensors for optimal system orientation measurement

- Investigation of the energy efficiency of photovoltaic solar power plant of 2.08 kW_P;
- Investigation of the energy efficiency of solar modules depending on their geographical orientation and the angle of inclination in real climate conditions;
- Investigation of the energy efficiency of a two-axis tracking PV system in real climate conditions;
- Study of the energy efficiency of PV solar modules depending on their soiling.

Solar Energy Laboratory at the ASARS has realized the following international projects: *Renewable energy sources as a model of sustainable development of the countries of West Balkans* (UNESCO project, 2010–2011), *Influence of Energy Efficiency of Solar Energy on Economic and Sustainable Development for the Western Balkans region* (UNESCO project, 2012–2013), and *The influence of renewable energy sources to the protection of the environment in the West Balkan Countries* (UNESCO project, 2014–2015).

In addition, *Solar Energy Laboratory* at the ASARS has realized a project *Investigation of the energy efficiency of photovoltaic solar power plant of 2.08 kWP*, financed by the Ministry of Science and Technology of the Republic of Srpska.

Results obtained at the ASARS were published in following Refs. [7–15].

1.7 PV Solar Power Plants in the Republic of Srpska

In the Republic of Srpska, there are several companies involved in the design and installation of solar systems: *Etmax Ltd.* (Banja Luka), *Bemind* (Banja Luka), *Koming* (Gradiska), *Klenik* (Gradiska), *Topling* (Prnjavor), *Pavlovic Mont* (Banja Luka), etc. Solar cells and solar modules are not produced in the Republic of Srpska, while solar modules, battery charging regulators, accumulators, and inverters are sold by several private companies.

In the Republic of Srpska, there are a total of 42 electricity producers with the status of privileged and temporarily privileged electricity producers using a PV system of up to 250 kW. Total installed power of privileged electricity producers in the Republic of Srpska on 25.9.2017 was 4.69361 MW.

Large PV solar power plants

Until now, in the Republic of Srpska, no PV solar power plant with a power exceeding 250 kW has been installed or put into operation.

Small PV solar power plants

So far, 43 PV solar power plants up to 250 kW have been installed in RS (Figs. 15, 16, 17, 18, 19, 20, 21, 22, 23 and 24; Table 5).

Fig. 15 PV solar power plant *Novakovic-besjeda* power of 249 kW_P Vijacani, Prnjavor (2014) [16]

Fig. 16 PV solar power plant *Tesla1* and *Tesla2* at Tesla Ltd. power of 2 × 120 kW_P, Modriča (2014)

Fig. 17 PV solar power plant *Neutron—MBM Ltd.* power of 180 kW_P, Bijeljina (2014/15) [17]

2 Lighting in the Republic of Srpska

2.1 Early Development of Lighting

The first light bulbs in the Trapists monastery of *Marija Zvijezda* in Deli Basin village and Banja Luka were illuminated on March 27, 1899. Thanks to electric lighting, the monastery *Marija Zvijezda* and *Delibašino selo* looked like a fairy late in the night.

Fig. 18 PV solar power plant *Oil refinery Modriča* power of 110 kW_P, Modriča (2013) [18]

Fig. 19 PV solar power plant *Madra Ltd.* power of 2 × 50 kW_P, Celinac (2014)

At the same time, with the construction of the *Delibašino selo* hydropower plant, Banja Luka started building a city network for the illuminated city center and households. Thanks to the electricity from the hydro power plant *Delibašino selo* in Banja Luka, in 1902, the city center, two railway stations, individual premises of the city administration, households near the power plant, etc., were illuminated.

Fig. 20 PV solar power plant *MI Trivas Ltd.* power of 50 kW_P, Prnjavor (2014/15) [19]

Fig. 21 PV solar power plant *Verano Motors* power of 48 kW_P, Banja Luka (2013)

2.2 Development of the Electric Lighting

Modern lighting in the Republic of Srpska is similar to lighting in Serbia. Lately, LED lighting has being increasingly used in households and other places in the Republic of Srpska. In some city streets in Banja Luka, classic lighting is replaced with LED bulbs (Figs. 25 and 26).

Fig. 22 PV solar power plant *Fratelo trade A.D.* power of 45 kW_P, Banja Luka (2014)

Fig. 23 PV solar power plant *BLC 1* power of 20 kW_P, (2012) and *BLC2* power of 10 kW_P, Banja Luka (2014)

Fig. 24 PV solar power plant power of 6.75 kW_P on the administrative building of the Electro power industry company of RS, Trebinje (2013)

Table 5 Small PV solar power plants in the Republic of Srpska

No.	PV solar power plant name	Community	Location	Total output power (kW)	Installed in (year)	Voltage level (kV)	Annual output (kWh)
1	MSE Solar1	Kozarska Dubica	Kozarska Dubica	49.92	2014	0.4	55,000
2	MSE Borik	Banja Luka	Banja Luka	12.50	2014	0.4	18,000
3	MSE Prnjavor 1	Prnjavor	Prnjavor	27.00	2014	0.4	29,793
4	MSE Fratelo 1	Banja Luka	Banja Luka	43.00	2014	0.4	55,000
5	MSE “BLC”	Banja Luka	Banja Luka	20.00	2014	0.4	26,000
6	MSE Glamočani	Banja Luka	Glamočani	10.00	2014	0.4	12,000
7	MSE Verano	Banja Luka	Banja Luka	48.00	2015	0.4	60,000
8	MSE Elektro Doboj-Teslić	Teslić	Teslić	41.76	2014	0.4	51,865
9	MSE 100 kW Oil refinery “Modriča”	Modriča	Modriča	110.25	2014		121,415
10	MSE Crnjelovo1	Bijeljina	Gornje Crnjelovo	10.12	2014	0.4	11,800
11	MSE Santing 1	Pale	Pale	49.68	2014	0.4	57,542

(continued)

Table 5 (continued)

No.	PV solar power plant name	Community	Location	Total output power (kW)	Installed in (year)	Voltage level (kV)	Annual output (kWh)
12	MSE Santing 2	Pale	Pale	49.68	2014	0.4	57,542
13	MSE Solar2	Kozarska Dubica	Kozarska Dubica	49.68	2014	0.4	60,000
14	MSE Solar 3	Kozarska Dubica	Kozarska Dubica	49.50	2014	0.4	60,000
15	MSE Woll	Teslić	Gomjenica	9.12	2015	0.4	12,000
16	MSE Martić	Derventa	Derventa	8.39	2014	0.4	7000
17	MSE Tesla 1	Modriča	Donji Kladari	122.40	2014	0.4	140,000
18	MSE Tesla 2	Modriča	Donji Kladari	122.40	2014	0.4	140,000
19	MSE Podgrab 1	Pale	Podgrab	249.50	2014	0.4	336,825
20	MCE Džungla	Doboj	Doboj	188.50	2014	0.4	74,000
21	MSE Neutron I	Bijeljina	Bijeljina	180.00	2015	0.4	188,616
22	MSE Herteks	Modriča	Modriča	10.00	2015	0.4	5164
23	MSE Solar 1 - Bileća	Bileća	Bileća	249.90	2015	0.4	362,420
24	MSE Madra 1	Čelinac	Čelinac	49.50	2015	0.4	57,000
25	MSE Madra 2	Čelinac	Čelinac	49.50	2015	0.4	57,000
26	MSE Novaković	Prnjavor	Prnjavor	247.86	2015	0.4	300,000
27	MSE Fratelo 2	Banja Luka	Banja Luka	105.00	2015	0.4	120.000
28	MSE BMB Delta	Gradiška	Gradiška	49.60	2015	0.4	63,875
29	MSE TE Ugljevik (in area TE Ugljevik)	Ugljevik	Ugljevik	240.00	2015	0.4	300,000
30	MSE 059 Bileća	Bileća	Bileća	37.00	2015	0.4	50,315
31	MSE Turmenti 1	Trebinje	Turmenti	248.40	2015	0.4	355,043
32	MSE Turmenti 2	Trebinje	Turmenti	248.40	2015	0.4	355,043
33	MSE Turmenti 3	Trebinje	Turmenti	248.40	2016	0.4	355,043

(continued)

Table 5 (continued)

No.	PV solar power plant name	Community	Location	Total output power (kW)	Installed in (year)	Voltage level (kV)	Annual output (kWh)
34	MSE Turmenti 4	Trebinje	Turmenti	248.40	2016	0.4	355,043
35	MSE Atlantik	Banja Luka	Banja Luka	150.00	2016	0.4	190,000
36	MSE Trivas	Prnjavor	Prnjavor	48.78	2015	0.4	355,043
37	MSE Tesla 3	Modriča	Donji Kladari	75.00	2016	0.4	95,000
38	MSE Solar 4	Kozarska Dubica	Kozarska Dubica	49.50	2016	0.4	62,000
39	MSE Solar 5	Kozarska Dubica	Kozarska Dubica	49.83	2016	0.4	62,000
40	MSE Derventa 1	Derventa	Gornji Cerani	248.40	2017	10	364,000
41	MSE Podromanija 2	Sokolac	Podromanija	250.00	2017	0.4	262,900
42	MSE Podromanija 3	Sokolac	Podromanija	250.00	2017	10	262,900
43	MSE Ramići	Banja Luka	Banja Luka	156.00	2017	0.4	198,875

Fig. 25 Banja Luka Town hall administrative building [20]

Fig. 26 Gospodska street in Banja Luka [21]

References

1. Pavlović TM, Tripanagnostopoulos Y, Mirjanić Lj. D, Milosavljevć DD (2015) SOLAR ENERGY in Serbia, Greece and the Republic of Srpska. Academy of Sciences and Arts of the Republic of Srpska, Banja Luka
2. Encyclopedia of the Republic of Srpska (2017) Academy of Sciences and Arts of the Republic of Srpska, Banja Luka (in serbian)
3. http://re.ec.europa.eu/pvgis/
4. https://www.rtrs.tv/_FOTO/nwz/0516/051677.jpg
5. Pavlović MT, Mirjanić LjD, Milosavljević DD (2018) Electric power industry in Serbia and the Republic of Srpska. Academy of Sciences and Arts of the Republic of Srpska, Banja Luka (in serbian)
6. Pavlović MT, Milosavljević DD, Mirjanić LjD (2013) Renewable energy sources. Academy of Sciences and Arts of the Republic of Srpska, Banja Luka
7. Pavlović T et al (2013) Assessments and perspectives of PV solar power engineering in the Republic of Srpska (Bosnia and Herzegovina). Renew Sust Energy Rev 18:119–133
8. Milosavljević DD et al (2016) Photovoltaic solar plants in the Republic of Srpska—current state and perspectives. Renew Sust Energy Rev 62:546–560
9. Milosavljević DD et al (2015) Energy efficiency of PV solar plant in real climate conditions in Banja Luka. Therm Sci 19(2):S331–S338
10. Pavlović T et al (2011) Application of solar cells of different materials in PV solar plants of 1 MW in Banjaluka. Contemp Mater (Renewable energy sources) II-2:155–163
11. Pavlović T et al (2011) Analyses of PV systems of 1 kW electricity generation in Bosnia and Herzegovina. Contemp Mater (Renewable energy sources) II-2:123–138
12. Pavlović T et al (2012) Assessment of the possibilities of building integrated PV systems of 1 kW electricity generation in Banja Luka. Contemp Mater III-2:167–176
13. Mirjanić D et al (2015) Investigation of energy efficiency of polycrystalline silicon solar modules in relation to their geographical orientation and tilt angle. Contemp Mater (Renewable energy sources) VI-2:87–94
14. Cekić N et al (2015) Application of solar cells in contemporary architecture. Contemp Mater (Renewable energy sources) VI-2:104–114

15. Mirjanić LjD, Pavlović MT, Milosavljević DD (2015) Contemporary materials for photovoltaic solar energy conversion. In: Proceedings of 3rd International Conference "New Functional Materials and High Technology" NFMaHT-2015, Russian Academy of Sciences, Academy of Sciences and Arts of the Republic of Srpska, G.A. Krestov Institute of Solution Chemistry of Russian Academy of Sciences (ISC-RAS), 29–30, June 2015, Tivat, Montenegro, pp 7–17
16. https://www.google.com/search?q=PV+solar+plant+Novakovic-besjeda+power+of+249+kWP+Vijacani,+Prnjavor+(2014)&source=lnms&tbm=isch&sa=X&ved=0ahUKEwjljNOTpL_kAhVv2aYKHQ8MB0UQ_AUIEigB&biw=1760&bih=848&dpr=1.09#imgrc=4X7msuhuk919fM:
17. https://encrypted-tbn0.gstatic.com/images?q=tbn:ANd9GcRPLj1mbd0IlmyACBAT92Sg-hsvjGZNRIo6dUQzukeGth0k7Tk9
18. https://ars.els-cdn.com/content/image/1-s2.0-S1364032116301101-gr5.sml
19. https://encrypted-tbn0.gstatic.com/images?q=tbn:ANd9GcSRmopWMweQavsUpp-6jeOYTXK1wGp5z7pIk0DXi7eTAXLYaGVH
20. https://encrypted-tbn0.gstatic.com/images?q=tbn:ANd9GcRHzh1egCKyVCEwcQUqJNm2GZBtINa3ckpeWokYdbuhG9GSNF_a
21. https://encrypted-tbn0.gstatic.com/images?q=tbn:ANd9GcSkzR0iSwFQ1Jg9G9YsE367WKsTjQG0BaOByTa1JRE6CjnNzG5uMA

Index

T. Pavlovic (ed.), *The Sun and Photovoltaic Technologies*,
Green Energy and Technology, https://doi.org/10.1007/978-3-030-22403-5

T

U

V

W

MIX
Papier aus verantwortungsvollen Quellen
Paper from responsible sources
FSC® C105338

If you have any concerns about our products,
you can contact us on
ProductSafety@springernature.com

In case Publisher is established outside the EU,
the EU authorized representative is:
Springer Nature Customer Service Center GmbH
Europaplatz 3, 69115 Heidelberg, Germany

Printed by Libri Plureos GmbH
in Hamburg, Germany